Differential Equations with Applications to Mathematical Physics

This is volume 192 in
MATHEMATICS IN SCIENCE AND ENGINEERING
Edited by William F. Ames, *Georgia Institute of Technology*

A list of recent titles in this series appears at the end of this volume.

Differential Equations with Applications to Mathematical Physics

Edited by

W. F. Ames
E. M. Harrell II
J. V. Herod

SCHOOL OF MATHEMATICS
GEORGIA INSTITUTE OF TECHNOLOGY
ATLANTA, GEORGIA

ACADEMIC PRESS, INC.

Harcourt Brace Jovanovich, Publishers

Boston San Diego New York
London Sydney Tokyo Toronto

ACADEMIC PRESS, INC.
1250 Sixth Avenue, San Diego, CA 92101-4311

United Kingdom edition published by
ACADEMIC PRESS LIMITED
24–28 Oval Road, London NW1 7DX

Library of Congress Cataloging-in-Publication Data

Differential equations with applications to mathematical physics /
    edited by W.F. Ames, E.M. Harrell II, J.V. Herod.
        p.   cm. — (Mathematics in science and engineering ; v. 192)
    Includes bibliographical references and index.
    ISBN 0-12-056740-7 (acid-free)
    1. Differential equations.   2. Mathematical physics.   I. Ames,
William F.   II. Harrell, Evans M.   III. Herod, J. V., 1937-   .
IV. Series.
QA377.D63   1993
515'.35—dc20   92-35875
                                                                    CIP

Printed in the United States of America
92 93 94 95   EB   9 8 7 6 5 4 3 2 1

Contents

Preface

Since the days of Newton, Leibniz, Euler and Laplace, mathematical physics has been inseparably bound to differential equations. Physical and engineering problems continue to provide very important models for mathematicians studying differential equations, as well as valuable intuition as to the solutions and properties. In recent years, advances in computation and in nonlinear functional analysis have brought rigorous theory closer to realistic applications, and a mathematical physicist must now be quite knowledgeable in these areas.

In this volume we have selected several articles on the forefront of research in differential equations and mathematical physics. We have made an effort to ensure that the articles are readable as well as topical, and have been fortunate to include as contributions many luminaries of the field as well as several young mathematicians doing creative and important work. Some of the articles are closely tied to work presented at the International Conference on Differential Equations and Mathematical Physics, a large conference which the editors organized in March, 1992, with the support and sponsorship of the National Science Foundation, the Institute for Mathematics and its Applications, the Georgia Tech Foundation, and IMACS. Other articles were submitted and selected later after a refereeing process, to ensure coherence of this volume. The topics on which this volume focuses are: nonlinear differential and integral equations, semiclassical quantum mechanics, spectral and scattering theory, and symmetry analysis.

These Editors believe that this volume comprises a useful chapter in the life of our disciplines and we leave in the care of our readers the final evaluation.

The high quality of the format of this volume is primarily due to the efforts of Annette Rohrs. The Editors are very much indebted to her.

W. F. Ames, E. M. Harrell II, J. V. Herod
Atlanta, Georgia, USA

An Elementary Model of Dynamical Tunneling

J. Asch
Technische Universität, Berlin, Germany
P. Duclos
Centre de Physique Théorique, Marseille, France
and Phymat, Université de Toulon et du Var, La Garde, France

Abstract

In the scattering of a quantum particle by the potential $V(x) := (1+x^2)^{-1}$, we derive bounds on the scattering amplitudes for energies E greater than the top of the potential bump. The bounds are of the form $cte\ exp - \hbar^{-1}s(k,k')$, where $s(k,k')$ is the classical action of the relevant instanton on the energy shell $E = k^2 = k'^2$. The method is designed to suit as much as possible the n-dimensional case but applied here only to the case $n = 1$.

1 Introduction

It is well known that a quantum particle is in general scattered in all directions by a potential bump even if its energy is greater than the top of this bump. May be less known is that this phenomenon could be considered as a manifestation of tunneling. The purpose of this exposé is twofold: to show how one may treat such a problem with tunneling methods and to actually give estimates of semiclassical type on the scattering amplitudes.

Differential Equations with
Applications to Mathematical
Physics

After a very active period of studying tunneling through potential barrier (in the configuration space) there is nowadays a growing interest for tunneling in phase space (see e.g. [1], [2, and ref. therein], [4], [10]). It is natural to ask whether the configuration space techniques can be applied or extended to this new field of interest. To this end we propose the study of a simple model: the reflection of a one dimensional quantum particle above a potential barrier. This problem was studied by several authors: [5], [6], [7], [8]. The results which are more or less complete were derived by O.D.E. methods. Our aim here is to present a new method based on functional analytic tools created in the study of tunneling in the configuration space. The hope is that this method can be applied to n dimensional situations.

In section 2 we introduce our model and explain its tunneling features. In section 3 we present the estimate on the reflection coefficient of our model and the method that we use; finally we end up by some concluding remarks in section 4.

2   The Model

2.1   The Dynamical Tunneling Model

A one dimensional quantum particle in an exterior potential V is described by the Schrödinger operator ($\hbar$ is the Planck constant)

$$H := V + H_0 \ , \ H_0 := -\hbar^2\Delta \text{ on } L^2(\mathbf{R}) =: \mathcal{H},$$

and the corresponding classical Hamiltonian reads: $h(p,q) := V(q) + p^2$. We further restrict the model by fixing V and the energy E as:

$$V(x) := (1 + x^2)^{-1} \text{ and } E > V(0) =: v_0. \tag{1}$$

If one considers scattering experiments with energies E above the barrier top we know that a quantum particle sent from the left will undergo a reflection when crossing the region where the potential barrier is maximum, whereas the classical one is totally transmitted to the right.

If we look at the phase space trajectories of the classical hamiltonian h, we see that the energy surface for a given E greater than v_0 has two disconnected components corresponding to the two possible movements, the one from the left to the right and the other one from the right to the left. We interpret the capacity of jumping from one connected component of the energy shell to the other one as tunneling, much in the same way as for the case of an energy E below the barrier top v_0. In this latter case the two components of the energy shell are separated by a classically forbidden region due to the potential barrier whereas for the case of E above the barrier top, the classically forbidden region must be read along the momentum axis. Accordingly one speaks of a *dynamical barrier* between the two disjoint phase space trajectories on the energy shell which in turn motivates the terminology *dynamical tunneling* to mean the corresponding tunneling process.

To study this reflection we shall estimate the off diagonal terms of the *on (energy) shell transition matrix*: $T(E) := (2i\pi)^{-1}(1 - S(E))$, where $S(E)$ stands for the scattering matrix at energy E. $S(E)$ and $T(E)$ act on $L^2(\{-\sqrt{E}, \sqrt{E}\}) \simeq \mathbf{C}^2$ and the quantity we are interested in, i.e. the reflection coefficient, is

$$r := T(E)(-\sqrt{E}, \sqrt{E}).$$

2.2 Tunneling and Complex Classical Trajectories

An equivalent way to define the matrix $T(E)$ is to solve the equation $-\hbar^2\psi'' + (V - E)\psi = 0$ with the following boundary conditions

$$\psi(x) \sim t \, exp(i\hbar^{-1}\sqrt{E}x) \quad as \quad x \to \infty$$

$$\psi(x) \sim exp(i\hbar^{-1}\sqrt{E}x) + r \, exp(-i\hbar^{-1}\sqrt{E}x) \quad as \quad x \to -\infty;$$

t, the other entry of $T(E)$, is usually called the *transmission coefficient*. To solve the Schrödinger equation one may use the method of characteristics: $\psi(x, \hbar) := a(x, \hbar)exp(-i\hbar^{-1}s(x))$, which leads to the equivalent system

$$s'^2 := E - V \quad and \quad -\hbar^2 a'' - i\hbar(as')' - i\hbar s'a' = 0. \quad (2)$$

Obviously the phase s has two determinations on $\mathbf{R}$ which asymptotic forms at $\pm\infty$ are respectively $\pm\sqrt{E}x$ and $\mp\sqrt{E}x$. So there is no way to obtain a term like $exp(-i\hbar^{-1}\sqrt{E}x)$ in ψ starting with the determination $\sqrt{E}x$ of s at $+\infty$. The remedy, as well known, consists in allowing the variable x to be complex so that turning around the complex turning points of $E-V$ will exchange the two determinations of s. Of course the phase s will become complex during this escapade on the complex energy surface which will cause an exponentially small damping factor for the component of ψ on $exp(-i\hbar^{-1}\sqrt{E}x)$.

As one can see from (2.5), s is nothing but the action of the solution of our classical hamiltonian at energy E. Hence by allowing the classical particle to wander on the complex energy surface $h(p,q) = E$, it becomes able to jump between the two real components of this surface. Thus tunneling in quantum mechanics between two regions of the phase space is intimately related to the existence of classical trajectories linking these two regions on the complex energy surface. Such trajectories are usually called *instantons*.

According to the above discussion we can predict the exponentially small damping factor in r. The shortest way to join the two components of the energy shell is described by the instanton: $\gamma_+(p) = (p, V^{-1}(E - p^2)) = (p, i(1 + (p^2 - E)^{-1})^{1/2})$ for p running in $(-\sqrt{E - v_0}, \sqrt{E - v_0})$. The imaginary part of the corresponding action is

$$\hbar d_\star := Im \int_{\gamma_+} q(p)dp = \int_{-\sqrt{E-1}}^{\sqrt{E-1}} (1 + \frac{1}{p^2 - E})^{1/2}dp.$$

We show in section 3 that r decays at least like $d_\star^2 exp - d_\star$ in the large energy limit. Notice that $\hbar d_\star$ is usually given rather like

$$\hbar d_\star = Im \int_{-q_\star}^{q_\star} \sqrt{E - V(it)}dt$$

which corresponds to a parametrisation of γ_+ in terms of the position q, $\pm q_\star$ being the complex turning points.

3 The Main Theorem

3.1 The Basic Formula for the Reflection Coefficient

We shall use the *off (energy) shell transition operator* defined by:

$$T : \mathbf{C} \setminus \mathbf{R}_+ \to \mathcal{L}(\mathcal{H}), \quad T(z) := V - VR(z)V$$

where $R(z) := (H - z)^{-1}$ denotes the resolvent of H; similarly $R_0(z) := (H_0 - z)^{-1}$.

With our potential V, it is standard to show that $\widehat{T}(E + i\varepsilon)$ has a limit in $\mathcal{L}(\widehat{\mathcal{H}}^{-1}, \widehat{\mathcal{H}}^1)$ as ε goes to zero from above where $\widehat{T}(z)$ denotes the Fourier transform of $T(z)$ and $\widehat{\mathcal{H}}^n$ the domain of $\widehat{V}^{-\frac{n}{2}}$ equipped with its graph norm. Notice that $\widehat{\mathcal{H}}^1$ is just the Sobolev space $\mathbf{H}^1(\mathbf{R})$. The Fourier transform we use in this exposé is the one which exchanges x and $-i\hbar \partial_x$. Moreover if one introduces the trace operators

$$\tau_\pm : \mathbf{H}^1(\mathbf{R}) \to \mathbf{C} , \quad \tau_\pm(u) := u(\pm\sqrt{E}),$$

the operator $\tau_- \widehat{T}(E + i0)\tau_+^\star$ makes sense and one has:

$$r := T(E)(-\sqrt{E}, \sqrt{E}) = \tau_- \widehat{T}(E + i0)\tau_+^\star. \tag{3}$$

A key formula for our method is

$$T(z) = (V^{-1} + R_0(z))^{-1}, \quad z \in \mathbf{C} \setminus \mathbf{R}_+$$

which is valid first for z such that $\|VR_0(z)\| < 1$ and then for all z in $\mathbf{C} \setminus \mathbf{R}_+$ by analyticity. Then if we introduce the family of operators

$$A(z) := V^{-1} + R_0(z) \quad \text{so that} \quad \widehat{A}(z) = -\hbar^2 \Delta + 1 + \frac{1}{x^2 - z}$$

we see that the reflection coefficient is nothing but the Green function of $\widehat{A}(E + i0)$ evaluated at $\mp\sqrt{E}$ with zero value of its spectral parameter

$$r = \tau_-(\widehat{A}(E + i0) - 0)^{-1}\tau_+^\star. \tag{4}$$

3.2   The Operator $\widehat{A}(E+i0)$ and the Dynamical Barrier

A convenient way to study $\widehat{A}(E+i0)$ is to use the sectorial form [24, p. 310] associated to $\widehat{A}(z)$ for z in $\mathbf{C} \setminus \mathbf{R}_+$:

$$t_z[u] := \hbar^2 \|u'\|^2 + \|u\|^2 + (\frac{1}{x^2-z}u, u), \quad \mathcal{D}(t_z) := \widehat{\mathcal{H}}^1 = \mathbf{H}^1(\mathbf{R}).$$

Since for each z in $\mathbf{C} \setminus \mathbf{R}_+$, $(x^2 - z)^{-1}$ is bounded, t_z is obviously closed and sectorial and moreover $\widehat{A}(z)$ is a type A analytic family of m-sectorial operators [24, p. 375].

Let $W(x) := 1 + \frac{1}{x^2-E}$, then the following lemma is nothing but a rephrasing of the limiting absorption principle with an Agmon potential.

Lemma 1. *As ϵ goes to zero from above the operator $\widehat{A}(E+i\epsilon)$, $E > 0$, converges in $\mathcal{L}(\widehat{\mathcal{H}}^1, \widehat{\mathcal{H}}^{-1})$ to the m-sectorial operator associated to the form defined on $\widehat{\mathcal{H}}^1$ by:*

$$t_{E+i0}[u] := \hbar^2 \|u'\|^2 + (Wu, u) + i\pi |u(-\sqrt{E})|^2 + i\pi |u(\sqrt{E})|^2.$$

Notice that W in the above formula must be understood in the sense of its Cauchy principal value. The operator $\widehat{A}(E+i0)$ can be represented symbolically by

$$\widehat{A}(E+i0) = -\hbar^2\Delta + W + i\pi\delta(x^2 - E).$$

Its real part is a Schrödinger Hamiltonian which exhibits for E greater than $v_0 = 1$ two potential wells in the vicinity of $\pm\sqrt{E}$ separated by a potential barrier. W plays the role of an *effective potential* for our auxiliary non selfadjoint Schrödinger operator $\widehat{A}(E+i0)$.

Thus the Green function of $\widehat{A}(E+i0)$ evaluated at $\pm\sqrt{E}$ must contain an exponentially small overall factor due to tunneling through this potential barrier. This potential barrier ω_+ is actually the dynamical barrier we were speaking of in section 2.

3.3   Estimate of the Reflection Coefficient

We have shown in section 3.1 that the estimate of the reflection coefficient r is reduced to the one of the Green function of $\widehat{A}(E+i0)$

evaluated at $\pm\sqrt{E}$. As was argued in section 3.2, $\pm\sqrt{E}$ being separated by the dynamical barrier ω_+ we expect an exponentially small behavior of r in the size of ω_+. To prove it we resort to our familiar methods developed in the context of tunneling in configuration space (see e.g. [10, and ref. therein]).

As usual we define the auxiliary function

$$\rho(x) := d(-\sqrt{E}, x) \text{ if } x \geq -\sqrt{E} \text{ and } 0 \text{ otherwise}$$

where d denotes the pseudo-distance in the Agmon metric $ds^2 = \hbar^{-2}\omega_+(x)dx^2$ and $\omega_+(x) := W_+(x)$ if $x^2 < E$ and 0 otherwise. Since $\exp\rho(-\sqrt{E})$ equals 1 one gets: $r = \tau_- e^{-\rho}\widehat{A}(E+i0)^{-1}e^\rho\tau_+^\star e^{-\rho(\sqrt{E})}$ $= e^{-d_\star}\tau_-\widehat{A}_\rho(E+i0)^{-1}\tau_+^\star$, where $d_\star$ is the diameter of the dynamical barrier in the Agmon metric,

$$d_\star := d(-\sqrt{E}, \sqrt{E}) = \hbar^{-1}\int_{-\sqrt{E-1}}^{\sqrt{E-1}} \sqrt{1 + \frac{1}{x^2 - E}}\, dx,$$

and $\widehat{A}_\rho$ denotes the *boosted* operator: $\widehat{A}_\rho(E+i0) := e^{-\rho}\widehat{A}(E+i0)e^\rho$.

Thus it remains to find a suitable bound on the Green function $\tau_-\widehat{A}_\rho(E+i0)^{-1}\tau_+^\star$. We shall do it as follows. Using the standard bound: $\|\tau_\pm(-\Delta+1)^{-\frac{1}{2}}\| \leq 1$, we are led to estimate $\widehat{A}_\rho(E+i0)^{-1}$ as an operator from $\widehat{\mathcal{H}}^{-1}$ into $\widehat{\mathcal{H}}^1$. One possible way is to find a lower bound on the real part of $\widehat{A}_\rho(E+i0)$ as an operator from $\widehat{\mathcal{H}}^1$ to $\widehat{\mathcal{H}}^{-1}$:

$$Re\widehat{A}_\rho(E+i0) = -\hbar^2\Delta + W - \hbar^2\rho'^2 \geq \frac{\hbar^2}{2E}(-\Delta+1). \qquad (5)$$

This last estimate will be explained in the next subsection. Due to the method we are using, it will be valid only in the large energy limit and more precisely for values $(\hbar, E)$ in the following domain:

$$\nu := \left\{(\hbar, E) \in \mathbf{R}_+ \times \mathbf{R}_+ , \ E > \max\left\{(C_1\hbar^{-4}, C_2\hbar^2\right\}\right\}, \qquad (6)$$

where $C_1 := 121$ and $C_2 := 3$. So we have proven the

Theorem 2. *For every* $(\hbar, E)$ *in the domain* ν *defined above one has*

$$|r| \leq \frac{2E}{\hbar^2}\exp -d_\star.$$

3.4   Estimate of the Boosted Resolvent

To show (15) it is sufficient to obtain a lower bound on the real part of $\widehat{A}_\rho$ of the type $\gamma(-\Delta + 1)$ with γ strictly positive. Let $\omega_\rho :=$ $W - \omega_+ =: \omega_0 + \omega_1$ be a splitting of the potential part of $Re\,\widehat{A}_\rho$ so that ω_1 contains the Cauchy principal part of W:

$$\omega_1(x) := \begin{cases} \omega_\rho(x) & \text{if } x^2 < E \\ -\omega_\rho(\pm 2\sqrt{E} - x) & \text{if } \pm x > \sqrt{E}. \end{cases} \tag{7}$$

Then with $0 < \alpha^2 < 1$, one has

$$Re\,\widehat{A}_\rho(E + i0) \geq -(1 - \alpha^2)\hbar^2\Delta + \omega_1 - \alpha^2\hbar^2\Delta + \tilde{\omega}_0 \tag{8}$$

where we have estimated ω_0 from below by the square well potential: $\tilde{\omega}_0(x) := 1$ if $x^2 > E$ and $\tilde{\omega}_0(x) := 0$ otherwise. This allows to estimate from below the second Schrödinger operators on the r.h.s of (8) by $C(\alpha\hbar, E) := \alpha^2\hbar^2\pi^2 E^{-1}(1 - \alpha\hbar E^{-1/2})$ under the condition

$$\frac{\alpha^2\hbar^2\pi^2}{E} \leq 1. \tag{9}$$

For the first Schrödinger operators on the r.h.s. of (8) we use the following inequality:

$$|(\omega_1 u, u)| \leq 2E^{-3/4}\|u'\|^{3/2}\|u\|^{1/2} \tag{10}$$

to deduce with $\|u\| = 1$ that:

$$(Re\,\widehat{A}_\rho u, u) \geq (1 - \alpha^2)\hbar^2\|u'\|^2 + 2E^{-3/4}\|u'\|^{3/2} + C(\alpha\hbar, E). \tag{11}$$

To derive (10) we have used Sobolev inequalities. Choosing for the moment $\gamma := C(\alpha\hbar, E)/2$ and fixing α by $\alpha^2 := 4(\pi^2 + 4)^{-1}$ it remains to check that for $(\hbar, E)$ in the domain ν defined in (6) one has: $ax^2 + bx^{3/2} + c \geq 0$ for every non-negative x, with $a := (1 - \alpha^2)\hbar^2 - C(\alpha\hbar, E)/2$, $b := 2E^{-3/4}$ and $c := C(\alpha\hbar, E)/2$. Finally we are allowed take a smaller but better looking $\gamma := \frac{\hbar^2}{2E}$ since due to (9) $C(\alpha\hbar, E) \geq \hbar^2 E^{-1}$. Hence we have proven the statement contained in (15) and (6).

4 Concluding Remarks

In addition to the explanation of section 2.2 one can also understand tunneling as a transition between different subspaces of the Hilbert space of physical states. For example in our model, the quantum reflection is a transition between the two subspaces $Ran\widehat{\chi}_\pm$ where $\widehat{\chi}_\pm$ are the sharp characteristic functions of $\pm(\sqrt{E-1},\sqrt{E})$. Therefore all the processes exhibiting non-adiabatic transitions may be called dynamical tunneling as well.

The adiabatic method has been used extensively in the study of the quantum reflection coefficient by transforming the Schrödinger equation into a system of two coupled first order equations, see [6], [7]. More recently in [11] the exact asymptotics of the reflection coefficient has been given in the true adiabatic case. At the time we are writing these lines T. Ramon has announced the same kind of result for the quantum reflection; his method using exact complex WKB method combined with micro analysis techniques is an adaptation of the one developed in [12] for the study of the asymptotics of the gaps of one dimensional crystals.

Both of these two results show that our upper bound has at least the correct exponential behaviour. If one wants to consider higher dimension problems, the hope to be able to derive exact asymptotics on the scattering amplitude is small because of the complicated structure of the caustics and singularities of the underlying classical Hamiltonian system. But deriving upper bounds for a suitable range of the parameters in the spirit of [10] should be possible with the method presented here.

Acknowledgments

One of us, P.D., has greatly benefitted during the progress of this work from the hospitality of the Bibos at the University of Bielefeld (RFA) and of discussions with D. Testard who was visiting Bibos at that time.

Bibliography

[1] M. Wilkinson, *Tunneling between tori in phase space*, Physica **21D**, 1986, p. 341–354.

[2] B. Helffer, and J. Sjöstrand, *Semiclassical analysis for the Harper's equation III. Cantor structure of the spectrum*, Mém. Soc. Math. France, **39**, 1989.

[3] A. Martinez, *Estimates on complex interactions in phase space*, Prep. 92-5, Lab. Anal. Geom. Applic., Université Paris-Nord, 1992.

[4] A. Barelli, and R. Fleckinger, *Semiclassical analysis of Harper-like models*, Prep. Centre de Physique Théorique, Marseille, 1992.

[5] N. Fröman, and P. O. Fröman, *JWKB Approximation, contribution to the Theory*, North holland Amsterdam 1965.

[6] M. V. Berry, *Semiclassical weak reflexions above analytic and non-analytic potential barriers*, J. Phys. A: Math. Gen., **15**, 1982 p. 3693–3704.

[7] J. T. Hwang, and P. Pechukas, *The adiabatic theorem in the complex plane and the semiclassical calculations of the Non-adiabatic transition Amplitudes*, Journ. Chem. Phys. **67**, 1977, p. 4640–4653.

[8] G. Benettin, L. Chiercha, and F. Fassó, *Exponential estimates on the one-dimensional Schrödinger equation with bounded analytic potential*, Ann. Inst. Henri Poincaré, **51**(1), 1989, p. 45–66.

[9] T. Kato, *Perturbation theory for linear operators*, Berlin Heidelberg, New York, Springer, 1966.

[10] Ph. Briet, J. M. Combes, and P. Duclos, *Spectral stability under tunneling*, Commun. Math. Phys. **126**, 1989, p. 133–156.

[11] A. Joye, H. Kunz, and Ch. E. Pfister, *Exponential decay and geometric aspect of transition probabilities in the adiabatic limit,* Ann. Phys. **208**, 1986, p. 299–332.

[12] T. Ramon, *Equation de Hille avec potentiel méromorphe,* to appear in the Bull. Soc. Math. France.

Discrete Schrödinger Operators with Potentials Generated by Substitutions

Jean Bellissard
Laboratoire de Physique Quantique, Université Paul Sabatier,
118 Route de Narbonne, Toulouse, France
Anton Bovier
Institüt für Angewandte Analysis und Stochastik,
Hausvogteiplatz 5-7, Berlin, Germany
Jean-Michel Ghez
Centre de Physique Théorique, Luminy Case 907, Marseille,
France and Phymat, Université de Toulon et du Var, La Garde,
France

Abstract

In the framework of the theoretical study of one-dimensional quasi-crystals, we present some general and particular results about the gap labelling and the singular continuity of the spectrum of Schrödinger operators of the type $H_V \psi_n = \psi_{n+1} + \psi_{n-1} + v_n \psi_n$, where $(v_n)_{n \in \mathbf{Z}}$ is an aperiodic sequence generated by a substitution.

1 Introduction

The quasi-crystals, discovered in 1984 [1], are studied in one dimension by means of tight-binding models, described by discrete

Differential Equations with
Applications to Mathematical
Physics

Schrödinger operators of the type

$$H_V \psi_n = \psi_{n+1} + \psi_{n-1} + v_n \psi_n \quad (\psi_n)_{n \in \mathbf{Z}} \in \ell^2(\mathbf{Z}) \qquad (1)$$

where $(v_n)_{n \in \mathbf{Z}}$ is a quasi-periodic sequence. A very interesting case, both mathematically and physically, is that of a sequence $(v_n)_{n \in \mathbf{Z}}$ generated by a substitution [2] (see sect. 2 for a definition). This is a rule which allows to construct words from a given alphabet or, from a physical point of view, a quasi-crystal from elementary pieces of a tiling of the space.

Such operators are in general expected to have a singular continuous spectrum, supported by a Cantor set of zero Lebesgue measure. This has been already proven for the Fibonacci [3], [4], [5] and Thue-Morse [6], [7] sequences. We show here how to obtain the same result for the period-doubling sequence [7].

In all these cases, the method which is used is that of transfer matrices. It can be summarized as follows: one writes the Schrödinger equation in matrix form:

$$\begin{pmatrix} \psi_{n+1} \\ \psi_n \end{pmatrix} = P_n \begin{pmatrix} \psi_n \\ \psi_{n-1} \end{pmatrix} \text{where} P_n = \begin{pmatrix} E - v_n & -1 \\ 1 & 0 \end{pmatrix}, \qquad (2)$$

defines the transfer matrices as products of the form $\prod_{k=n}^{0} P_k$ and deduces the spectral properties of H_V from those of their traces.

This method was first developed in the Floquet theory of periodic Schrödinger operators [8] and recently generalized to the Anderson model [9] and then to the quasi-periodic case [10] and in particular to quasi-crystals [11], [12]. These last models exhibit Cantor spectra, which gaps are labelled by a set of rational numbers, depending of the particular example one considers, their opening being studied in details, for instance for the Mathieu equation [13] or the Kohmoto model [14], [15].

The program, still in progress, which results are described in this lecture, is the investigation of the particular class of one-dimensional substitution Schrödinger operators. A *substitution* is a map ξ from a finite alphabet A to the set of words on A. A *substitution sequence* or *automatic sequence* is a ξ-invariant infinite word u [2]. A substitution

Schrödinger operator is an operator of type (1) defined by a sequence $(v_n)_{n \in \mathbf{Z}}$ obtained by assigning numerical values to each letter of u.

In this case, the substitution rule implies a recurrence relation between the transfer matrices, which itself gives a recurrence relation on their traces, called the "trace map" [16]. Then one proves that the spectrum of H_V is obtained as the set of stable conditions of this dynamical system, which also coincides with the set of zero Lyapunov exponents of H_V. Finally, a general result of Kotani implies that the spectrum is singular continuous and supported on a Cantor set of zero Lebesgue measure. This has been done for the Fibonacci [5], Thue-Morse [7] and period-doubling [7] sequences. In the last two cases, a detailed study of the trace map allows also to compute the labelling and the opening mode of the spectral gaps [6], [7].

Now, one is naturally led to try to generalize these results to a large class of substitutions. For primitive substitutions, an easy way of computing the label of the gaps is obtained - and applied to some examples - [17] combining the K-theory of $C^\star$-algebras [18], [19], [20] and the general theory of substitution dynamical systems [2] (there are only perturbative conjectures for their real opening [21]).

The second expected common feature of substitution Schrödinger operators, that is the singular continuity of their spectrum, can also be obtained, by extending to a general situation the analysis of the trace map. Indeed, for primitive substitutions which trace map satisfies a simple supplementary hypothesis, two of us proved this result recently and applied it to the same examples as before [22].

The plan of this contribution is the following. In section 2, we define what are substitution hamiltonians and we show how K-theory of $C^\star$-algebras provides with a general gap labelling theorem for such operators. In section 3, we apply the method of transfer matrices to the case of the period-doubling sequence, namely we prove that the spectrum is singular continuous and has a zero Lebesgue measure and we study the labelling and opening of the spectral gaps. In section 4, we generalize the singular continuity of their spectrum to a rather large class of substitutions.

2 Gap Labelling Theorem [17]

We show in this section how K-theory of C^*-algebras provides with a simple way of computing the values of the integrated density of states in the gaps of the spectrum of a substitution hamiltonian.

We first summarize some basic definitions on substitutions [2].

Given a finite alphabet A, a substitution ξ is a map from A to $A^* = \bigcup_{k \geq 1} A^k$. ξ induces in a natural way a map from $A^{\mathbf{N}}$ to $A^{\mathbf{N}}$, which admits a fixed point u if it satisfies the conditions:

(C1) there is a letter 0 in A such that the word $\xi(0)$ begins with 0;

(C2) for any $\beta \in A$, the length of $\xi^n(\beta)$ tends to infinity as $n \to \infty$.

We say that a Schrödinger operator H_V of type (1) is *generated* by ξ if $v_n = \pm V$ following the $n - th$ letter of $u = \xi^\infty(0)$. For example, the period-doubling substitution defined by $\xi(a) = ab$, $\xi(b) = aa$ has a fixed point given by $u = \xi^\infty(a) = abaaabab...$ Assigning the values V to v_0, $-V$ to v_1, V to v_2, v_3 and v_4, $-V$ to v_5... and completing by symmetry for negative n, we obtain the period-doubling hamiltonian.

The *integrated density of states* (IDS) $\mathcal{N}(E)$ of H_V is the number per unit length of eigenvalues of H_V smaller than E in the infinite length limit. A gap labelling theorem consists in the determination of the set of values that the IDS takes in the spectral gaps of H_V. We prove it for *primitive* substitutions, that is substitutions ξ such that there is a k such that for any α and β in A, $\xi^k(\alpha)$ contains β.

For $\ell = 1, 2$, the matrices $M_\ell(\xi)$ of a substitution ξ are defined by putting $M_{\ell,ij}$ equal to the number of times the letter i occurs in the image of the letter j by ξ_ℓ, where $\xi_1 = \xi$ and ξ_2 is defined on the alphabet of the words of length 2 appearing in the $\xi(\alpha\beta)$ by setting $\xi_2(w_0 w_1) = (y_0 y_1)(y_1 y_2)...(y_{|\xi(w_0)|-1} y_{|\xi(w_0)|})$ if $\xi(w_0 w_1) = y_0 y_1 \cdots y_{|\xi(w_0 w_1)|-1}$. If ξ is primitive, the Perron-Frobenius theorem implies that M_1 and M_2 have a strictly positive simple maximal eigenvalue θ (the same for both), which corresponding eigenvectors v_ℓ, normalized such that the sums of their components equal 1, can be chosen strictly positive [2].

Now we can state our gap labelling theorem:

THEOREM 2.1 : *Let H_V be a 1D discrete Schrödinger operator of type (1) generated by a primitive substitution on a finite alphabet.*

Then the values of the integrated density of states of H_V on the spectral gaps in [0,1] belong to the $\mathbf{Z}$-module generated by the density of words in the sequence u, which is equal to the $\mathbf{Z}[\theta^{-1}]$-module generated by the components of the normalized eigenvectors v_1 and v_2 with the maximal eigenvalue θ of the substitution matrices M_1 and M_2.

The proof of theorem 2.1 is divided in four steps.

Step 1: Shubin's formula: $\mathcal{N}(E) = \tau \{\chi(H \leq E)\}$, the trace per unit length τ of the projector $\chi(H \leq E)$ in the infinite length limit.

Step 2: Abstract gap labelling theorem 1: Let $\mathcal{A}_{H_V}$ be the $C^\star$-algebra of H_V, that is the $C^\star$-algebra generated by the translates of H_V. Shubin's formula, together with general results about the K-theory of $C^\star$-algebras (referenced in [17]), implies the
Abstract gap labelling theorem 1: *The values of $\mathcal{N}(E)$ in the spectral gaps of H_V belong to the countable set $[0, \tau(1)] \cap \tau_\star(K_0(\mathcal{A}_{H_V}))$, where $\tau_\star$ is the group homomorphism $K_0(\mathcal{A}_{H_V}) \to \mathbf{R}$ induced by τ.*

Step 3 : Abstract gap labelling theorem 2: Let T be the two-sided shift on $A^{\mathbf{Z}}$, Ω the closure of the orbit of u by T in $A^{\mathbf{Z}}$ ((Ω, T) is called the *hull* of u) and μ the unique (by primitivity [2]) T-invariant ergodic probability measure on Ω. The study of the K-theory of $\mathcal{C}(\Omega)$ leads to the
Abstract gap labelling theorem 2: $\tau^\star(K_0(\mathcal{A}_{H_V})) = \mu(\mathcal{C}(\Omega, \mathbf{Z}))$.

Step 4: Computation of μ: Every function in $\mathcal{C}(\Omega, \mathbf{Z})$ is an integral linear combination of characteristic functions of cylinders $[B]$ in Ω (B being a word in u). Since the $\mu([B])$ are of the form $\frac{1}{\theta^n}$ times (integral linear combination of the components of v_1 and v_2) [2], our gap labelling theorem is proved, putting together the results of these four steps.

3   The Period-Doubling Hamiltonian [7]

The period-doubling sequence (see sect. 2) defines two sequences of unimodular transfer matrices $(T_E^{(n)}(a))_{n \in \mathbf{N}}$ and $(T_E^{(n)}(b))_{n \in \mathbf{N}}$, corresponding to the two numerical sequences associated to $\xi^\infty(a)$ and $\xi^\infty(b)$. The substitution rule implies a recurrence relation between their traces x_n and y_n:

$$\begin{cases} x_{n+1} = x_n y_n - 2 \\ y_{n+1} = x_n^2 - 2 \end{cases} \tag{3}$$

with initial conditions $x_0 = E - V, y_0 = E + V$.

The unstable set of (3) is defined as $\mathcal{U} = \{(x_0, y_0) \in \mathbf{R}^2 \ s.t. \ \exists N > 0 \ s.t. \ |x_n| > 2 \ \forall n > N\}$. The identification of the set $\mathcal{E}(\mathcal{U})^c = \{E \ s.t. \ (E-V, E+V) \in \mathcal{U}^c\}$ of stable initial conditions of (3), and also of the set $\mathcal{O}_V$ of zero Lyapunov exponents $\gamma(E) = \lim_{n \to \infty} \frac{1}{n} Ln \|T_E^{(n)}\|$ of H_V, with the spectrum of H_V gives us its properties. We need first the following more convenient description of $\mathcal{U}$:

Lemma: $\mathcal{U} = \cup_{n \geq 0} \left\{ (x_0, y_0) \ s.t. \ (x_n, y_n) \in D_\pm^0 \right\}$, where
$$D_\pm^0 = \{(x, y) \ s.t. \ \pm > 2, y > 2\}$$

3.1   Cantor Spectrum of H_V

THEOREM 3.1 : *The spectrum of H_V is purely singular continuous and supported on a Cantor set of zero Lebesgue measure.*

Our method is similar to those of [4] and [5]. First, by a general result based on Floquet theory [6], $\sigma(H_V) \subset (\text{int } \mathcal{E}(\mathcal{U}))^c$. Then we use the lemma to prove an exponential upper bound for the norm of $T_E^{(n)}$, for $E \in \mathcal{E}(\mathcal{U})^c$, which implies that $\mathcal{E}(\mathcal{U})^c \subset \mathcal{O}_V$. Finally, the general fact that $(\sigma(H_V))^c \subset \mathcal{O}_V^c$ [23] allows to write the following sequence of inclusions, $\mathcal{E}(\mathcal{U})$ being open in our case:

$$\sigma(H_V) \subset \mathcal{E}(\mathcal{U})^c \subset \mathcal{O}_V \subset \sigma(H_V) \tag{4}$$

Therefore $\sigma(H_V) = \mathcal{E}(\mathcal{U})^c = \mathcal{O}_V$. Now $|\mathcal{O}_V| = 0$. This is obtained in two steps. First, let Ω be the hull of the period-doubling sequence, $\gamma_\omega(E)$ the Lyapunov exponent of the hamiltonian $H_V(\omega)$ generated by $\omega \in \Omega$, μ the unique T-invariant ergodic probability measure on Ω and $\gamma_\mu(E) = \int \mu(d\omega) \gamma_\omega(E)$ the mean Lyapunov exponent (see sect. 2). By Kotani [24], the set $\mathcal{O}_\mu = \{E \ s.t. \ \gamma_\mu(E) = 0\}$

has zero Lebesgue measure. Then, to complete the proof of theorem 3.1, we have to show that $|\mathcal{O}_{\mu\Delta}\mathcal{O}_\omega| = 0 \ \forall \omega \in \Omega$. This is achieved by using a lemma of Herman [25] to extend to substitution potentials a proof of Avron and Simon [26] about almost periodic potentials.

Finally, $|\sigma(H_V)| = 0$. Since we can prove that H_V has no eigenvalues and no generalized eigenfunctions tending to zero at infinity, this implies theorem 3.1.

Remark 1: $|\mathcal{O}_V| = 0$ is a general result for primitive substitutions, used in sect. 4 to extend theorem 3.1 to a large class of substitutions.

3.2   Labelling and Opening of the Gaps

Let $\tau_\pm$ be the two inverses of the trace map (3) and $\tau_\omega = \tau_{\omega_n}...\tau_{\omega_0}$ if $\omega = (\omega_0,...,\omega_n)$ and $\omega_i = \pm 1, i = 0,...,n$. Since $\sigma(H_V) = \mathcal{E}(\mathcal{U})^c$, the lemma implies that

$$[\sigma(H_V)]^c = \left\{E \ s.t. \ \exists \ \omega \ s.t. \ (E - V, E + V) \in \tau_\omega(D_\pm^\infty)\right\}, \quad (5)$$

where $D_\pm^\infty = \tau_\pm^\infty(D_\pm^0)$

This gives the two families of spectral gaps constructed from $D_\pm^\infty$:

THEOREM 3.2 :

   i) The gaps at the points $\tau_\omega(0,0)$ open linearly, with opening angle of order $2^{-|\omega|}$, and are labelled by $\mathcal{N}(E) = \frac{k}{2^n}$;

   ii) The gaps at the points $\tau_\omega(-1,-1)$ open exponentially, with width of order $e^{\frac{-3Ln2}{2V}}V^{Ln2}$, and are labelled by $\mathcal{N}(E) = \frac{k}{3.2^n}$.

Remark 2: These values of $\mathcal{N}(E)$ come for the formula for the free laplacian: $\mathcal{N}(E) = \frac{1}{\pi}\arccos(-E/2)$

Remark 3: Similar results were obtained for the Thue-Morse sequence defined by $\xi(a) = ab, \xi(b) = ba$, with the difference that the gaps labelled by purely dyadic $\mathcal{N}(E)$ (except 1/2) remain closed, due to the symmetry of the potential [6].

4   Singularity of the Spectrum [22]

We have seen in section 2 that a general gap labelling theorem can be proven for substitution hamiltonians H_V. Here, we show how, under a simple supplementary hypothesis, which can be verified algorithmically, the second general result, that is the singularity of the spectrum of H_V, has been very recently generalized by two of us [22]. This is achieved by extending the analysis of the stable set of the trace map performed for the period-doubling sequence.

We start with a primitive substitution ξ defined on a finite alphabet A. For $\omega \in A^{\mathbf{N}}$, let $x^{(n)}(\omega)$ be the trace of the transfer matrix associated to ω. By construction, there is a finite alphabet B, including A, such that the trace map of ξ, that is the map $(f_{\beta_i})_{i \leq |B|}$ defined by $x^{(n+1)}(\beta_i) = f_{\beta_i}\left(x^{(n)}(\beta_i), ..., x^{(n)}(\beta_{|B|})\right)$, is a dynamical system on $\mathbf{R}^{|B|}$ [27]. It is clear that the essential role in the vanishing of the Lyapunov exponent is played by the dominant terms in the f_{β_i}. Therefore its crucial property is the existence for each i of a unique monomial of highest degree $\tilde{f}_{\beta_i}$, called the *reduced trace map*, and of the associated substitution Φ on B. Actually, defining a *semi-primitive* substitution as a substitution satisfying:

i) $\exists C \subset B$ s.t. $\Phi|_C$ is a primitive substitution from C to $C^\star$;

ii) $\exists k$ s.t. $\forall \beta \in B$, $\Phi^k(\beta)$ contains at least one letter from C,

we can prove:

THEOREM 4.1 : *Let H_V be a 1D discrete Schrödinger operator generated by a primitive substitution ξ on a finite alphabet. Assume that there is a trace map such that the substitution Φ associated to its reduced trace map is semi-primitive and also that there is a finite k s.t. $\xi^k(0)$ contains the word $\beta\beta$ for some $\beta \in B$. Then the spectrum of H_V is singular and supported on a set of zero Lebesgue measure.*

The proof of theorem 4.1 can be summarized as follows: Let $\tilde{\mathcal{U}} \subset \mathcal{U}$ be the open "generalized" unstable set of ξ (see [22] for a precise definition). Generalizing the proof of theorem 3.1, we use the crucial fact that, for primitive ξ, the lengths of the words $|\xi^n \alpha|$ ($\alpha \in A$) grow with n exponentially fast with the same rate θ^n, where θ is the Perron-Frobenius eigenvalue of the substitution matrix [2], [17], to show that, for semi-primitive Φ, $\mathcal{E}(\tilde{\mathcal{U}})^c \subset \mathcal{O}_V$.

As in sect. 2, this implies the following sequence of inclusions:

$$\mathcal{E}(\tilde{\mathcal{U}})^c \subset \mathcal{O}_V \subset \sigma(H_V) \subset (Int(\mathcal{E}(U))^c \subset \mathcal{E}(\tilde{\mathcal{U}})^c \tag{6}$$

and thus $\sigma(H_V) = \mathcal{O}_V$, which concludes the proof of theorem 4.1 (see Remark 1 after the proof of theorem 3.1).

Remark 4: If we assume that $\xi^k(0)$ begins with the word $\beta\beta$, we can prove that H_V has no eigenvalues and therefore that the spectrum of H_V is singular continuous and supported on a Cantor set of zero Lebesgue measure.

Bibliography

[1] D. Shechtman, I. Blech, D. Gratias and J.V. Cahn, Phys. Rev. Lett. **53**, 1984, p.1951-1953.

[2] M. Queffélec, *Substitution dynamical systems. Spectral analysis*, Lecture Notes in Mathematics, vol. 1294, Berlin, Heidelberg, New York, Springer, 1987.

[3] M. Casdagli, Commun. Math. Phys. **107**, 1986, p. 295-318.

[4] A. Sütö, J. Stat. Phys. **56**, 1989, p. 525-531.

[5] J. Bellissard, B. Iochum, E. Scoppola and D. Testard, Commun. Math. Phys. **125**, 1989, p. 527-543.

[6] J. Bellissard, in *Number theory and physics*, J.-M. Luck, P. Moussa and M. Waldschmidt, Eds., Springer proceedings in physics, vol. 47, Berlin, Heidelberg, New York, Springer, 1990, p. 140-150.

[7] J. Bellissard, A. Bovier and J.-M. Ghez, Commun. Math. Phys. **135**, 1991, p. 379-399.

[8] L. Brillouin, J. Phys. Radium **7**, 1926, p. 353-368.

[9] H. Kunz and B. Souillard, Commun. Math. Phys. **78**, 1980, p.201-246.

[10] J. Bellissard, R. Lima and D. Testard, Commun. Math. Phys. **88**, 1983, p. 207-234.

[11] M. Kohmoto, L.P. Kadanoff and C. Tang, Phys. Rev. Lett. **50**, 1983, p. 1870-1872.

[12] S. Ostlund, R. Pandit, D. Rand, H.J. Schnellnhuber and E.D. Siggia, Phys. Rev. Lett. **50**, 1983, p. 1873-1876.

[13] J. Bellissard and B. Simon, J. Funct. Anal. **48**, 1982, p. 408-419.

[14] C. Sire and R. Mosseri, J. de Physique **50**, 1989, p. 3447-3461.

[15] C. Sire and R. Mosseri, J. de Physique **51**, 1990, p. 1569-1583.

[16] J.-P. Allouche and J. Peyrière, C. R. Acad. Sci. Paris **302**, No 18, serie II, 1986, p. 1135-1136.

[17] J. Bellissard, A. Bovier and J.-M. Ghez, Rev. Math. Phys. **4**, 1992, p. 1-37.

[18] J. Bellissard, R. Lima and D. Testard, in *Mathematics+Physics. Lectures on recent results*, vol. 1, L. Streit, Ed., Singapore, Philadelphia, World Scientific 1985, p. 1-64.

[19] J. Bellissard, in *Statistical mechanics and field theory*, T.C. Dorlas, M.N. Hugenholtz and M. Winnink, Eds., Lecture Notes in Physics, vol. 257, Berlin, Heidelberg, New York, Springer 1986, p. 99-156.

[20] J. Bellissard, in *From number theory to physics*, M. Waldschmidt, P. Moussa, J.-M. Luck and C. Itzykson, Eds., Berlin, Heidelberg, New York, Springer, 1992, p. 538-630.

[21] J.-M. Luck, Phys. Rev. **B39**, 1989, p. 5834-5849.

[22] A. Bovier and J.-M. Ghez, *Spectral properties of one dimensional Schrödinger operators with potentials generated by substitutions*, Preprint CPT-92/2705.

[23] F. Martinelli and E. Scoppola, Rivista del Nuovo Cimento **10**, 1987.

[24] S. Kotani, Rev. Math. Phys. **1**, 1990, p. 129-133.

[25] M. Herman, Comment. Math. Helvetici **58**, 1983, p. 453-502.

[26] J. Avron and B. Simon, Duke Math. J. **50**, 1983, p. 369-391.

[27] M. Kolár and F. Nori, Phys. Rev. **B42**, 1990, p. 1062-1065.

Wave Packets Localized on Closed Classical Trajectories

S. De Bièvre, J.C. Houard and M. Irac-Astaud
L.P.T.M. Université Paris 7

Abstract

In the classical limit eigenfunctions of Hamiltonians tend to localize in phase space on energy surfaces if the system is ergodic, or on invariant tori for completely integrable systems. In cases when the energy levels are highly degenerate, one may hope to construct eigenstates that localize on lower dimensional flow invariant manifolds such as closed orbits. This is known to be true for the Kepler problem. We establish the same result for n-dimensional harmonic oscillators. The construction generalizes to yield states well-localized on closed orbits of more general Hamiltonians.

1 Introduction

Let H_0 be a C^∞ Hamiltonian on phase space $\mathbb{R}^{2n} = T^*\mathbb{R}^n$. Let $\gamma : t \in [0,T] \to \gamma(t) \equiv (q(t),p(t)) \in \mathbb{R}^{2n}$ be a periodic solution $(\gamma(0) = \gamma(T))$ of the corresponding Hamiltonian equations of motion. We shall write $E_0 = H_0(q(t),p(t))$. We then consider

$$< >_\gamma: f_0 \in C^\infty(\mathbb{R}^{2n}) \to < f_0 >_\gamma \equiv \frac{1}{T} \int_0^T dt f_0(q(t),p(t)). \quad (1.1)$$

This defines a classical state, i.e. a probability measure on phase space, which is concentrated on γ and flow invariant in the sense

Differential Equations with
Applications to Mathematical
Physics

25

that

$$< f_0 \circ \phi_t >_\gamma = < f_0 >_\gamma, \quad \forall t \in \mathbb{R}, \tag{1.2}$$

where we wrote ϕ_t for the flow defined by H_0.

Consider then self-adjoint operators $H(\hbar), F(\hbar)$ on $L^2(\mathbb{R}^n)$, which have H_0, respectively f_0, as their principal Weyl symbols. The eigenstates $\psi_\hbar$ of $H(\hbar)$ satisfy the quantum equivalent of (1.2), i.e.

$$< \psi_\hbar, e^{\frac{itH(\hbar)}{\hbar}} F(\hbar) e^{-\frac{itH(\hbar)}{\hbar}} \psi_\hbar > = < \psi_\hbar, F(\hbar)\psi_\hbar > . \tag{1.3}$$

It is then natural to ask whether it is possible to construct a family of eigenstates

$$H(\hbar)\psi_\hbar = E(\hbar)\psi_\hbar, \tag{1.4a}$$

with

$$E(\hbar) \to E_0 \quad \text{as} \quad \hbar \to 0 \tag{1.4b}$$

and such that

$$< \psi_\hbar, F(\hbar)\psi_\hbar > \to \frac{1}{T} \int_0^T dt f_0(q(t), p(t)), \tag{1.5}$$

for all $F(\hbar)$ as above.

In general, this is impossible. Indeed, as a first example, think of the double symmetric potential well. In that case, all eigenstates satisfy $| \psi_\hbar(x) |^2 = | \psi_\hbar(-x) |^2$. Hence they can never concentrate on a classical trajectory in one of the two wells in the limit $\hbar \to 0$. More generally, consider the case when H_0 is completely integrable. The classical limit of energy eigenstates for such systems has been studied extensively in the literature [8] [1]. Let $T^*\mathbb{R}^n = \mathbb{R}^n \times \mathbb{R}^{n*}$ be the classical phase space and $\vec{P}_0 : T^*\mathbb{R}^n \to \mathbb{R}^n$ n commuting constants of the motion for the Hamiltonian H_0, i.e.

$$\{P_0^i, P_0^j\} = 0 \tag{1.6a}$$

and

$$H_0 = P_0^1. \tag{1.6b}$$

In the corresponding quantum system, one has self-adjoint operators $P_i(\hbar)$ having P_i^0 as their principal Weyl symbol. They form a complete set of commuting observables on the Hilbert space $L^2(\mathbb{R}^n)$. As a result, fixing their eigenvalues $\lambda_i(\hbar)$ determines a unique eigenstate of the quantum Hamiltonian $H(\hbar)$ and one expects that, as $\hbar \to 0$, this eigenstate concentrates - in phase space - uniformly on the corresponding classical torus $\vec{P}^{-1}(\vec{\lambda}_0)$. This is indeed established in [1], under suitable conditions on H_0. The results in [1] lead one to conclude that non-degenerate eigenstates of $H(\hbar)$, which are automatically eigenstates of all the $P_i(\hbar)$, cannot in general be expected to satisfy (1.5). In fact, one expects that (1.4)-(1.5) can only be satisfied if $H(\hbar)$ admits highly degenerate eigenspaces so that one can construct many eigenstates of $H(\hbar)$ that are not simultaneously eigenstates of the other $P_i(\hbar)$.

There are two known examples where (1.4)-(1.5) can be satisfied for all the classical closed trajectories. They are the hydrogen atom [3] and the isotropic harmonic oscillator [2]. In both cases the method of construction is based on group-theoretical arguments using the hidden symmetries of the problem.

In section 2, we construct eigenstates of the anisotropic harmonic oscillator satisfying (1.5). Symmetry arguments cannot be used in this case, but instead we propose a very natural construction using coherent states.

Since the requirement that $\psi_\hbar$ is an eigenstate is in general incompatible with (1.5), it is customary to replace it by the weaker condition

$$\| (H(\hbar) - E(\hbar))\psi(\hbar) \| = O(\hbar^N) \qquad (1.7)$$

for some $N \in \mathbb{N}$. One then says that $\psi_\hbar$ is a quasimode. Quasimodes localized on closed classical trajectories were constructed by Ralston [6] for a class of partial differential operators under certain natural stability conditions on γ which determine N and supposing $\dot{q}(t) \neq 0, \forall t \in [0, T]$.

In section 3 we show how our construction of section 2 can be generalized very simply to construct states satisfying (1.5) and hence (1.7) with $N = 1$, without any stability conditions on γ. In the absence of stability requirements, one can probably not hope to do better than this. While this work was in progress, we learned of recent results of Paul and Uribe [5], who use the same construction to prove (1.7) for all N in the case where $n = 1$ and $H(\hbar)$ is an ordinary differential operator with polynomial coefficients.

2   The Anisotropic Oscillator

Let

$$H = \sum_{i=1}^{n} \frac{P_i^2}{2m_i} + \frac{1}{2} m_i \omega_i^2 Q_i^2 \qquad (2.1)$$

be the usual harmonic oscillator Hamiltonian on $L^2(\mathbb{R}^n)$. Its spectrum is given by

$$E_m = \hbar(\omega_1 m_1 + \omega_2 m_2 + ... + \omega_n m_n + \frac{1}{2}(\omega_1 + ... + \omega_n)). \qquad (2.2)$$

The corresponding classical system, with Hamiltonian function

$$H_0(q,p) = \sum_{i=1}^{n} \frac{1}{2m_i} p_i^2 + \frac{1}{2} m_i \omega_i^2 q_i^2 \qquad (2.3)$$

always admits closed classical trajectories. If all ω_i are two by two incommensurate, the only such trajectories are the ones in which only one mode of the oscillator is excited. If, on the other hand, $\omega_{i_1}, \omega_{i_2}, ..., \omega_{i_k} (k \leq n)$ are two by two commensurate, the others being incommensurate, then all trajectories in which only the degrees of freedom $i_1, ..., i_k$ are excited, will be periodic. They then have a common period, which is the least common multiple of the $T_{i_j} = \frac{2\pi}{\omega_{i_j}}$.

Let us now fix a closed trajectory

$$\gamma : t \in [0, T] \to (q(t), p(t)) \in \mathbb{R}^{2n} \qquad (2.4)$$

of the Hamiltonian in (2.3). We shall write

$$E_0 = H_0(q(t), p(t)) \qquad (2.5)$$

for the corresponding energy. In the rest of this section, we construct an $\hbar$-dependent sequence of eigenfunctions of H, all with energy E_0, concentrating on γ as $\hbar \to 0$ in the sense explained in section 1.

First, we briefly recall the definition of coherent states. We define

$$< x \mid q,p > = [\frac{detK}{(\pi\hbar)^n}] \exp(i\frac{p \cdot x}{\hbar} - \frac{1}{\hbar} < (x - q), K(x - q) > -\frac{i}{2}\frac{p \cdot q}{\hbar})$$

$$(2.6a)$$

where K is the matrix

$$K_{ij} = m_i\omega_i\delta_{ij}. \qquad (2.6b)$$

It is then well known that

$$e^{-i\frac{sH}{\hbar}} \mid q(t),p(t) > = e^{-i\frac{|\omega|s}{2}} \mid q(t + s),p(t + s) > \qquad (2.7a)$$

where we introduced the notation

$$\mid \omega \mid = \sum_{i=1}^{n}\omega_i. \qquad (2.7b)$$

The coherent state $\mid q,p >$ being optimally localized around the phase space point (q,p), it is natural to construct a state

$$\mid \gamma > = \frac{1}{T}\int_0^T dt\ a(t)\ e^{i\frac{\phi(t)}{\hbar}} \mid q(t),p(t) > \qquad (2.8)$$

which is a superposition of coherent states localized on points of the trajectory γ. Taking

$$\phi(t) = E_0t \qquad (2.9a)$$

$$a(t) = \exp -i\frac{\mid \omega \mid t}{2} \qquad (2.9b)$$

it is easily verified that, $\forall s \in \mathbb{R}$,

$$e^{-i\frac{sH}{\hbar}} \mid \gamma > = e^{-i\frac{sE_0}{\hbar}} \mid \gamma > \qquad (2.10a)$$

provided $\exists n \in \mathbb{Z}$ so that

$$[E_0 - \hbar\frac{\mid \omega \mid}{2}] = \frac{2\pi n\hbar}{T}. \qquad (2.10b)$$

Furthermore one verifies readily that $\mid \gamma >$ in (2.11) is identically zero, unless $\exists m_{i_1},...,m_{i_k} \in \mathbb{N}$ so that

$$\frac{2\pi}{T}n = m_{i_1}\omega_{i_1} + ... + m_{i_k}\omega_{i_k}, \qquad (2.11)$$

where we recall that $i_1, ..., i_k$ label the degrees of freedom of the oscillator that are excited on the trajectory γ. We conclude that $| \gamma >$ in (2.8)-(2.9) is an eigenstate of H with eigenvalue E_0 provided $\hbar$ is chosen so that (2.10b) and (2.11) are satisfied.

In the next section, we prove a general result which implies that $| \gamma >$ in (2.8), after normalization, satisfies (1.5).

3   Localized Wave Packets

Let γ be a closed C^∞ curve in $T^*\mathbb{R}^n$. We construct

$$| \gamma >= \frac{1}{T} \int_0^T dt \, a(t) \, e^{\frac{i}{2\hbar} \int_0^t (pdq - qdp)} \, | q(t), p(t) > \qquad (3.1)$$

where $a(t)$ is a C^∞ function on $[0, T]$. Now let $f_\hbar$ be a strongly $\hbar$-admissible symbol on $T^*\mathbb{R}^n$ and $F(\hbar)$ the corresponding Weyl-quantized operator, i.e.

$$F(\hbar) = Op_\hbar^W f_\hbar. \qquad (3.2)$$

For precise definitions of "strongly admissible" and $Op_\hbar^W$ we refer to [7]. Let us just say that $f_\hbar(q, p)$ depends smoothly on $\hbar$ and on (q, p), is polynomially bounded in (q, p) for each $\hbar$ and has an asymptotic expansion

$$f_\hbar \sim \sum_{n=0}^\infty f_n \, \hbar^n \qquad (3.3)$$

where each f_n is again C^∞ and polynomially bounded. Formally, for ψ in the Schwartz space

$$Op_\hbar^W(f_\hbar)\psi(x) = \int \int e^{\frac{i}{\hbar}<(x-y), p>} f_\hbar(\frac{1}{2}(x + y), p)\psi(y)\frac{dydp}{(2\pi\hbar)^n}. \qquad (3.4)$$

We then have the following result.

Theorem 3.1 *For $\hbar$ such that*

$$a(t)e^{\frac{i}{2\hbar} \int_0^t (pdq - qdp)} \qquad (3.5)$$

is periodic with period T and provided

$$| a(t) |^2 = (< \dot{q}(t), K\dot{q}(t) > + < \dot{p}(t), K^{-1}\dot{p}(t) >)^{\frac{1}{2}}, \qquad (3.6)$$

we have that

$$\frac{< \gamma \mid F(\hbar) \mid \gamma >}{< \gamma \mid \gamma >} = \frac{1}{T} \int_0^T dt\, f_0(q(t), p(t)) + O(\hbar). \qquad (3.7)$$

Sketch of the proof: We first remark that ([7], Proposition II.56)

$$\begin{aligned}
< \gamma \mid F(\hbar) \mid \gamma > &= Tr F(\hbar) \mid \gamma >< \gamma \mid \\
&= \int \frac{dq\, dp}{(2\pi\hbar)^n} f_\hbar(q, p) W_\gamma(q, p; \hbar) \qquad (3.8)
\end{aligned}$$

where W_γ is the Weyl symbol of $\mid \gamma >< \gamma \mid$. The latter can be written

$$W_\gamma(q, p; \hbar) = \frac{1}{T^2} \int_0^t dt \int_0^t dt'\, a(t)\overline{a(t')} e^{\frac{i}{\hbar}(\phi(t) - \phi(t'))} W_{t,t'}(q, p; \hbar), \qquad (3.9a)$$

where

$$\phi(t) = \frac{1}{2} \int_0^t (p\, dq - q\, dp). \qquad (3.9b)$$

and $W_{t,t'}(q, p; \hbar)$ is the Weyl symbol of $\mid q(t), p(t) >< q(t'), p(t') \mid$. Computing the latter explicitly and inserting the result into (3.8), one gets

$$< \gamma \mid F(\hbar) \mid \gamma > = \int_0^T dt \int_0^T dt' \int dq\, dp\, k(q, p; \hbar) e^{\frac{i}{\hbar}\psi_t(t', q, p)} \qquad (3.10a)$$

with

$$k(q, p, \hbar) = \frac{2^n}{T^2 (2\pi\hbar)^n} a(t)\overline{a}(t') f_\hbar(q, p). \qquad (3.10b)$$

The phase ψ_t is a smooth function of (t', q, p) having a unique critical point at $t' = t, q = q(t), p = p(t)$. Applying a stationary phase argument ([4], Theorem 7.7.5), the result then follows.

   The claim made at the end of section 2 is now an easy consequence of the above result.

Bibliography

[1] A. M. Charbonnel, *Contribution à l'étude du spectre conjoint de systèmes d'opérateurs pseudo-différentiels qui commutent,* Thèse de doctorat d'Etat, Université de Nantes, 1989 and *Comportement semi-classique des systémes ergodiques,* Ann. Inst. H. Poinc. A, **56**, 2 (1992) 187-214.

[2] S. De Bièvre, *Oscillator eigenstates concentrated on classical trajectories,* J. Phys. A: Math. Gen. **25** (1992) 3399-3418.

[3] J. Gay, D. Delande, and A. Bommier, *Atomic quantum states with maximum localization on classical elliptical orbits,* Phys. Rev. A, Brief Reports, **39**, 12 (1989), 6587-6590.

[4] L. Hörmander, *The analysis of linear partial differential operators,* Vol. 1, Springer-Verlag, Berlin 1983.

[5] T. Paul and A. Uribe, *A construction of quasimodes using coherent states,* preprint April 1992.

[6] J. V. Ralston, *On the construction of quasimodes associated with stable periodic orbits,* Commun. Math. Phys. **51** (1976) 219-24.

[7] D. Robert, *Autour de l'approximation semi-classique,* Progress in Mathematics, Vol. 68, Birkhaüser, Boston 1987.

[8] A. Voros, *Développements semi-classiques,* Thèse de Doctorat d'Etat, Université de Paris Sud, 1977, and *Semi-classical approximations,* Ann. Inst. H. Poinc. A, **24**, 1 (1976) 31-90.

Lower Bounds on Eigenfunctions and the First Eigenvalue Gap

R. M. Brown* and P. D. Hislop[†]
Department of Mathematics
University of Kentucky
Lexington, KY 40506-0027 USA

A. Martinez
Département de Mathématiques
Université de Paris-Nord
Av. J.-B. Clemente
93430 Villetaneuse France

Abstract

We give upper and lower bounds on the difference $\Delta E \equiv E_2 - E_1$, of the first two Dirichlet eigenvalues for a dumbbell region in $\mathbf{R}^n$. These bounds are exponentially small in the diameter ε of the straight tube connecting two identical bounded cavities as $\varepsilon \to 0$. The proof relies on a lower bound for the first Dirichlet eigenfunction for one cavity with a thin tube attached.

*Supported in part by NSF DMS 91-03046.
†Supported in part by NSF DMS 91-06479 and INT 90-15895.

Differential Equations with
Applications to Mathematical
Physics

1   Introduction

The purpose of this note is to discuss some recent results on lower bounds for eigenvalue differences for Dirichlet Laplacians on domains. We present an alternative proof of one of the main results of [2]. The problem we consider here is the following. Let $\mathcal{C} \subset \mathbf{R}^n$ be a bounded domain and let $T(\varepsilon)$ be a tube of diameter $\varepsilon > 0$ described as follows. Let $D_1 \subset \mathbf{R}^{n-1}$ be a bounded, connected region containing the origin. We assume ∂D_1 is smooth, see [2] for more general situations. For $\varepsilon > 0$, let $D_\varepsilon \equiv \varepsilon D_1$ be the scaled cross-section of the tube $T(\varepsilon) \equiv D_\varepsilon \times (-\delta, \ell + \delta)$, for some $\delta > 0$ small and independent of ε. We choose coordinates $(x', x_n) \in \mathbf{R}^{n-1} \times \mathbf{R} = \mathbf{R}^n$ such that $(0, 0) \in \partial\mathcal{C}$. We take R to be the reflection of the half-space $x_n < \ell/2$ in the $x_n = \ell/2$ plane, to obtain a symmetric dumbbell region with $\mathcal{C}_1 \equiv \mathcal{C}$ and $\mathcal{C}_2 \equiv R\mathcal{C}_1$, defined by $\Omega(\varepsilon) \equiv \mathcal{C}_1 \cup T(\varepsilon) \cup \mathcal{C}_2$. That is, $\Omega(\varepsilon)$ consists of two symmetric cavities (with respect to $x_n = \ell/2$) joined by a straight tube of diameter ε. Note that $(0, \ell) \in \partial\mathcal{C}_2$.

   Let $P(\varepsilon) \equiv -\Delta_{\Omega(\varepsilon)}$ be the Dirichlet Laplacian on $\Omega(\varepsilon)$. Let $0 < E_1(\varepsilon) < E_2(\varepsilon) \le \ldots$ be the Dirichlet eigenvalues and define $\Delta E(\varepsilon) \equiv E_2(\varepsilon) - E_1(\varepsilon)$. We refer to this difference as the splitting of the first two Dirichlet eigenvalues. Our goal is to bound $\Delta E(\varepsilon)$ from above and from below in terms of the tube diameter ε and the tube length ℓ. Note that when $\varepsilon = 0$, the two cavities are identical and disjoint. We also have that $-\Delta_{\Omega(\varepsilon)} \to -\Delta_{\mathcal{C}_1} \oplus -\Delta_{\mathcal{C}_2}$ in an appropriate sense as $\varepsilon \to 0$. For the limit operator $\Delta E = 0$, i.e. the first eigenvalue is doubly degenerate. Let α^2 be the first Dirichlet eigenvalue of D_1. By scaling, $\left(\frac{\alpha}{\varepsilon}\right)^2$ is the first Dirichlet eigenvalue of D_ε. For the case of a straight tube, as described above, our main result is the following.

THEOREM 1.1 *Let $\Omega(\varepsilon) \subset \mathbf{R}^n$ be a symmetric dumbbell region with a straight tube of length ℓ. Let $\Delta E(\varepsilon) \equiv E_2(\varepsilon) - E_1(\varepsilon)$ be the difference of the first two Dirichlet eigenvalues. For any $\tilde{\ell} < \ell$ there exists $\varepsilon_0 > 0$ and constants $C_1, C_2 > 0$ such that for $\varepsilon < \varepsilon_0$*

$$C_1 \varepsilon^{n+6} e^{-\alpha\ell/\varepsilon} \le \Delta E(\varepsilon) \le C_2 e^{-\alpha\tilde{\ell}/\varepsilon} \tag{1}$$

We will sketch the proof of Theorem 1.1 in the following two sections. The upper bound in (1.1) relies on L^2-exponential decay estimates on the Dirichlet eigenfunctions for the region $\mathcal{C}(\varepsilon) \equiv Int\overline{\mathcal{C}} \cup T(\varepsilon)$, i.e. one cavity with a tube attached. These estimates were obtained using Agmon-type [1] positivity arguments in [6]. We note here that for the nth Dirichlet eigenfunction $u_{n,\varepsilon}$ those estimates can be improved to give

$$\|u_{n,\varepsilon}\|_{L^2(T(\varepsilon))} \le C_0 \varepsilon^{\frac{1}{2}} e^{-\alpha(\ell-\delta)/\varepsilon} \tag{2}$$

for any $\delta > 0$ in the case of a straight tube with cross-section D_ε. For this, it suffices simply to replace the weight ρ in [6] with $\rho(x_n) = x_n\sqrt{\left(\frac{\alpha}{\varepsilon}\right)^2 - E_n(\varepsilon)}$, $x_n \in]0, \ell[$.

Here we concentrate on the lower bound in (1.1). It depends upon a lower bound on the first Dirichlet eigenfunction (which is nonnegative) in $\mathcal{C}(\varepsilon)$. In [2], we obtain a lower bound using a Harnack inequality and a comparison principle for parabolic equations. Here, we give a different proof which results in an L^2-lower bound for $u_\varepsilon \equiv u_{1,\varepsilon}$ in the tube.

THEOREM 1.2 *Let u_ε be the first normalized Dirichlet eigenfunction on $\mathcal{C}(\varepsilon) \subset \mathbf{R}^n$. For all $x_n^0 \in]0, \ell[\; \exists C_0 = C_0(x_n^0) > 0$ such that for any $\delta > 0$ and for all $\varepsilon > 0$ small enough*

$$\|u_\varepsilon\|_{L^2(D_\varepsilon \times [x_n^0, \ell])} \ge C_0 \varepsilon^{1+\frac{n+5}{2}+\delta} e^{-\alpha x_n^0/\varepsilon}, \tag{3}$$

where α^2 is the first Dirichlet eigenvalue for $D_1 \subset \mathbf{R}^{n-1}$.

We mention that more general results are given in [2]. Although lower bounds on the splitting are well-known for the Schrödinger operator $-h^2\Delta + V$ on $\mathbf{R}^n$ in the semi-classical regime (see, for example, [9] and references therein) not that much is known for the Dirichlet Laplacian on bounded domains. One such result is due to Singer, Wong, Yau and Yau [12]. If Ω is a bounded convex domain with diameter d and $D \equiv \max\{\delta \mid B(\delta, x) \subset \Omega\}$, then they prove

$$\frac{1}{4}\pi d^{-2} \le \Delta E \le 4\pi^2 n D^{-2}$$

(which is a special case of a more general result).

In section 2 of this note, we derive Theorem 1.1 from Theorem 1.2. This derivation is rather well-known (see [5], [10],[11]) so we simply sketch the proof. In section 3, we prove Theorem 1.2.

2   Bounds on ΔE : Proof of Theorem 1.1

We sketch the derivation of Theorem 1.1 given Theorem 1.2. We use the method of Helffer and Sjöstrand [5] which reduces the estimation of ΔE for $\Omega(\varepsilon)$ to that of estimating the first Dirichlet eigenfunction for $\mathcal{C}(\varepsilon)$ in the tube (see also [10], [11]). Let $M_i \equiv Int\overline{\mathcal{C}_i \cup T(\varepsilon)}$, $i = 1, 2$, be the left and right cavities with the tube attached, respectively. We consider the Dirichlet Laplacian $P_i \equiv -\Delta_i$ on $L^2(M_i)$, $i = 1, 2$. Let $\phi_1^{(i)}$ be the first Dirichlet eigenfunction for P_i with eigenvalue $E_0(\varepsilon)$. Let $\chi_i \in C^\infty(\mathbf{R}^n)$ denote cut-off functions such that $\nabla\chi_1$ is supported in $\{(x', x_n) \mid \ell - \eta \le x_n \le \ell\}$ and $\nabla\chi_2$ is supported in $\{(x', x_n) \mid 0 \le x_n \le \eta\}$ for some $\eta > 0$ small, and such that χ_i is identically one on the rest of M_i. Then $\psi_i \equiv \chi_i \phi_1^{(i)} \in D(P(\varepsilon))$ and

$$P(\varepsilon)\psi_i = E_0(\varepsilon)\psi_i - (2\nabla\chi_i \cdot \nabla\psi_1^{(i)} + (\Delta\chi_i)\phi_1^{(i)}), \qquad (4)$$

for $i = 1, 2$. Let E be the subspace of $L^2(\Omega(\varepsilon))$ spanned by $\{\psi_1, \psi_2\}$. Let u_i be the first two eigenfunctions of $P(\varepsilon)$ on $L^2(\Omega(\varepsilon))$ and let F be the subspace spanned by these eigenfunctions. Since the error terms in (2.4) are localized far from the cavities where $\phi_1^{(i)}$ are small, E should be a good approximation to F. To quantify this statement, we need the following result of [6] (modified as described in section 1).

PROPOSITION 2.1 *For $\beta = 0, 1$, for all $\mathcal{K} > 0$, there exist constants $C_{\beta,\mathcal{K}}$, $\tilde{C}_{\beta,\mathcal{K}} > 0$ and an $\varepsilon_0 > 0$ such that for $\varepsilon < \varepsilon_0$ and $i = 1, 2$,*

$$\|e^{(1-\mathcal{K})\alpha x_n^{(i)}/\varepsilon} \partial^\beta \phi_1^{(i)}\|_{L^2(T(\varepsilon))} \le \tilde{C}_{\beta,\mathcal{K}} \varepsilon^{-C_{\beta,\mathcal{K}}} \qquad (5)$$

where $x_n^{(i)} = x_n$ and $x_n^{(2)} = x_n - \ell$ and $C_{0,\mathcal{K}} = 0$, $C_{1,\mathcal{K}} = 1$.

We conclude from Proposition 2.1 that

$$\|\chi_i \partial^\beta \phi_1^{(i)}\| = \mathcal{O}\left(e^{-\tilde{\ell}\alpha/\varepsilon}\right) \tag{6}$$

for any $\tilde{\ell} < \ell$. Consequently, following Helffer-Sjöstrand [5] we easily obtain

PROPOSITION 2.2 *Let* $\Pi_0 : L^2(\Omega(\varepsilon)) \to E$ *be the projection onto* E *along* $F^\perp$. *For* $\tilde{\ell} < \ell$, *the matrix* $\Pi_0 P(\varepsilon)|_E$, *in the basis* $\{\psi_1, \psi_2\}$ *for* E, *has the form*

$$\begin{pmatrix} E_0 & 0 \\ 0 & E_0 \end{pmatrix} + (W_{ij}) + \mathcal{O}(e^{-2\tilde{\ell}\alpha/\varepsilon}) \tag{7}$$

where, for $1 \leq i, j \leq 2$, $W_{ii} = 0$ *and*

$$W_{ij} = \int_{\Omega(\varepsilon)} \chi_i \left(\phi_1^{(j)} \nabla \phi_1^{(i)} - \phi_1^{(i)} \nabla \phi_1^{(j)} \right) \cdot \nabla \chi_j. \tag{8}$$

Furthermore, $W_{ij} = \mathcal{O}(e^{-\alpha \tilde{\ell}/\varepsilon})$.

We analyze the interaction matrix (W_{ij}) in the usual manner (see [10], [11], [2]) and omit the details here. We mention only that we use the symmetry of the eigenfunctions and the Poincaré inequality for D_ε.

PROPOSITION 2.3 *For any* $\tilde{\ell} < \ell$ *and all* ε *sufficiently small,*

$$\Delta E(\varepsilon) \geq 4 \left(\left(\frac{\alpha}{\varepsilon}\right)^2 - E_0(\varepsilon) \right) \int_0^{\ell/2} dx_n \int_{D_E} dx' \phi_1^{(2)}(x)^2 + \mathcal{O}(e^{-2\alpha\tilde{\ell}/\varepsilon}). \tag{9}$$

Proof of Theorem 1.1 (given Theorem 1.2)

1) Lower bound. We obtain directly from (3) (using symmetry, $\phi_1^{(1)} = R\phi_1^{(2)}$) that

$$\int_0^{\ell/2} dx_n \int_{D_E} dx' \phi_1^{(2)}(x)^2 \geq C_0 \varepsilon^{2\left(1 + \frac{n+5}{2} + \delta\right)} e^{-\alpha\ell/\varepsilon}, \tag{10}$$

so the result follows from (2.11) and the fact that $|\, E_0(\varepsilon) - E_0\,| < C$, where E_0 is the first Dirichlet eigenvalue for $\mathcal{C}$.

2) Upper bound. From (2.8), it is easy to find an upper bound on ΔE :

$$\Delta E \;\leq\; 4 \int_0^{\ell/2} dx_n \int_{D_E} dx' \left\{ | \, \partial_n \phi_1^{(2)} \, |^2 + E_0(\varepsilon) \, | \, \phi_1^{(2)} \, |^2 \right\}$$
$$+ \mathcal{O}(e^{-2\alpha\tilde{\ell}/\varepsilon}), \tag{11}$$

for any $\tilde{\ell} < \ell$. Consequently, from Proposition 2.1, we obtain

$$\Delta E \leq C_2 e^{-\alpha\tilde{\ell}/\varepsilon}, \tag{12}$$

for any $\tilde{\ell} < \ell$. This proves the theorem.   □

3 Proof of Theorem 1.2: Lower Bounds for Straight Tubes

We sketch the proof of Theorem 1.2 in this section. We refer to [2] for a different proof and more general results. Our goal is to derive an L^2-lower bound for the first Dirichlet eigenfunction restricted to a small tube. Let $\mathcal{C}$ be a bounded, connected, open region in $\mathbf{R}^n$ with a C^2-boundary (this can be relaxed, see [2]). We choose coordinates so $0 \in \partial\mathcal{C}$. Let $D_1 \subset \mathbf{R}^{n-1}$ be an open connected bounded region with smooth boundary and define $D_\varepsilon \equiv \varepsilon D_1$, a scaled cross-section. Let the tube $\tilde{T}(\varepsilon)$ be defined by $D_\varepsilon \times [-\delta, \ell]$, for $\delta > 0$ small so that $\{(x', -\delta) \mid x' \in D_\varepsilon\} \subset \mathcal{C}$. We also define $T(\varepsilon) \equiv \mathcal{C}^c \cap \tilde{T}(\varepsilon)$ and set $\mu = \max\{x_n \mid (x', x_n) \in \partial\mathcal{C} \cap \tilde{T}(\varepsilon)\}$, the first point of contact, along the x_n-axis from $x_n = \ell$, of the tube $\tilde{T}(\varepsilon)$ with $\partial\mathcal{C}$. We also require an obvious transversality condition: $\nu(0) \cdot \hat{x}_n > 0$, where $\nu(p)$, $p \in \partial\mathcal{C}$, is the outward normal. We define $\mathcal{C}(\varepsilon) \equiv Int\overline{\mathcal{C} \cup T(\varepsilon)}$, the cavity with a small tube attached.

We need notation for several operators associated with these regions

1) $-\Delta$ is the Dirichlet Laplacian on $\mathcal{C}(\varepsilon)$ with first eigenfunction
   u_ε : $-\Delta u_\varepsilon = E_1(\varepsilon)u_\varepsilon$;

2) $-\Delta_\mathcal{C}$ is the Dirichlet Laplacian on $\mathcal{C}$ with first eigenfunction
   u_0 : $-\Delta_\mathcal{C} u_0 = E_1 u_0$;

3) $-\Delta_{x'}$ is the Dirichlet Laplacian on D_1 with eigenfunctions b_p and $-\Delta_{x'}b_p = \alpha_p^2 b_p$, $p = 1, 2, \ldots$.

Note that if $-\Delta_{x',\varepsilon}$ denotes the Dirichlet Laplacian on D_ε, then the corresponding eigenvalues are $(\alpha_p/\varepsilon)^2$ and the eigenfunctions are $b_{p,\varepsilon}(x') = \varepsilon^{-\left(\frac{n-1}{2}\right)} b_p(x'/\varepsilon)$.

We prove Theorem 1.2 by contradiction. We suppose $\exists\, x_n^0 \in]0, \ell[$, a constant $C_0 > 0$ and a sequence $\varepsilon_n \to 0$ such that, for each $\varepsilon = \varepsilon_n$,

$$\|u_\varepsilon\|_{L^2(D_\varepsilon \times [x_n^0, \ell])} \le C_0 \varepsilon^{N_1} e^{-\alpha x_n^0/\varepsilon} \tag{13}$$

where $N_1 = 1 + (n+5)/2 + \delta$, for any $\delta > 0$. We propagate this estimate back to a neighborhood of zero in $\partial\mathcal{C}$. There, we compare u_ε with u_0. We conclude that in an ε-neighborhood of zero in $\mathcal{C}$, $B(0, \varepsilon\eta) \cap \mathcal{C}$, η sufficiently small,

$$\|u_0\|_{L^2(B(0,\varepsilon\eta)\cap\mathcal{C})} \le C_0 \varepsilon^{\frac{n}{2}+1+\delta} \tag{14}$$

On the other hand, we have the following special case of a lemma of Hopf (see [4], section 3.2).

LEMMA 3.1 *Suppose L is a uniformly elliptic operator on Ω and $Lu \ge 0$ on Ω with $u(x_0) = 0$ for some $x_0 \in \partial\Omega$. Suppose $\partial\Omega$ is sufficiently smooth (C^2 suffices), u is continuous at x_0, and $u(x) < 0$ on Ω. Then the outer normal derivative of u at x_0 satisfies a strict inequality:*

$$\frac{\partial u}{\partial\nu}(x_0) > 0.$$

We apply this lemma to $L \equiv \Delta_\mathcal{C} + E$ and $u = -u_0$ on $\mathcal{C}$. Since u_0 is the first Dirichlet eigenfunction, u satisfies the hypotheses of the lemma. We conclude from the positivity of the normal derivative at x_0 that $\exists\, C_0 > 0$ such that for η sufficiently small

$$u_0(x) \ge C_0 d(x, \partial\mathcal{C}), \ \forall\, x \in B(0, \varepsilon\eta) \cap \mathcal{C}.$$

Consequently, we conclude that for ε sufficiently small

$$C_0 \varepsilon^{1+\frac{n}{2}} \le \|u_0\|_{L^2(B(0,\varepsilon\eta)\cap\mathcal{C})}. \tag{15}$$

This contradicts (14).

We now prove estimate (14) in 4 steps. In the first, we obtain some a priori estimates on u_ε following from (13). Next, in the second, we expand u_ε in $T(\varepsilon)$ in the eigenfunctions $b_{p,\varepsilon}$ of $-\Delta_{x',\varepsilon}$ and show by ODE techniques that (13) implies that u_ε is small near $x_n = \mu$. We use Harnack and other inequalities in step 3 to extend these estimates for u_ε and ∇u_ε into a neighborhood of zero in $\mathcal{C}$ away from the corners $\partial \mathcal{C} \cap \partial T(\varepsilon)$. Finally, we compare u_0 and u_ε in such a region and derive (14).

Step 1

We begin with some a priori estimates on u_ε in $T(\varepsilon)$. Recall from [6] that $E_1(\varepsilon) \to E_1$ as $\varepsilon \to 0$.

LEMMA 3.2 *For each* $\alpha \in \mathbf{N}^n \; \exists \; N_\alpha \geq 0$ *such that* $|\partial^\alpha u_\varepsilon(x)| = \mathcal{O}(\varepsilon^{-N_\alpha})$ *for* $x \in D_\varepsilon \times]\mu, \ell[$.

PROOF These estimates follow from the Sobolev embedding theorem for $T(\varepsilon)$ and a scaling argument.   □

LEMMA 3.3 *For* $k = 0, 1$, $\|\partial_n^k u_\varepsilon\|_{H^1([x_n^0+\varepsilon,\ell]\times D_\varepsilon)} \leq C_0 \varepsilon^{N_1-k-1} e^{-\alpha x_n^0/\varepsilon}$.

PROOF Let $\chi_\varepsilon \in C^\infty$, $\chi_\varepsilon \geq 0$, be s.t. $\chi_\varepsilon|[x_n^0+\varepsilon, \ell] = 1$ and $supp\, \chi_\varepsilon \subset [x_n^0, \infty)$. Then $\chi_\varepsilon^{(k)}(x) = \mathcal{O}(\varepsilon^{-k}) \forall\, k$. We consider

$$\Delta(\chi_\varepsilon u_\varepsilon) = -E_1(\varepsilon)\chi_\varepsilon u_\varepsilon + 2\nabla\chi_\varepsilon\nabla u_\varepsilon + (\Delta\chi_\varepsilon)u_\varepsilon \tag{16}$$

which implies

$$\begin{aligned}
\|\nabla(\chi_\varepsilon u_\varepsilon)\|^2 &= E_1(\varepsilon)\|\chi_\varepsilon u_\varepsilon\|^2 - 2\langle\chi_\varepsilon\partial_n u_\varepsilon, (\partial_n\chi_\varepsilon)u\rangle \\
&\quad - \langle u_\varepsilon, (\partial_n^2\chi_\varepsilon)u_\varepsilon\rangle \\
&= E_1(\varepsilon)\|\chi_\varepsilon u_\varepsilon\|^2 - 2\langle\partial_n(\chi_\varepsilon u_\varepsilon), (\partial_n\chi_\varepsilon)u_\varepsilon\rangle \\
&\quad + 2\|(\partial_n\chi_\varepsilon)u_\varepsilon\|^2 - \langle u_\varepsilon, (\partial_n^2\chi_\varepsilon)u_\varepsilon\rangle .
\end{aligned}$$

It follows from this and (13) that

$$\|\nabla u_\varepsilon\|_{L^2(D_\varepsilon\times[x_n^0+\varepsilon,\ell])} = \mathcal{O}(\varepsilon^{N_1-1}e^{-\alpha x_n^0/\varepsilon}), \tag{17}$$

from which the result for $k = 0$ is evident. For $k = 1$, it suffices to estimate

$$\|\nabla u_\varepsilon\|_{L^2(D_\varepsilon \times [x_n^0 + \varepsilon, \ell])}^2 \leq C_1 \|\chi_\varepsilon u_\varepsilon\|^2 + C_2 \|(\partial_n \chi_\varepsilon)\partial_n u_\varepsilon\|^2 \\ + C_3 \|(\partial_n^2 \chi_\varepsilon)u_\varepsilon\|^2,$$

which follows from (13) and (17). □

Step 2

In this part of the proof, we use the assumption (13) to obtain estimates on $\partial^\beta u_\varepsilon$ in a small cylindrical region near $x_n = \mu$. We use ODE techniques to estimate the coefficients occurring in the expansion of u_ε in $T(\varepsilon)$ in the eigenfunctions of $-\Delta_{x'}$.

For all x with $x_n > \mu$, we expand $u_\varepsilon(x)$ as

$$u_\varepsilon(x) = \sum_{p \geq 1} < u_\varepsilon(\cdot, x_n),\ b_{p,\varepsilon} >_{D_\varepsilon}\ b_{p,\varepsilon}(x'). \tag{18}$$

The coefficient $B_{p,\varepsilon}(x_n) \equiv< u_\varepsilon(\cdot, x_n),\ b_{p,\varepsilon} >_{D_\varepsilon}$ satisfies the ODE

$$\frac{d^2}{dx_n^2} B_{p,\varepsilon}(x_n) = \left(\left(\frac{\alpha_p}{\varepsilon}\right)^2 - E_1(\varepsilon)\right) B_{p,\varepsilon}(x_n),$$

for $x_n \geq \mu$, so we obtain

$$B_{p,\varepsilon}(x_n) = \alpha_\varepsilon e^{\gamma_{p,\varepsilon}(x_n - \mu)/\varepsilon} + B_\varepsilon e^{-\gamma_{p,\varepsilon}(x_n - \mu)/\varepsilon} \tag{19}$$

where

$$\gamma_{p,\varepsilon} = [\alpha_p^2 - \varepsilon^2 E_1(\varepsilon)]^{\frac{1}{2}}.$$

Evaluating $B_{p,\varepsilon}$ at $x_n = \mu$, we obtain

$$2\alpha_\varepsilon = B_{p,\varepsilon}(\mu) + \varepsilon \gamma_{p,\varepsilon}^{-1} B_{p,\varepsilon}'(\mu) \tag{20}$$

$$2\beta_\varepsilon = B_{p,\varepsilon}(\mu) - \varepsilon \gamma_{p,\varepsilon}^{-1} B_{p,\varepsilon}'(\mu) \tag{21}$$

where

$$B_{p,\varepsilon}'(\mu) \equiv< \partial u_\varepsilon(\cdot, \mu)/\partial x_n,\ b_{p,\varepsilon} > .$$

It follows from Lemma 3.2 that for some $N \geq 0$,

$$|\alpha_\varepsilon| + |\beta_\varepsilon| = \mathcal{O}(\varepsilon^{-N}) \tag{22}$$

The Dirichlet boundary condition $u(x', \ell) = 0$ and (22) imply that

$$\alpha_\varepsilon = \mathcal{O}(e^{-2\alpha_p/\varepsilon}\varepsilon^{-N}). \tag{23}$$

Next, we express $B_{p,\varepsilon}(x_n)$ in terms of $B_{p,\varepsilon}(\mu)$ plus a small remainder. From (19), we have

$$B_{p,\varepsilon}(x_n) = B_{p,\varepsilon}(\mu)e^{-\gamma_{p,\varepsilon}(x_n-\mu)/\varepsilon} + r_\varepsilon(x_n)$$

where

$$r_\varepsilon(x_n) = e^{-\gamma_{p,\varepsilon}(x_n-\mu)/\varepsilon}(B'_{p,\varepsilon}(\mu)\varepsilon\gamma_{p,\varepsilon}^{-1} - \beta_\varepsilon) + \alpha_\varepsilon e^{\gamma_{p,\varepsilon}(x_n-\mu)/\varepsilon}.$$

Evaluating this at $x_n = \mu$ and using (23) we find

$$r_\varepsilon = \mathcal{O}(\varepsilon^{-N}e^{-\alpha_p/\varepsilon}),$$

so that

$$B_{p,\varepsilon}(x_n) = B_{p,\varepsilon}(\mu)e^{-\gamma_{p,\varepsilon}(x_n-\mu)/\varepsilon} + \mathcal{O}(\varepsilon^{-N}e^{-\alpha_p/\varepsilon}). \tag{24}$$

Since $\{b_{p,\varepsilon}\}$ is a complete orthonormal set for $L^2(D_\varepsilon)$, we compute for $x_n \geq \mu$

$$
\begin{aligned}
\|u_\varepsilon(\cdot, x_n)\|^2_{L^2(D_\varepsilon)} &= \sum_{p \geq 1} |B_{p,\varepsilon}(x_n)|^2 \\
&= \sum_{p \geq 1} \Bigg\{ |<u_\varepsilon(\cdot, \mu), b_{p,\varepsilon}>_{D_\varepsilon}|^2 e^{-2\gamma_{p,\varepsilon}\cdot(x_n-\mu)/\varepsilon} \\
&\quad + \mathcal{O}\left(\varepsilon^{-N}\alpha_p^{\frac{n}{2}}e^{-\alpha_p/\varepsilon-\alpha_p x_n/\varepsilon}\right) \Bigg\},
\end{aligned}
\tag{25}
$$

where we used the fact that $\mu = \mathcal{O}(\varepsilon)$ and

$$|B_{p,\varepsilon}(\mu)| \leq K_0 \alpha_p^{\frac{n}{2}}, \tag{26}$$

as follows, for example, from the formula (35) below. We extract the $p = 1$ term from (26). Note that by Weyl's law, $\alpha_p = \mathcal{O}\left(p^{\frac{2}{n-1}}\right)$. We obtain

$$\|u_\varepsilon(\cdot, x_n)\|_{L^2(D_\varepsilon)} = \; < u_\varepsilon(\cdot, \mu), b_{1,\varepsilon} >_{D_\varepsilon} e^{-\gamma_{1,\varepsilon}(x_n - \mu)/\varepsilon} \tag{27}$$
$$+ \mathcal{O}(e^{-\alpha_1 x_n/\varepsilon})$$

for $x_n > \mu + \delta$. The error term depends on δ, but is uniform in ε, even if $\delta = \mathcal{O}(\varepsilon)$.

We combine this result with the hypothesis (13). For $t > 0$, $x' \in D_\varepsilon$,

$$u_\varepsilon(x', x_n^0 + \varepsilon) = \int_t^{x_n^0 + \varepsilon} \frac{\partial u_\varepsilon}{\partial x_n}(x', x_n) dx_n + u_\varepsilon(x', t)$$

and, upon integrating over $t \in [x_n^0 + \varepsilon, \ell]$, we get

$$u_\varepsilon(x', x_n^0 + \varepsilon) \leq C \left[\int_{x_n^0 + \varepsilon}^\ell \left(\left| \frac{\partial u_\varepsilon}{\partial x_n}(x', x_n) \right|^2 + |u_\varepsilon(x', x_n)|^2 \right) dx_n \right].$$

This implies

$$\|u_\varepsilon(\cdot, x_n^0 + \varepsilon)\|_{L^2(D_\varepsilon)} \leq C\|u_\varepsilon\|_{H^1(D_\varepsilon \times [x_n^0 + \varepsilon, \ell])}. \tag{28}$$

From the hypothesis (13) and the expansion (27), we conclude that

$$< u_\varepsilon(\cdot, \mu), b_{1,\varepsilon} >_{D_\varepsilon} = \mathcal{O}(\varepsilon^{N_1 - 1}). \tag{29}$$

In [13], Davies gives a general estimate on boundary behavior of eigenfunctions provided the boundary is smooth. Applied to the present situation, we have the following: $\exists\, C > 0$ s.t.

$$\sup_{x' \in D_\varepsilon} \frac{|b_{p,\varepsilon}(x')|}{b_{1,\varepsilon}(x')} \leq C_0 \alpha_p^{\frac{n}{2}}. \tag{30}$$

Using this result together with (29) gives

$$< u_\varepsilon(\cdot, \mu), b_{p,\varepsilon} >_{D_\varepsilon} = \mathcal{O}\left(\varepsilon^{N_1 - 1} \alpha_p^{\frac{n}{2}}\right)$$

and, combining this with (24) gives

$$< u_\varepsilon(\cdot, \mu + c\varepsilon), b_{p,\varepsilon} >_{D_\varepsilon} = \mathcal{O}\left(\alpha_p^{\frac{n}{2}} e^{-\alpha_p/\varepsilon} \varepsilon^{N_1 - 1}\right). \tag{31}$$

Again, using the fact that $\{b_{p,\varepsilon}\}$ is an orthonormal basis for $L^2(D_\varepsilon)$ and the above estimate on α_p, we obtain

$$\|u_\varepsilon(\cdot, \mu + c\varepsilon)\|_{L^2(D_\varepsilon)} = \mathcal{O}(\varepsilon^{N_1 - 1}) \qquad (32)$$

For $\kappa > 2$, we consider u_ε in a cylinder near $x_n = \mu$ of the form $D_{\theta\varepsilon} \times [\mu + \varepsilon/\kappa, \ \mu + \varepsilon\kappa]$, for $\theta \in]0, \frac{1}{2}[$. From a local boundedness theorem for $W^{1,2}$-solutions (see [4], Theorem 8.17), we have

$$\sup_{D_{\theta\varepsilon} \times [\mu + \varepsilon/(2+\kappa), \mu + \varepsilon\kappa]} u_\varepsilon \le \varepsilon^{-\frac{n}{2}} C_0 \|u_\varepsilon\|_{L^2(D_{2\varepsilon\theta} \times I_\kappa)}, \qquad (33)$$

where $C_0 = C_0(\theta, \mu, \kappa)$ and $I_\kappa \equiv [\mu + \varepsilon/(2 + \kappa), \mu + \varepsilon(\kappa + \theta)]$. This result, together with (32), yields

$$\sup_{D_{\theta\varepsilon} \times [\mu + \varepsilon/(2+\kappa), \mu + \varepsilon\kappa]} u_\varepsilon = \mathcal{O}\left(\varepsilon^{N_1 - \frac{n+1}{2}}\right). \qquad (34)$$

Step 3

We extend estimate (34) to an ε-neighborhood of 0 in $\mathcal{C}(\varepsilon)$. For a constant $C > 0$, define

$$A_{C\varepsilon} \equiv B(0, C\varepsilon) \cap \mathcal{C}(\varepsilon).$$

We note the following well-known bound on Dirichlet eigenfunctions ϕ_κ for a bounded domain,

$$|\phi_\kappa(x)| \le e^{1/2}(\lambda_\kappa/4\pi)^{n/4}, \qquad (35)$$

which is proved, for example, in [2]. This allows us to derive the bound

$$\|u_\varepsilon\|_{L^2(A_{C\varepsilon})} = \mathcal{O}(\varepsilon^{n/2}) \qquad (36)$$

which, for $n > 2$, is stronger than the bound which follows from the Poincaré inequality. We first use the Harnack inequality in the interior of $A_{C\varepsilon}$. For $C_0 > 0$, we define

$$B_{C_0\varepsilon} \equiv A_{C\varepsilon} \setminus \{x | d(x, \partial\mathcal{C}(\varepsilon)) < C_0\varepsilon\}.$$

Estimate (34) and a version of the Harnack inequality due Jerison [7] (see also [2]) yield

$$\sup_{B_{C_0\varepsilon}} u_\varepsilon = \mathcal{O}\left(\varepsilon^{N_1 - \frac{n+1}{2}}\right). \tag{37}$$

To extend (37) from $B_{C_0\varepsilon}$ to $A_{C\varepsilon}$ we have to use the boundary Harnack inequality developed for non-negative solutions to parabolic equations of the form $Lu = Au - \partial_t u = 0$, where A is an elliptic operator (see [3] for a discussion). This estimate applied to u_ε states that $\exists\, C_1 > 0$ depending only on the Lipschitz character of $\partial \mathcal{C}(\varepsilon)$ such that for any $x \in \partial \mathcal{C}(\varepsilon)$,

$$\sup_{B(x, C_0\varepsilon) \cap \overline{\mathcal{C}(\varepsilon)}} u_\varepsilon \leq C_1 u_\varepsilon(x + C_0\varepsilon). \tag{38}$$

This immediately implies

$$\sup_{A_{C\varepsilon}} u_\varepsilon = \mathcal{O}\left(\varepsilon^{N_1 - \frac{n+1}{2}}\right). \tag{39}$$

We next obtain an L^2-estimate for ∇u_ε in $A_{C\varepsilon} \setminus \Sigma_{\varepsilon/C}$, where

$$\Sigma_{\varepsilon/C} \equiv \left\{ x \in \mathcal{C}(\varepsilon) | d(x, \partial \mathcal{C} \cap \partial T(\varepsilon)) > \frac{1}{C}\varepsilon \right\}$$

It is known that the gradient is poorly behaved near the corners. We must assume that the boundary set $\partial \mathcal{C}(\varepsilon) \cap \left(A_{C\varepsilon} \setminus \Sigma_{\varepsilon/C} \right)$ is C^2. We need the following lemma, which is a version of a Caccioppoli inequality (see [14], for example, for a proof).

LEMMA 3.4 *Let $u \in (H^2 \cap H_0^1)(\Omega)$. Let $r > 0, x \in \overline{\Omega}$, and $\eta \geq 0, \eta \in C^\infty$, be a smooth cut-off function such that $\eta | B(x, r) = 1$, supp $\eta \subset B(x, 2r)$. Then*

$$\int_{B(x,r) \cap \Omega} |\nabla u|^2 \leq C \left(\varepsilon^{-1} \int_{B(x,2r) \cap \Omega} |\nabla u|^2 u^2 + \int_{B(x,2r) \cap \Omega} \eta^2 u \Delta u \right)$$

for any $\varepsilon > 0$.

To continue the proof of the theorem, we apply, the lemma to u_ε. From estimate (39) and the fact that $\Delta u_\varepsilon = \mathcal{O}(1)$, we obtain a boundary estimate for $r = \frac{1}{C}\varepsilon, x \in \partial\mathcal{C}(\varepsilon) \cap \left(\overline{A_{C\varepsilon} \setminus \sum_{\varepsilon/C}}\right)$:

$$\int_{B(x,\varepsilon/C)\cap\mathcal{C}(\varepsilon)} |\nabla u_\varepsilon|^2 = \mathcal{O}(\varepsilon^{2N_1-2}). \tag{40}$$

Finally, we obtain an interior L^2-estimate on ∇u_ε as in (40) by choosing x in Lemma 3.4 such that $B(x, 2r) \subset A_{C\varepsilon}$ (so $r = \mathcal{O}(\varepsilon)$ as above). These results yield

$$\int_{A_{C\varepsilon}\setminus\sum_{\varepsilon/C}} |\nabla u_\varepsilon|^2 = \mathcal{O}(\varepsilon^{2N_1-2}). \tag{41}$$

Step 4

We now relate estimate (39) on u_ε and the L^2-estimate (41) on ∇u_ε to u_0. Let $\chi \in C_0^\infty(B(0,2))$ be a smooth cut-off function such that $\chi \geq 0, \chi|B(0,1) = 1$. Define a function in $\mathcal{C}$ by

$$\widetilde{u}_\varepsilon(x) = (1 - \chi(x/C_0\varepsilon))u_\varepsilon(x),$$

where $C_0 > 0$ is chosen such that $T(\varepsilon) \cap \partial\mathcal{C} \subset B\left(0, \frac{C_0\varepsilon}{2}\right)$. We have that $\widetilde{u}_\varepsilon \in H^2(\mathcal{C}) \cap H_0^1(\mathcal{C})$ and hence

$$- \Delta_\mathcal{C}\widetilde{u}_\varepsilon = E_1(\varepsilon)\widetilde{u}_\varepsilon + r_\varepsilon, \tag{42}$$

which follows by a simple calculation. The remainder r_ε has the form

$$r_\varepsilon = 2(C_0\varepsilon)^{-1}\chi\nabla\chi.\nabla u_\varepsilon + (C_0\varepsilon)^{-2}(\Delta\chi)u_\varepsilon$$

and, due to the support of χ' and estimates (39) and (41), the remainder r_ε satisfies

$$\|r_\varepsilon\|_{L^2(\mathcal{C})} = \mathcal{O}(\varepsilon^{N_1-5/2}). \tag{43}$$

Let Γ be a simple closed contour about E_1 independent of ε. We take ε small enough so $E_1(\varepsilon)$ lies inside Γ. By a simple calculation based on (42), we have for $z \in \Gamma$

$$(-\Delta_\mathcal{C} - z)^{-1}\widetilde{u}_\varepsilon = (E_1(\varepsilon) - z)^{-1}\widetilde{u}_\varepsilon - \widetilde{r}_\varepsilon, \tag{44}$$

where

$$\tilde{r}_\varepsilon = (E_1(\varepsilon) - z)^{-1}(-\Delta_{\mathcal{C}} - z)^{-1} r_\varepsilon,$$

and

$$\|\tilde{r}_\varepsilon\|_{L^2(\mathcal{C})} = \mathcal{O}(\varepsilon^{N_1 - 5/2}). \tag{45}$$

Since E_1 is a simple eigenvalue, the integral of (44) along Γ and estimate (45) yield

$$\tilde{u}_\varepsilon = <u_\varepsilon, u_0> u_0 + \mathcal{O}(\varepsilon^{N_1 - 5/2}). \tag{46}$$

Note that $|\mathcal{C}\backslash\{x | \chi(x) = 1\}| = \mathcal{O}(\varepsilon^n)$ so

$$\|\tilde{u}_\varepsilon\|_{L^2(\mathcal{C})} = 1 + \mathcal{O}(\varepsilon^{N_1 - 5/2}). \tag{47}$$

These two results, (46) and (47), imply that

$$\tilde{u}_\varepsilon = u_0 + \mathcal{O}(\varepsilon^{N_1 - 5/2})$$

in $L^2(\mathcal{C})$ and for each $\varepsilon = \varepsilon_n \to 0$ as in (13). We take $x \in \{x \in \mathcal{C} | d(x, 0) < C_0\varepsilon\}$, so there $\tilde{u}_\varepsilon = 0$ and

$$u_0(x) = \mathcal{O}(\varepsilon^{N_1 - 5/2}). \tag{48}$$

Since u_0 is independent of ε, this estimate holds for all ε sufficiently small. We now recall from Lemma 3.1 that $u_0(x) \geq C_0 d(x, \partial\mathcal{C})$, for $C_0 > 0$ independent of ε. This lower bound and (48) imply

$$C_1\varepsilon^{n+2} \leq \int_{B(0, C\varepsilon) \cap \mathcal{C}} u_0^2 = \mathcal{O}(\varepsilon^{2N_1 - 5}).$$

Since $N_1 = 1 + \frac{n+5}{2} + \delta$, we obtain a contradiction for ε sufficiently small. This concludes the proof. □

Bibliography

[1] S. Agmon, *Lectures on Exponential Decay of Solutions of Second-Order Elliptic Equations*, Princeton University Press, Princeton, N.J., 1982.

[2] R. M. Brown, P. D. Hislop, A. Martinez, *Lower bounds on the interaction between cavities connected by a thin tube*, submitted to Duke Math. J.

[3] E. B. Fabes, N. Garofalo and S. Salsa, *A Backward Harnack Inequality and Fatou Theorem for Non-negative Solutions of Parabolic Equations*, Ill. J. Math. **30**, 1986, p. 536-565.

[4] D. Gilbarg and N. S. Trudinger, *Elliptic Partial Differential Equations of Second Order*, Springer-Verlag, Berlin, 1983.

[5] B. Helffer and J. Sjöstrand, *Multiple Wells in the Semi-classical Limit*, I., Commun. Partial Diff. Eqns., **9**, 1984, p. 337-408.

[6] P. D. Hislop and A. Martinez, *Scattering resonances of a Helmholtz resonator*, Ind. Univ. Math. J., **40**, 1991, p. 767-788.

[7] D. S. Jerison, *The first nodal set of a convex domain*, Preprint, 1991.

[8] C. E. Kenig and J. Pipher, *The h-path distribution of the lifetime of conditioned Brownian motion for non-smooth domains*, Probab. Th. Rel. Fields, **82**, 1989, p. 615-623.

[9] W. Kirsch and B. Simon, *Comparison theorems for the gap of Schrödinger operators*, J. Func. Anal., **75**, 1987, p. 396-410.

[10] A. Martinez, *Estimations de l'effet Tunnel pour le Double Puits*, I, J. Math. Pure et Appl. **66**, 1987, p. 195-215.

[11] A. Martinez and M. Rouleux, *Effet Tunnel entre Puits dégénères*, Commun. Partial Diff. Eqns. **13**, 1988, p. 1157-1187.

[12] I. M. Singer, B. Wong, S. T. Yau and S. S. T. Yau, *An estimate of the gap of the first two eigenvalues in the Schrödinger operator*, Ann. Scuola Norm. Sup. Pisa, **12**, 1985, p. 319-333.

[13] E. B. Davies, *Heat Kernels and Spectral Theory*, Cambridge University Press, Cambridge, 1989.

[14] R. M. Brown, P. D. Hislop, A. Martinez, *Eigenvalues and Resonances for Domains with Tubes: Neumann Boundary Conditions*, preprint 1992.

Nonlinear Volterra Integral Equations and the Apéry Identities

P. J. Bushell
Mathematics Division
University of Sussex
Brighton, Sussex, U.K.

W. Okrasiński
Institute of Mathematics
University of Wroclaw
Poland

1 Introduction

The study of nonlinear diffusion and free boundary value problems frequently leads to a Volterra integral equation of the form

$$u(x) = \int_0^x k(x - s)g(u(s))ds \qquad (1.1)$$

either by consideration of a special case or by choice of similarity variables. For such applications see, for example, Keller (1981) and the many references given by Okrasiński (1989).

We suppose that for some $c > 0$:

(g) g is an increasing absolutely continuous function on $[0, c]$, $g(0) = 0$ and $u/g(u) \to 0$ as $u \to 0^+$;

(k) k is a monotone absolutely continuous integrable function on $(0, d]$ with $k(x) > 0$ for $0 < x \le d$.

Equation (1.1) has the trivial solution $u = 0$, but since g does not satisfy a Lipschitz condition in $[0, c]$, there may be other nontrivial

Differential Equations with
Applications to Mathematical
Physics

solutions, that is, a solution u with $u > 0$ in $(0, d]$ for some $d > 0$. These are the physically interesting solutions.

For the important special case

$$u(x) = \int_0^x (x - s)^{\alpha - 1} g(u(s)) ds \qquad (\alpha > 0) \qquad (1.2)$$

we have the following result.

Theorem 1 *Let*

$$I(\alpha) = \int_{0+}^c \left(\frac{u}{g(u)} \right)^{1/\alpha} \frac{du}{u}.$$

Then there exists a nontrivial solution of (1.2) *if, and only if,* $I(\alpha) < \infty$.

This condition was discovered by Gripenberg (1981), the hypotheses relaxed by Okrasiński (1990) and Gripenberg (1990), and a simpler and more general approach provided by Bushell and Okrasiński (1990) and (1992).

Thus a non–trivial solution of (1.2) exists if $g(u) = u^{1/p}$ with $p > 1$ and if $g(u) = \left(\ln \frac{1}{u} \right)^\beta$ with $\beta > \alpha$, but there is only the trivial solution $u = 0$ for (1.2) if $g(u) = u$ or if $g(u) = \left(\ln \frac{1}{u} \right)^\beta$ with $\beta \leq \alpha$.

To generalize Theorem 1 we use the comparison method for positive integral operators, that is, under very general conditions, if

$$T_i u(x) = \int_0^x k_i(x, s, u(s)) ds \qquad \text{for } i = 1, 2,$$

where

$$k_1(x, s, u) \leq k_2(x, s, u),$$

and there exists a nontrivial solution to the equation $u \leq T_1 u$, then the same is true for the equation $u = T_2 u$ (see Gripenberg (1981) or Zeidler (1986)).

The comparison equations are found using inequalities established with the help of an identity due to Apéry (1953).

Let $K(x) = \int_0^x k(s)ds$ and K^{-1} denote the inverse function to K. Let

$$I_1 = \int_{0+}^{c} [g(s)k \circ K^{-1}(s/g(s))]^{-1}ds$$

and let

$$I_2 = \int_{0+}^{c} [g'(s)/g(s)]K^{-1}(s/g(s))ds.$$

Theorem 2 (Necessary conditions). *Let g and k satisfy conditions (g) and (k), and suppose that equation (1.1) has a nontrivial solution u in $[0, d]$ with $c = u(d)$.*

(i) If k is increasing and $\ln k$ is concave, then $I_1 < \infty$.

(ii) If k is decreasing $I_2 < \infty$.

Theorem 3 (Sufficient conditions). *Let g and k satisfy conditions (g) and (k). Then equation (1.1) has a nontrivial solution in $[0, d]$ with $d > 0$ if*
either   (i) k is increasing and $I_2 < \infty$,
or   (ii) k is decreasing, $\ln k$ is convex and $I_1 < \infty$.

The proofs of these results are given in Bushell and Okrasiński (1992).

Following the remark in Bushell and Okrasiński (1989) we can suppose that the nontrivial solution is nondecreasing.

2   The Apéry Identity and Steffensen Inequalities

An elementary calculation verifies the following version of Apéry's identity:

$$
\begin{aligned}
\int_a^x f(s)h(s)ds \;=\; & \int_a^\lambda f(s)\phi(s)ds \\
& + \int_a^\lambda [f(\lambda) - f(s)][\phi(s) - h(s)]ds \\
& + \int_\lambda^x [f(s) - f(\lambda)]h(s)ds \\
& + f(\lambda)\left[\int_a^x h(s)ds - \int_a^\lambda \phi(s)ds\right]. \quad (2.1)
\end{aligned}
$$

It is easy to deduce the Steffensen inequalities:

Lemma 2.1 *Suppose that* $0 < h(s) \leq h(x)$ *for* $a < s \leq x$ *and let*

$$\lambda = a + \int_a^x [h(s)/h(x)]ds.$$

(i) If f is increasing,

$$\int_a^x f(s)h(s)ds \geq h(x) \int_a^\lambda f(s)ds. \tag{2.2}$$

(ii) If f is decreasing

$$\int_a^x f(s)h(s)ds \leq h(x) \int_a^\lambda f(s)ds. \tag{2.3}$$

The full details of the proofs of Theorems 2 and 3 are somewhat lengthy, but the main idea can be illustrated easily. Suppose that we can find a nontrivial solution v to the equation

$$\int_0^x g(v(s))ds = g(v(x))K^{-1}(v(x)/g(v(x))).$$

Then, from Lemma 2.1 (i),

$$
\begin{aligned}
\int_0^x k(x-s)g(v(s))ds &= \int_0^x k(s)g(v(x-s))ds \\
&\geq g(v(x))K\left(\int_0^x [g(v(s))/g(v(x))]ds\right) \\
&= v(x),
\end{aligned}
$$

and the existence of a non–trivial solution to equation (1.1) follows from the classical comparison theorem.

3   Power Nonlinearity

The function $g(u) = u^{1/p}$ $(p > 1)$ is of particular interest in applications. In this case, if $\frac{1}{p} + \frac{1}{p} = 1$, then

$$I_1 = qK^{-1}(c^{1/q})$$

and

$$I_2 = \frac{q}{p} \int_{0+}^{c^{1/q}} K^{-1}(t) \frac{dt}{t}.$$

If $k(0) = k^{(1)}(0) = \cdots = k^{(n-1)}(0)$ and $k^{(n)}(0) > 0$, it is easy to establish the a priori bounds on a solution,

$$m x^{(n+1)q} \le u(x) \le M x^{(n+1)q}$$

and existence and uniqueness of a nontrivial solution follows using weighted metric fixed point methods as in Askhabov and Betilgiriev (1990) or projective metrics as in Bushell and Okrasiński (1989).

Kernels such as $k_1(x) = \exp(-1/x^\alpha)$ and $k_2(x) = \exp(-\exp(1/x^\alpha))$ are not covered by the theorems given above. However, Okrasiński (1991) has shown that nontrivial solutions exist for k_2 with $0 < \alpha < 1$ but do not exist if $\alpha \ge 1$. Very different conditions which apply to these extremely flat kernels have been given recently by Szwarc (1992).

4   Estimates and Bounds for Solutions

A second identity due to Apéry leads to a simple proof of further inequalities due to Steffensen. The identity is as follows:

$$
\begin{aligned}
\int_a^x f(s)h(s)ds &= \int_{x-\lambda}^x f(s)\phi(s)ds - \int_a^{x-\lambda} [f(x-\lambda) - f(s)]h(s)ds \\
&\quad - \int_{x-\lambda}^x [f(s) - f(x-\lambda)][\phi(s) - h(s)]ds \\
&\quad - f(x-\lambda)\left\{ \int_{x-\lambda}^x \phi(s)ds - \int_a^x h(s)ds \right\}.
\end{aligned}
$$

Lemma 4.1 *Suppose that $0 < h(s) \le h(x)$ for $a < s \le x$ and let $\lambda = \int_a^x [h(s)/h(x)]ds$.*
(i) If f is increasing

$$\int_a^x f(s)h(s)ds \le h(x) \int_{x-\lambda}^x f(s)ds.$$

(ii) If f is decreasing

$$\int_a^x f(s)h(s)ds \geq h(x) \int_{x-\lambda}^x f(s)ds.$$

The inequalities in Lemmas 2.1 and 4.1 provide bounds on solutions to equation (1.1).

Example. Bernis and McLeod (1991) consider a fourth order nonlinear diffusion equation. Using similarity solutions they reduce the problem to an equation of our type. An important step in their analysis is the establishment of a lower bound for a solution of the equation

$$u(x) \geq (k/6) \int_b^x (x - t)^3 u(t)^{1/m} dt$$

with $k > 0$, $m > 1$ and $x \geq b > 0$.

Consider the slightly more general problem

$$u(x) \geq \alpha A \int_b^x (x - t)^{\alpha-1} u(t)^{1/m} dt$$

with $A > 0$ and $\alpha, m > 1$. Assuming, as usual, that the solution u is non–decreasing, it follows from Lemma 4.1 (ii) that

$$u(x) \geq \alpha A u(x)^{1/m} \int_{x-\lambda}^x (x - t)^{\alpha-1} dt$$

and hence that

$$u(x)^{(m-1+\alpha)/\alpha m} \geq A^{1/\alpha} \int_b^x u(s)^{1/m} ds = w(x), \quad \text{say.}$$

From the last inequality it follows easily that

$$w'(x) \geq A^{1/\alpha} w(x)^{m/(m-1+\alpha)}$$

and hence that

$$u(x) \geq \left(\frac{m - 1}{m - 1 + \alpha}\right)^{m\alpha/m-1} \{A^{1/\alpha}(x - b)\}^{m\alpha/m-1}.$$

When $\alpha = 4$ we obtain

$$u(x) \geq \left(\frac{m - 1}{m + 3}\right)^{4m/m-1} A^{m/m-1}(x - b)^{4m/m-1},$$

which is a constant multiple of the function found by Bernis and McLeod.

Bibliography

[1] R. Apéry, *Une inégalité sur les fonctions de variable réelle,* Atti del Quarto Congresso dell' Unione Matematica Italiana 1951, vol. 2, Roma 1953, 3–4.

[2] G. Gripenberg, *Unique solutions of some Volterra intergal equations,* Math. Scand. **48**, 1981, 59–63.

[3] G. Gripenberg, *On the uniqueness of solutions of Volterra equations,* J. Integral Equations and Applications **2**, 1990, 421–430.

[4] P. J. Bushell and W. Okrasiński, *Uniqueness of solutions for a class of nonlinear Volterra integral equations with convolution kernel,* Math. Proc. Camb. Phil. Soc. **106**, 1989, 547–552.

[5] P. J. Bushell and W. Okrasiński, *Nonlinear Volterra integral equations with convolution kernel,* J. London Math. Soc. **41**, 1990, 503–510.

[6] P. J. Bushell and W. Okrasiński, *Nonlinear Volterra equations and the Apéry identities,* J. London Math. Soc., 1992.

[7] S. N. Askhabov and M. A. Betilgiriev, *Nonlinear convolution type equations,* Seminar Analysis, Operator Equations and Numerical Analysis, 1989/90, Karl–Weierstrass–Institut für Mathematik, Berlin, 1990, 1–30.

[8] W. Okrasiński, *Remarks on nontrivial solutions to a Volterra integral equation,* Math. Methods Appl. Sci. **13**, 1990, 273–279.

[9] W. Okrasiński, *Nonlinear Volterra equations and plysical applications,* Extracta Mathematicae 4, 1989, 51–80.

[10] W. Okrasiński, *Nontrivial solutions for a class of nonlinear Volterra equations with convolution kernel,* J. Integral Equations and Applications **3**, 1991, 399–409.

[11] R. Szwarc, *Nonlinear integral inequalities of the Volterra type,* Math. Proc. Camb. Phil. Soc., 1992, 599–608.

[12] E. Zeidler, *Nonlinear Functional Analysis and Its Applications I*, Springer, New York, 1986.

[13] F. Bernis and J. B. McLeod, *Similarity solutions of a higher order nonlinear diffusion equation*, Nonlinear Analysis **17**, 1991.

[14] J. J. Keller, *Propagation of simple nonlinear waves in gas filled tubes with friction*, Z. Angew, Math. Phys. **32**, 1981, 170–181.

Connections Between Quantum Dynamics and Spectral Properties of Time-Evolution Operators

Jean-Michel Combes
Centre de Physique Théorique, Luminy Case 907, 13288
Marseille, France and Phymat, Université de Toulon et du Var,
B.P. 132, 83957 La Garde, France

Abstract

Lower bounds for time averages of mean square displacement are discussed in terms of the Hausdorff dimension of the spectrum.

1 Introduction

A few decades ago D. Ruelle [10] stated the first general result relating space-time behaviour for solutions of the Schrödinger equation with the spectral type of the corresponding quantum Hamiltonian. Equipped with technical refinements this result became the well-known RAGE theorem [11] which supports the conventional wisdom that continuous spectrum manifests itself in the time decay of local space averages whereas point spectrum implies localisation in configuration space. Variants of this theorem have been proved by Enss and Veselic for time periodic forces [3] and Jauslin and Lebowitz for quasi periodic time dependent forces [8]; here point spectrum of the Floquet operator (or more generally the quasi-energy operator) is

Differential Equations with
Applications to Mathematical
Physics

Floquet operator (or more generally the quasi-energy operator) is
related to quantum stability whereas a continuous spectrum implies
unbounded growth of the energy. The need for quantitative refinements of these general connections appeared in the last ten years with
the investigation of models in solid state physics exhibiting "extraor-
dinary spectra" (in the terminology of Avron and Simon [1]) like
dense point spectrum or continuous singular spectrum supported on
Cantor sets. Such models reveal in addition "unusual" dynamical behaviours as opposed to what is "usual" for well-behaved potentials;
to be more precise let

$$\langle A(t) \rangle = \langle \psi_t, A\psi_t \rangle \tag{1}$$

with ψ_t the solution of the Schrödinger equation with $\psi_{t=0} = \psi_0$ and
A a self-adjoint operator such that $\psi_t \in \mathcal{D}(A) \ \forall t \in \mathbf{R}$ if ψ_0 does. For
$A = |X|^2$, the mean square displacement, one thinks of "ordinary"
dynamics as either the localisation regime where $< |X|^2(t) > < C \ \forall t$
or the ballistic regime where $|X|^2(t) > \sim Ct^2(t \to \infty)$, which are
supposed to correspond respectively to discrete or absolutely continuous spectrum from our experience of well-behaved (locally and
asymptotically) potentials (although this has no general mathematical ground). On the other hand it is well-known that intermediate
behaviours between these two exist. For example in the hierarchical
models considered by Jona-Lasinio et al. [9] one has $< |X|^2(t) > \sim$
$C(Logt)^\beta$ for some $\beta > 0$; but more generally the importance of
these intermediate regimes is due to the relation between the "diffusion constant":

$$D_{\psi_0} = \lim_{t \to +\infty} \frac{1}{t} < |X|^2(t) > \tag{2}$$

and static conductivity. Without going into the details of this connection (see e.g. [4], [15]) let us just mention that one electron models of
metals (resp. insulators) should have $0 < D_{\psi_0} < \infty$ (resp. $D_{\psi_0} = 0$).
Thus in particular diffusive behaviour is the rule in models of con-
ducting media and one would like to know which type of spectra is
responsible for this. Clearly connections between these "extraordi-
nary" spectral and dynamical properties go beyond the mere RAGE
theorems which might not even provide the right intuition. For example think of the naive conjecture that $D_{\psi_0} = 0$ corresponds to

a pure point spectral measure for ψ_0; although it is correct that $< |X|^2(t) > < C \ \forall t$ implies that ψ_0 has no continuous component the only general result about the converse is a recent one by B. Simon [12] stating that in this last case $\lim\limits_{t \to \infty} \frac{1}{t^2} < |X|^2(t) > = 0$ which is far from the expected answer $D_{\psi_0} = 0$. One of the reasons why our intuition might be misleading is that unusual spectra like dense pure point or singular continuous are very unstable. As shown e.g. by Simon and Wolff [14] and Howland [6] even a rank one perturbation with arbitrary small norm can induce a transition from one type to the other. On the other hand one does not expect that the dynamics should be strongly affected by such perturbations. Thus if one believes in this last argument any "extraordinary" dynamics produced by some singular continuous spectral measure should also show up with some pure point measure obtained from the first one by a small perturbation; in other words Simon's result might be optimal!

The interest into such questions is not limited to the choice $A = |X|^2$; when considering external time-periodic forces it is natural to let A be the internal energy operator. Then one considers $< A(nT) >$, $n \in \mathbf{Z}$, where T is the period so that $A(nT) = F^n A F^{-n}$ with F the Floquet operator. Boundedness of $< A(nT) >$ is related to quantum stability and this problem has attracted considerable interest recently in connection with quantum chaos since classically chaos manifests itself through a diffusive growth of energy. It would be of course of primary interest to have criteria allowing to deduce such a diffusive growth from spectral properties of the Floquet operator (conditions for F to have pure point spectrum will be discussed by J. Howland [7] in this conference).

It turns out that the first step towards a refined RAGE theorem obtained by I. Guarneri [5] was motivated in fact by the investigation of dynamical localization for the kicked rotator. This problem is one particular aspect of quantum diffusion on a one dimensional lattice; Guarneri provides arguments, both heuristic and rigorous, to connect time asymptotic regimes with what he calls "spectra of peculiar type". More precisely he obtains remarkable lower bounds on $< A(t) >$ in terms of the lattice dimension d, counting function for A and Hausdorff dimension of the support of spectral measures with respect to the evolution operator over one period of time. These results will be described in §2 below; they imply in the case $A = |X|^2 (X \in \mathbf{Z}^d)$ that

$$\frac{1}{N} \sum_{n=0}^{N} < |X|^2(n) > \geq C N^{2\alpha/d}/(\log N)^{2/d} \qquad (3)$$

where α is the dimension of the spectral measure for ψ_0 (see def. below) and the time period is chosen equal to one. Forgetting about the logarithmic term, which seems to be a technically irrelevant consequence of Guarneri's method, we notice that for $d = 1$ the time behaviour is at least ballistic for the absolutely continuous spectrum ($\alpha = 1$) and localized for point spectrum whereas diffusion requires $\alpha \leq \frac{1}{2}$. This is no more true for $d \geq 2$ and diffusive behaviour does not seem anymore incompatible with absolutely continuous spectrum. One might think that this is due to the fact that Guarneri considers only lattice dynamics and Floquet spectrum instead of the Hamiltonian spectrum as in the RAGE theorem. Surprisingly it appears that Guarneri's bounds can be extended to quantum dynamics on $\mathbf{R}^d$; this follows from recent results of R. Strichartz [13] about Fourier transform of α-dimensional measures which provide a substitute to Guarneri's Dirichlet like estimates for Fourier series; this will be described in §2 below.

2 Spectral Dimension and Quantum Diffusion

Let us first describe Guarneri's lower bounds [5] for the spreading of wave-packets in terms of the Hausdorff dimension of the spectrum. Consider time averages:

$$<< A >>_T = \frac{1}{N} \sum_{n=0}^{N} < F^n \psi_0, A F^n \psi_0 > \qquad (4)$$

where F is the evolution operator over an interval of time T_0, $T = NT_0$ and A is a self adjoint operator having a spectral decomposition

$$A = \sum_{k \in \mathbf{Z}^d} \phi(|k|)|e_k > < e_k| \tag{5}$$

where $||e_k|| = 1 \forall k \in \mathbf{Z}^d$, the function ϕ being positive non decreasing. (If T_0 is one period of some time periodic perturbation then F is just the Floquet operator.) For ordinary lattice dynamics of tight binding models one takes $A = |X|^2 (X \in \mathbf{Z}^d)$ so that $\phi(|k|) = |k|^2$ and $e_k(m) = \delta_{km} \forall m \in \mathbf{Z}^d$. For the kicked rotator the dynamics is given by periodic kicks and A is the kinetic energy $A = \sum_{k^2 \in \mathbf{Z}} k^2 |e_k > < e_k)$ where $e_k(\theta) = (2\pi)^{-\frac{1}{2}} e^{ik\theta}$ are the angular momentum eigenstates etc... The counting function for A is defined as:

$$\nu(x) = \# \left\{ k \in \mathbf{Z}^d, \phi(|k|) \le x \right\} \tag{6}$$

To state Guarneri's result one needs to make a very specific assumption about the spectral measure μ_{ψ_0} of the initial state ψ_0 with respect to the unitary operator F.

DEFINITION 2.1 *A positive measure μ on* **R** *is said to be locally uniformly α-dimensional if for some positive constant C*

$$\mu(B_r(\lambda)) \le C r^\alpha \tag{7}$$

for every ball $B_r(\lambda)$ of center λ and radius r, $0 < r \le 1$.

(The measures considered here as defined on the Borel sets of **R** *and are only assumed to be locally finite.)*

We refer to [13] for the properties of such measures. In particular one can show that they are absolutely continuous with respect to the α-dimensional Hausdorff measure μ_α and admits a Radon-Nikodym decomposition $\mu = \varphi d\mu_\alpha + \nu$ where ν is null with respect to μ_α in the sense that $\nu(B) = 0$ for any B such that $\mu_\alpha(B) < \infty$.

Examples of such measures have been constructed e.g. by Avron and Simon [1] in connection with their analysis of recurrent absolutely continuous spectrum.

We can now state Guarneri's main result:

PROPOSITION 2.1 *If μ_{ψ_0} is locally uniformly α dimensional then for sufficiently large T and for all $x > 0$*

$$<< A >>_T \geq x \left[1 - C_1 \nu(x) T^{-\alpha} \log T\right] \qquad (8)$$

for some constant C_1.

A remarkable consequence of (8) follows from its application to $A = |X|^2, X \in \mathbf{Z}^d$; here one has $\nu(x) \sim C x^{d/2}$ from which it immediately follows that for T large enough:

$$<< A >>_T \geq C\ T^{2\alpha/d}/(\log T)^{1/d} \qquad (9)$$

The basic ingredient in the derivation of (8) is the following inequality obtained by Guarneri by elementary Dirichlet like estimates:

$$<< P_k >>_T \leq C\ T^{-\alpha} \log T \qquad (10)$$

where $P_k = |e_k \times e_k|$. Such an inequality is in fact a weak form of a result of R. Strichartz [13] stating that if μ is a locally uniformly α-dimensional measure on $\mathbf{R}$ and $f \in L^2(d\mu)$ then the Fourier transform $\widehat{f d\mu}$ satisfies:

$$\sup_{T \geq 1} T^{\alpha - n} \int_0^T |\widehat{f d\mu}(t)|^2 dt \leq C||f||_2^2 \qquad (11)$$

This suggests to consider now time averages

$$<< A >>_T = \frac{1}{T} \int_0^T < A(t) > dt \qquad (12)$$

with $< A(t) >$ given by (1) and A of the form (5). Let $\mathcal{P}_0$ be the projection operator on the cyclic subspace generated by $\{\psi_t, t \in \mathbf{R}\}$; then $\forall k$:

$$\mathcal{P}_0 e_k = \int f_k(\lambda) dE_\lambda \psi_0 \qquad (13)$$

where E_λ is the spectral family for the Hamiltonian (or quasi-energy) operator.

Furthermore $f_k \in L^2(d\mu_{\psi_0})$ with

$$||f_k||_2^2 = ||\mathcal{P}_0 e_k||^2 \leq 1. \qquad (14)$$

Since $\widehat{f_k d\mu_{\psi_0}}(t) = <\psi_t, e_k>$ inequality (11) implies the stronger form of (10):

$$<< P_k >>_T \leq C\, T^{-\alpha} \tag{15}$$

From this it follows as in [5] that if μ_{ψ_0} is locally uniformly α dimensional then $\forall x > 0$ and $T \geq 1$:

$$<< A >>_T \geq x\left[1 - C_1\nu(x)T^{-\alpha}\right] \tag{16}$$

which is a generalized form of Guarneri's inequality (8).

There is an obvious difficulty if one wants to apply (16) to the investigation of quantum dynamics on $\mathbf{R}^d$ instead of $\mathbf{Z}^d$ as Guarneri did since then $A = |X|^2$ is obviously not of the form (5). This can be easily overcome if we make the extra assumption that ψ_0 has bounded energy; then one has for example:

PROPOSITION 2.2 *Let the quantum Hamiltonian have the form* $H = -\Delta + V$ *on* $L^2(\mathbf{R}^d)$ *where* V *is real and bounded below. Let* $\psi_0 \in \mathcal{D}(e^H)$ *be such that if* $\varphi_0 = e^H\psi_0$ *then* $d\mu_{\varphi_0}$ *is locally uniformly* α-*dimensional; then for* $T \geq 1$:

$$<< |X|^2 >>_T \geq C\, T^{2\alpha/d} \tag{17}$$

Let us mention briefly how one can obtain (17) from (16); one has:

$$< |X|^2(t) > \geq \sum_{k \in \mathbf{Z}^d} |k|^2 < \psi_t, \chi_k\psi_t > \tag{18}$$

where χ_k is the characteristic function of $\{X \in \mathbf{R}^d, k_j \leq X_j < k_j + 1, j = 1, \ldots, d\}$. Then write

$$\langle \psi_t, \chi_k\psi_t \rangle = \langle \varphi_t, A_k\varphi_t \rangle \tag{19}$$

with $\varphi_t = e^H\psi_t$ and $A_k = e^{-H}\chi_k e^{-H}$; using e.g. semi-group kernel inequalities it is easy to see that A_k is trace-class and denoting by $||A_k||_1$ it's trace norm one has

$$||A_k||_1 \leq C||e^\Delta\chi_k e^\Delta||_1 = ||e^\Delta\chi_0 e^\Delta||_1 < \infty \tag{20}$$

so that $||A_k||_1$ is uniformly bounded in $k \in \mathbf{Z}^d$.

Then (11) implies easily that:

$$\sup_{T \geq 1} << A_k >>_T \leq C \ T^{-\alpha}||A_k||_1 \qquad (21)$$

for some constant $C_0 < \infty$. Returning to (18) one has $\forall x \in \mathbf{N}$:

$$
\begin{aligned}
<< |X|^2 >>_T \ &\geq \ x \sum_{|n|^2 > x} |n|^2 << \chi_n >>_T \\
&\geq \ x \left[1 - \sum_{|n|^2 \leq x} |n|^2 << \chi_n >>_T \right] \\
&\geq \ x \left[1 - C \ x^{d/2} T^{-\alpha} \right]
\end{aligned}
$$

where we have used (19) and (21) in the last step; this gives (17).

Remark: The arguments developed above can also be used to derive directly inequalities like

$$<< (1 + |X|^2)^{-1} >>_T \leq C \ T^{-\delta} \qquad (22)$$

for any $\delta < \dfrac{2\alpha}{d}$ under the same assumptions as in Prop. 2.1; details will appear elsewhere [2].

3   Concluding Remarks

The dependence of bounds like (17) or (22) on space dimension, in particular in the case of absolutely continuous spectrum is somewhat unexpected from the conventional wisdom inherited from RAGE theorem. However this appears as natural if one thinks that in disordered media trajectories should look more like random walks rather than the well-behaved asymptotic straight lines of potential scattering models which motivated Ruelle's initial work. Of course there remains the question of whether these bounds are sharp; in particular Definition 4 of the dimension of a measure is somewhat ambiguous. As emphasized by Strichartz [13] α-dimensional measures with $0 < \alpha < 1$ need not exhibit any fractal behaviour; however if in the Radon-Nikodym decomposition of an α-dimensional measure with respect to Hausdorff measure μ_α the coefficient of μ_α is the characteristic function of a quasi-regular set then the upper-bound (21) has an associated related lower-bound. This is enough to get directly lower-bounds e.g. for $<< (1 + |X|^2)^{-\gamma} >>_T$ for $\gamma > \frac{d}{2}$.

Bibliography

[1] J. Avrom and B. Simon, *Transient and recurrent spectrum*, J. Func. Anal. **43**, 1981, 1.

[2] J.-M. Combes, *Lower bounds on quantum diffusion and spectral dimension*, to appear.

[3] V. Enss, K. Veselic, *Bound states and propagating states for time-dependent Hamiltonians*, Ann. Inst. H. Poincaré **39A**, 1983, 159.

[4] J. Frohlich and T. Spencer, *A rigorous approach to Anderson localization*, Phys. Rep. **103**, 1984, 9.

[5] I. Guarneri, *Spectral properties of quantum diffusion on discrete lattices*, Europh. Lett. **10(2)**, 1989, 95.

[6] J. Howland, *Perturbation theory of dense point spectra*, J. Funct. Anal. **74**, 1980, 52.

[7] J. Howland, Proceedings of this conference and references therein.

[8] H. R. Jauslin, J.-L. Lebowitz, *Spectral and stability aspects of quantum chaos*, Preprint Rutgers Univ., 1991.

[9] G. Jona-Lasinio, F. Martinelli and E. Scoppola, *Multiple tunnelings in d dimensions: a quantum particle in a hierarchical potential*, Ann. Inst. H. Poincaré, **42A**, 1985, 73.

[10] D. Ruelle, *A remark on bound states in potential scattering theory*, Nuovo Cimento, **61A**, 1969, 655.

[11] M. Reed and B. Simon, *Methods of modern mathematical physics III scattering theory*, Acad. Press, 1979.

[12] B. Simon, *Absence of ballistic motion*, Com. Math. Phys., **134**, 1990, 209.

[13] R. Strichartz, *Fourier asymptotics of fractal measures*, Jour. Funct. Anal., **89**, 1990, 154.

[14] B. Simon and T. Wolff, *Singular continuous spectrum under rank one perturbations and localization for random Hamiltonians*, Com. Pure Appl. Math., **39**, 1986, 75.

[15] D.J. Thouless, *Electrons in disordered systems and the theory of localization*, Phys. Rep., **13**, 1974, 93.

Quasilinear Reaction Diffusion Models For Exothermic Reaction

W. E. Fitzgibbon and C. B. Martin
Department of Mathematics
University of Houston
Houston, TX 77204-3476

In this note we discuss an idealized model of irreversible chemical reaction. Actual chemical reactions involve a large number of chemical species and many intermediate chemical reactions. For example, it is argued in [20] that the production of water by combination of molecular hydrogen and molecular oxygen is described by a reaction sequence involving eight chemical species and a minimum of sixteen reactions whereas a more complicated process such as methane oxidation involves twelve chemical species and twenty two reactions. In an effort to make such processes analytically and computationally tractable various idealized models have been put forth. The model which we consider describes an irreversible exothermic chemical of the form,

$$A + B \; \rightarrow \; 2B$$
$$2B \; \rightarrow \; \text{Products.}$$

This is the idealized two step reaction of Zeldovich [26] as formulated by Niioka [18]. Here it is assumed that the first reaction has a high activation energy and negligible heat release and that the second has negligible temperature dependence and high heat release. If we account for diffusion the partial differential equations modelling this

Differential Equations with
Applications to Mathematical
Physics

reaction sequence are of the form:

$$\partial u/\partial t - \nabla \cdot d_1(x, u, v, \theta)\nabla u = -uvf(\theta), \qquad (1)$$

$$\partial u/\partial t - \nabla \cdot d_2(x, u, v, \theta)\nabla v = uvf(\theta) - \mu v^2, \qquad (2)$$

$$\partial \theta/\partial t - \nabla \cdot d_3(x, u, v, \theta)\nabla \theta = \mu v^2, \qquad (3)$$

for $x \in \Omega, t > 0$. We impose homogeneous Neumann boundary conditions,

$$\partial u/\partial n = \partial v/\partial n = \partial \theta/\partial n = 0 \qquad (4)$$

for $x \in \partial\Omega, t \geq 0$ and require that the initial data

$$u(x, 0) = u_0(x) \quad v(x, 0) = v_0(x) \quad \theta(x, 0) = \theta_0(x) \qquad (5)$$

for $x \in \Omega$ be continuous and nonnegative on $\bar{\Omega}$. We stipulate Ω is a bounded region in $\mathbb{R}^n$ with smooth boundary $\partial\Omega$ such that Ω lies locally on one side of $\partial\Omega$. Admittedly only $n = 1, 2$, or 3 have any physical significance, however, assumption of arbitrary spatial dimension does not change our analysis.

We assume that there exists $a, b > 0$ so that

$$0 < a < \max\{d_1(\ldots), d_2(\ldots), d_3(\ldots)\} < b$$

and that each $d_i \in C^\infty(\bar{\Omega} \times \mathbb{R}^3_+)$. The nonlinear function $f(\ )$ represents a prototypical Arrhenius temperature dependence. It is nonnegative, monotone increasing, smooth and uniformly bounded. It has the form,

$$f(\theta) \begin{cases} = & 0 \quad \text{if } \theta \leq 0 \\ & Ke^{-E/\theta} \quad \text{if } \theta > 0, \end{cases}$$

where K and E are positive constants. Roughly speaking the variables u and v represent concentrations of A and B respectively with θ representing a nondimensional temperature of the reaction vessel. The homogeneous boundary conditions require that the vessel is insulated and that these chemical species remain confined to the vessel for all time.

Extensive treatments of chemical reaction kinetics may be found in [6, 8, 9, 14, 20]. In the note at hand we extend semilinear results appearing in [10] and argue that solutions are globally well posed and detail their asymptotic convergence. The mathematical literature on this type of reaction diffusion system is extensive and the interested reader is referred to [3, 4, 5, 7, 4, 11].

If $u : \Omega \to \mathbb{R}$ we shall denote the n–dimensional gradient by ∇u and its Euclidean norm by $|\nabla u|$. The space time cylinder $\Omega \times [0, t)$ will be denoted by Q_t with Q_∞ denoting $\Omega \times [0, \infty)$. We shall use the standard $Lp(\Omega)$ spaces ($p \geq 1$) whose norms will be denoted by $\| u \|_{p,\Omega}$ with the norm on $C(\bar{\Omega})$ being denoted by $\| u \|_{\infty,\Omega}$.

Our first result provides global existence of solutions and precompactness of their trajectories.

Theorem 1 *There exists an unique classical solution to (1 - 3) on Q_∞ such that $u(x, t) \geq 0$, $v(x, t) \geq 0$ and $\theta(x, t) \geq 0$, and a constant $M > 0$ so that*

$$\sup_{Q_\infty}\{u(x, t), v(x, t), \theta(x, t)\} < M. \tag{6}$$

Finally the trajectories $\Gamma(u_0, v_0, \theta_0) = \{u(,t), v(,t), \theta(,t) \mid t \geq 0\}$ are precompact in $C(\bar{\Omega})$.

Proof: Local existence, uniqueness and continuous dependence follow from arguments of abstract parabolic theory due to Amann, [2]. The non-negativity of solutions is a established by standard maximum principle arguments. If one can establish the existence of uniform a priori bounds for solution components on $[0, T_{\max})$ then Amann's continuation arguments', [2], yield global wellposednessed.

Adding the components and integrating on $Q_{T_{\max}}$ we immediately obtain

$$\int_\Omega (u(x, t) + v(x, t) + \theta(x, t))dx = \int_\Omega (u_0(x) + v_0(x) + \theta_0(x))dx \tag{7}$$

and the non-negativity of solutions yields the existence of uniform a priori L_1 bounds for the solution components on $[0, T_{\max})$. It is straightforward to observe that

$$\| u(,t) \|_{\infty,\Omega} \leq \| u_0() \|_{\infty,\Omega} \qquad \text{for } t \in [0, T_{\max}).$$

To obtain uniform a priori bounds for the second component we observe that,

$$f(\theta)uv - \mu v^2 = v[uf(\theta) - \mu v]$$
$$\leq v[uf(\theta)]$$

and for $(x,t) \in Q_{T_{\max}}$

$$u(x,t)f(\theta(x,t)) \leq K \parallel u_0 \parallel_{\infty,\Omega} .$$

It is now possible to utilize Moser-Alikakos iteration, cf. Alikakos, [1], to bootstrap the L_1 to a uniform L_∞ bound. This is a lengthy and complicated argument which will not be reproduced here. The reader is referred to [11] for the application of the argument to a similar system. Uniform a priori L_∞ bounds for $\theta(,)$ are produced in a similar manner. The existence of these bounds allows us to conclude that $T_{\max} = \infty$ and that (6) holds. Furthermore, the fore-mentioned work of Amann guarantees uniform a priori bounds imply precompactness of tractories.

If one knows that trajectories, $\Gamma(u_0, v_0, \theta_\infty)$ are precompact one may draw upon the powerful results of abstract LaSalle-Lyapunov theory, cf [15]. We have the following result:

Theorem 2 *If* $u_0(x), v_0(x), \theta_0(x) > 0$ *for* $v \in \bar{\Omega}$ *then the following are true*

$$\lim_{t \to \infty} \parallel u(,t) \parallel_{\infty,\Omega} = 0 \tag{8}$$

$$\lim_{t \to \infty} \parallel v(,t) \parallel_{\infty,\Omega} = 0 \tag{9}$$

$$\lim_{t \to \infty} \parallel \theta(,t) - w_0 \parallel_{\infty,\Omega} = 0 \tag{10}$$

where $w_0 = |\Omega|^{-1} \int_\Omega (u_0 + v_0 + \theta_0) dx$.

Proof: Let ω denote the ω-limit set for u_0, v_0, θ_0. By virtue of the precompactness of trajectories we know [15] that each trajectory $\Gamma(u_0, v_0, \theta_0)$ has a compact, connected forward invariant ω-limit set. Our verification of (8 - 10) subdivides into several parts. We first argue that if $(u^*, v^*, \theta^*) \in \omega(u_0, v_0, \theta_0)$ then $v^* = 0$ establishing (8).

Using $v^* = 0$ as initial data we establish that u^* is really a constant function c, and that this c is uniquely determined and that $c = 0$ establishing (1.6a). By similar techniques we show that θ^* is a constant function. Therefore if strictly positive initial data is chosen for this system the one-dimensional subspace of $\mathbb{R}^3$ of the form $\{(0,0,a) \mid a \in \mathbb{R}^+\}$ acts as a global attractor.

If we add the first two components and integrate on Q_t, we obtain

$$\int_\Omega (u(x,t) + v(x,t))dx + \int_0^t \int_\Omega Mv^2(x,s)dx \leq \int_\Omega (u_0 + v_0)dx, \quad (11)$$

Therefore, the improper integral $\int_0^\infty \parallel v(,s) \parallel_{2,\Omega}^2 ds$ exists. If we multiply (2) by $v()$ and integrate on Ω we obtain

$$\frac{1}{2}d/dt \parallel v(,t) \parallel_{2,\Omega}^2 + \int_\Omega d_2(x,u,v,\theta)|\nabla v|^2 dx$$

$$\leq K \parallel v(,t) \parallel_{2,\Omega}^2 \parallel u(,t) \parallel_{\infty,\Omega} + \mu \parallel v(,t) \parallel_{\infty,\Omega}^3$$

Consequently, there is a uniform upper bound for the quantity $d/dt(\parallel v(,t) \parallel_{2,\Omega}^2)$. This together with the finiteness of the improper integral and the boundedness of $\parallel v(,t) \parallel_{2,\Omega}$ imply that

$$\lim_{t \to \infty} \parallel v(,t) \parallel_{2,\Omega} = 0 \quad (12)$$

We can bootstrap the $L_2(\Omega)$ convergence of v to $L_\infty(\Omega)$ convergence by closely examining the estimates produced by the Moser-Alikakos iteration scheme. If we retrace the argument of Theorem 2 [11], we can construct a constant $N > 0$ so that,

$$\left[\int_\Omega v^{2k}dx \right]^{\frac{1}{2}k} \leq N \left[\int_\Omega v^2 \right]^{\frac{1}{2}} \quad (13)$$

and (9) follows immediately from (12) and (13) by taking the limit as $k \to \infty$.

To establish (8) we point out that trajectories $\Gamma(u_0,v_0,\theta_0)$ are precompact in $L_2(\Omega)$ as well as being precompact in $C(\bar{\Omega})$. We shall argue $u(,t)$ converges to an unique constant function in $L_2(\Omega)$ as $t \to \infty$. Therefore any convergent subsequence $u(,t_k)$ must also converge to this constant function in $L_\infty(\Omega)$. We thereby establish a constant function in the first spatial component of $\omega(u_0, v_0, \theta_0)$ in $L_\infty(\Omega)$. We then argue that this constant function must be zero. We point out that if $(u_*, v_*, \theta_*) \in \omega(u_0, v_0, \theta_0)$ then the previous argument insures that $v_* = 0$. If we multiply (1) by $u()$ and integrate on Q_t we observe that

$$\| u(,t) \|_{2,\Omega} \leq \| u_0 \|_{2,\Omega}$$

and we may observe that $\| u(,t) \|_{2,\Omega}$ is nonincreasing in t and bounded below. We let $r = \lim_{t \to \infty} \| u(,t) \|_{2,\Omega}$. It is clear that if $(u_*, 0, \theta_*) \in \omega(u_0, v_0, \theta_0)$ then $\| u_* \|_{2,\Omega} = r$. We solve the initial value problem (1-3) with initial data $(u_*, 0, \theta_*) \in \omega(u_0, v_0, \theta_0)$. Parabolic uniqueness implies the reaction terms decouple and solutions are given by $(u(,t), 0, \theta(,t))^T$ where u and θ satisfy,

$$\partial u/\partial t - \nabla \cdot d_1(u, 0, \theta)\nabla u = 0 \tag{14}$$

$$\partial\theta/\partial t - \nabla \cdot d_3(u, 0, \theta)\nabla\theta = 0 \tag{15}$$

with

$$\partial u/\partial n = \partial\theta/\partial n = 0 \quad \text{for } x \in \partial\Omega$$

and

$$u(x, 0) = u_*(x) \text{ and } \theta(x, 0) = \theta_*(x) \text{ for } x \in \Omega.$$

Forward invariance of $\omega(u_0, v_0, \theta_0)$ implies that for $t > 0$ $\|u(,t)\|_{2,\Omega} = r = \| u(x, 0) \|_{2,\Omega}$. Thus if we multiply (14) by $u(,)$ and integrate on Q_t we obtain

$$\int_0^t \int_\Omega d_1(u, 0, \theta)|\nabla u|^2 dx ds = 0$$

We thereby conclude that $\| |\nabla u| \|_{2,\Omega} = 0$ and deduce that $u(x, t) = u(t)$ is spatially homogeneous. However, $\| u(t) \|_{2,\Omega} = r$ and thus does not evolve in time. Moreover because $\lim_{t \to \infty} \| u(,t) \|_{2,\Omega} = r$ this constant is unique. Hence $\omega(u_0, v_0, \theta_0) = (c, 0, \theta_*)$. We now sketch the argument insuring $c = 0$.

We assume for the sake of contradiction that $\lim_{t \to \infty} u(,t) = c > 0$ in $C(\bar\Omega)$. The comparison principle implies that there exists an $\alpha > 0$ so that

$$f(\theta(x, t)) > \alpha > 0 \quad \text{for } (x, t) \in Q_\infty.$$

Consequently there exists a $t_1 > 0$ and a $\sigma > 0$ so that if $t > t_1$, then

$$u(x,t)f(\theta(x,t)) > \sigma.$$

Because $v(,t)$ converges to zero, there exists a $t_2 > 0$ so that $t > t_2$ implies

$$0 < \mu v(x,t) < \sigma/2.$$

Thus, if $t > \max\{t_1, t_2\}$ then

$$\partial v/\partial t - \nabla \cdot d_2 \nabla v > \sigma v/2$$

This inequality precludes convergence of $v(\ )$ to zero.

To establish (10) we first argue that there exists an $r > 0$ so that

$$r = \lim_{t \to \infty} \| \theta(,t) \|_{2,\Omega} \tag{16}$$

Toward this end we select $M > 0$ so that $\sup \| \theta(,t) \|_{\infty,\Omega} < M$ and set

$$y(x,t) = M - \theta(x,t).$$

We observe that $y(x,t) > 0$ for $(x,t) \in Q(\infty)$. Moreover

$$\partial y/\partial t - \nabla \cdot d_3 \nabla y = -\mu v^2$$

If we multiply the above equation by y and integrate on Ω to observe that

$$\frac{1}{2} d/dt(\| y(,t) \|_{2,\Omega}^2) + \int_\Omega d_3 |\nabla y|^2 dx \leq 0.$$

Consequently, $\| y(,t) \|_{2,\Omega}^2$ is nonincreasing and we are assured of the existence of $r_* = \lim_{t \to \infty} \| y(,t) \|_{2,\Omega}^2$ and we thereby deduce the existence of r satisfying (16). Thus if $\theta_* \in w(u_0, v_0, \theta_0)$ then $\| \theta_* \|_{2,\Omega} = r$. We then solve (1-5) with initial data $(0, 0, \theta_*) \in w(u_0, v_0, \theta_0)$ and argue that $d/dt(\| \theta(,t) \|^2) = d/dt(r) \equiv 0$. It is then not difficult to see that θ is constant and that the value of the constant is determined by

$$|\Omega|c = \int_\Omega (u_0 + v_0 + \theta_0) dx.$$

We point out that our methods apply equally well to the case of general quasilinear divergence from operators,

$$\partial \left(\sum_{j,k=1}^{n} d_i^{jk}(x,u,v,\theta)\partial()/\partial x_j \right) /\partial x_k.$$

From a physical point of view it is perhaps most important that the diffusivities are allowed to be nonlinear functions of the temperature. Our results agree with those obtained for semilinear models, [10] and we are lead to the conjecture that nonlinear diffusion does not effect the wellposedness or the longterm asymptotics. However, numerical experiments indicate that nonlinear diffusion does qualitatively effect the intermediate dynamics of the system.

Physically, our results are perhaps not too surprising. General principles of chemical thermodynamics postulate that closed balanced systems attract to constant steady states. In forthcoming work we shall treat quasilinear models with nonhomogeneous Robin boundary conditions. We point out that ideas contained herein will be central.

Bibliography

[1] N. Alikakos, *An application of the invariance principle to differential equations.* J. Diff Eqns., **33**(1979), 201-225.

[2] H. Amann, *Global existence and positivity for triangular quasilinear reaction diffusion systems.*, Preprint.

[3] J. Avrin, *Decay and boundedness results for a model of laminar flames with complex chemistry,* Proc. AMS, to appear.

[4] ———, *Qualitative theory of the Cauchy problem for a one step reaction model on bounded domains,* SIAM J. Math. Anal., to appear.

[5] ———, *Qualitative theory for a model of laminar flames with arbitrary non-negative initial data,* J. Diff. Equns. **84**(1990), 290-308.

[6] J. Bebernes and D. Eberly, *Mathematical Problems from Combustion Theory*, Applied Math. Sciences, **83**, Springer-Verlag, New York, 1989.

[7] H. Berestychki, B. Nicolaenko and B. Scheuer, *Traveling wave solutions to combustion models and their singular limits*. SIAM J. Math Anal. **6** (1985), 1207-1242.

[8] J. D. Buckmaster, *An introduction to combustion theory*, The Mathematics of Combustion (ed. J.D. Buckmaster,) *Frontiers in Applied Mathematics, 2*, SIAM, Philadelphia, 1985, 3-46.

[9] J. D. Buckmaster and G. Ludford, *Theory of Laminar Flames*, Cambridge University Press, Cambridge, 1982.

[10] W. E. Fitzgibbon and C. B. Martin "Semilinear parabolic systems modelling spatially inhomogeneous exothermic systems" JMAA (to appear).

[11] W. E. Fitzgibbon and C. B. Margin, *The longtime behavior of solutions to a quasilinear combustion model*, J. Nonlinear Anal (to appear).

[12] W. E. Fitzgibbon and J. J. Morgan, *A diffusive epidemic model on a bounded domain of arbitrary dimension*, Diff. and Int. Eqns, **1**, (1988), 125-132.

[13] W. E. Fitzgibbon, J. J. Morgan and S. S. Waggoner, *A quasilinear system modelling the spread of infectious disease*, Rocky Mount. J. of Math., (to appear).

[14] D. Frank-Kamenetskii, **Diffusion and Heat Transfer in Chemical Kinetics**, Plenum Press, New York, 1969.

[15] J. Hale, *Asymptotic Behavior of Dissipative Dynamical*, J. of Providence, R. I. 1988.

[16] O. A. Ladyzenskaja, V. A. Solonnikov and N.N. Ural'ceva, *Linear and Quasilinear Equations of Parabolic Type*, Translations of A.M.S., **23**, Providence, 1968.

[17] J. J. Morgan, *Boundedness and decay results for reaction diffusion systems*, SIAM J. Math. Anal., **21**(1990), 1172-1189.

[18] T. Niioka, *An asympotic analysis of branched-chain ignition of cold combustible gas by hot inert gas*, Combustion and Flame, **76**(1989), 143-149.

[19] J. Smoller, **Shock Waves and Reaction Diffusion Equations**, Springer-Verlag, Berlin,, 1983.

[20] D. B. Spalding, *The theory of flame phenomena with a chain reaction*, Phil. Trans. Roy. Soc. London, **A249**, 1-25.

[21] D. Terman, *Stability of planar wave solutions to a combustion model*, SIAM J. Math. Anal. **21**(1990) 1139-1171.

[22] Y. Zeldovich, A. Istratvos, N. Kidin and V. Librovich, *Flame propagation in tubes: hydrodynamics and stability*, Combustion Sci. Tech., **24**(1980), 1-13.

A Maximum Principle for Linear Cooperative Elliptic Systems

J. Fleckinger
Univ. Toulouse I

J. Hernandez
Univ. Autonoma Madrid

F.de Thélin
Univ. Toulouse III

Abstract

We give here some conditions for having a Maximum principle for cooperative systems with variable coefficients. They are stated in terms of the first eigenvalue for cooperative systems. This yields a necessary and sufficient condition in the case of a symmetric system.

1 Introduction

The Maximum Principle is a very important tool for many questions concerning partial differential equations, not only for proving existence and uniqueness of solutions, but also for studying their qualitative properties as positivity, symmetry,... (see e.g. [14]). In recent years there has been some progress concerning Maximum Principles for linear elliptic systems. The results in [14] for the cooperative case have been extended in [8], [9] (see also [15]), improving the sufficient conditions given in [14] and providing a necessary and sufficient condition in the constant coefficient case; this last result has been extended by the present authors to nonlinear problems involving the

Differential Equations with
Applications to Mathematical
Physics

p-Laplacian $\Delta_p u := \text{div}(|\nabla u|^{p-2} \nabla u), 1 < p < +\infty$, instead of Δ (see interesting Maximum Principle for non cooperative systems was given in [7], see [4] for a general presentation of these developments).

A closely related problem is the existence of principal eigenvalues (eigenvalues having positive eigenfunctions) for linear non cooperative systems; we mention in this direction the results in [2], [12], [1], [3], see also [5], [4].

In this short note, we give some conditions for having a Maximum Principle for cooperative systems with variable coefficients. They are stated in terms of the first eigenvalue for symmetric cooperative systems. This yields a necessary and sufficient condition in the case of a symmetric system.

2 The Symmetric Case

We study first the symmetric case. Let Ω be a smooth bounded domain in $I\!\!R^d$, we consider the following problem

$$(S) \begin{cases} -\Delta u_i = \sum_{j=1}^n a_{ij}(x)u_j + f_i \text{ in } \Omega \\ u_i = 0 \text{ on } \partial\Omega, \end{cases}$$

where the coefficients $a_{ij}(1 \leq i, j \leq n)$ are bounded and

$$a_{ij} \geq 0 \text{ for } i \neq j. \tag{1}$$

Such systems are called cooperative (or quasi-monotone). We assume that $f_i \in L^2(\Omega)$.

We say that (S) satisfies the Maximum Principle if $f_i \geq 0$ implies $u_i \geq 0, i = 1, \ldots, n$, for any solution $(u_1, \ldots, u_n)$. System (S) can also be written as

$$-\Delta U = AU + F \text{ in } \Omega, \qquad U = 0 \text{ on } \partial\Omega,$$

where U (resp. F) denotes a column matrix with elements u_i (resp. f_i) and $A = (a_{ij}) \in M_{n,n}$. We also consider the eigenvalue problem associated with (S): Find $(\lambda, U) \in \mathbb{C} \times (H_0^1(\Omega))^n$ such that

$$-\Delta U = AU + \lambda U \text{ in } \Omega, \tag{2}$$

in the distributional sense.

The usual spectral theory for linear operators with compact inverse can be applied here (see [6]).

We consider first the case where A is symmetric:

$$a_{ji} = a_{ij} \forall i, j = 1, \ldots, n. \tag{3}$$

We can introduce the bilinear form defined on $(H_0^1(\Omega))^n$ by

$$\mathcal{L}(U, V) = \int_\Omega [\sum_{i=1}^n \nabla u_i \cdot \nabla v_i - \sum_{i,j=1}^n a_{ij}(x) u_i v_j]. \tag{4}$$

It follows from (1) and (3) that $\mathcal{L}$ is continuous and coercive on $(H_0^1(\Omega))^n$; more precisely, there exist positive constants $c_i, i = 0, 1, 2$ such that:

$$\mathcal{L}(U, V) \leq c_0((U, V))$$
$$\mathcal{L}(U, U) + c_1(U, U) \geq c_2((U, U))$$

where (U, V) (resp. $((U, V))$) denotes the scalar product in $(L^2(\Omega))^n$ (resp. $(H_0^1(\Omega))^n$). Hence by applying the Riesz Theorem, we can define self-adjoint compact linear operator associated to (4) in the usual way. Therefore (2) admits an infinite sequence of real eigenvalues and the first one, which is simple, is given by the variational characterization

$$\lambda_1(S) = \inf\{\mathcal{L}(U, U)/(U, U); U \in (H_0^1(\Omega))^n\}. \tag{5}$$

The existence of an eigenvalue of (2) which is simple and has a positive eigenfunction has been studied (also for non necessarily symmetric systems) in [2], [12], [1], [3]; the main tools used there are the Maximum Principle and the Krein-Rutman Theorem. Here (symmetric case), the fact that principal eigenfunctions do not change sign follows from $\mathcal{L}(|U|, |U|) \leq \mathcal{L}(U, U)$, where $|U| = (|u_i|)$.

Theorem 1 *If (1) and (3) are satisfied, then (S) satisfies the Maximum Principle if and only if $\lambda_1(S) > 0$.*

Proof *The condition is necessary.* Consider the "principal eigenvector" $\Phi > 0$. We have

$$-\Delta(-\Phi) = A(-\Phi) + \lambda_1(S)(-\Phi) \text{ in } \Omega, \text{ and } \Phi = 0 \text{ on } \partial\Omega.$$

When $\lambda_1(S) \leq 0, \lambda_1(S)(-\Phi) \geq 0$ and (S) does not satisfy the Maximum Principle.

The condition is sufficient. Multiplying (S) by $u_i^- = \max(-u_i, 0)$ we get

$$\int_\Omega \nabla u_i \cdot \nabla u_i^- = \int_\Omega \sum_{j=1}^n a_{ij}(x) u_j u_i^- + \int_\Omega f_i u_i^-.$$

Hence using classical results by Stampacchia

$$\int_\Omega | \nabla u_i^- |^2 = -\int_\Omega \sum_{j=1}^n a_{ij}(x)(u_j^+ - u_j^-)u_i^- - \int_\Omega f_i u_i^-$$

$$\leq \int_\Omega \sum_{j=1}^n a_{ij}(x) u_j^- u_i^-.$$

By adding these inequalities, $\mathcal{L}(U^-, U^-) \leq 0$, so that, by (5),

$$\lambda_1(S).(U^-, U^-) \leq 0.$$

Since by hypothesis, $\lambda_1(S) > 0$, we obtain $(U^-, U^-) = 0$ and hence

It is very easy to check that the condition given in Theorem 1 coincides with the one in [8] in the constant coefficient case and $n = 2$. If one looks for positive solutions of

$$\begin{cases} -\Delta u = au + bv + \delta u \text{ in } \Omega \\ -\Delta v = bu + dv + \delta v \text{ in } \Omega \\ u = v = 0 \qquad \text{on } \partial\Omega \end{cases} \qquad (6)$$

of the form $(\alpha\varphi_1, \beta\varphi_1)$, where $(\lambda_1(-\Delta), \varphi_1)$ is the principal eigenpair associated with the Dirichlet Laplacian on Ω, one obtains the linear system

$$[\lambda_1(-\Delta) - \lambda - a]\alpha - b\beta = 0$$
$$-b\alpha + [\lambda_1(-\Delta) - \lambda - d]\beta = 0;$$

the first eigenvalue λ_1 of (6) which ensures that $\alpha > 0, \beta > 0$ is given by

$$\lambda_1 = \lambda_1(-\Delta) - \frac{a+d}{2} - \frac{1}{2}(\sqrt{(a-d)^2 + 4b^2}).$$

Now, it is easy to see that $\lambda_1 > 0$ if and only if

$$[\lambda_1(-\Delta) - a][\lambda_1(-\Delta) - d] > b^2,$$

which is the condition obtained in [8], [9].

3   The General Case

When A is not symmetric, we can introduce $\mathcal{L}(U,V)$ as above and still apply Lax-Milgram's Theorem in order to treat the eigenvalue problem (2); for doing this the symmetry of $\mathcal{L}$ is not required. But the associated compact linear operator is not self-adjoint, and the corresponding general theory as in [6] cannot be applied; in particular the variational characterization (5) is lost.

However, it is shown in [1] that there exists a unique principal eigenvalue (eigenvalue associated with a positive eigenfunction) by using a result of Krasnosel'skii ([13], Th. 2.5, p. 67). The results in [12], [3], concern classical solutions and cannot apply directly to the weak solutions of (2).

It is possible to obtain necessary and/or sufficient conditions for the Maximum Principle by considering symmetric systems associated to (S). Let us define the matrices

$$A^V := (a_{ij} V a_{ji}) \text{ and } A^\wedge := (a_{ij} \cdot a_{ji})$$

where

$$\mathrm{pvq} := \sup(p,q) \text{ and } p \cdot q := \inf(p,q),$$

and let us denote by S^V and $S^\wedge$ the associated (symmetric) systems.

Theorem 2 *If the Maximum Principle holds for (S), then $\lambda_1(S) > 0$ and $\lambda_1(S^\wedge) > 0$.*

Theorem 3 *If $\lambda_1(S^V) > 0$, then the Maximum Principle holds for (S) .*

Proof of Theorem 2. The proof of the first part of Theorem 2 is exactly the same as the proof of the first part in Theorem 1 (the condition is necessary). For proving the second part we adapt the same proof. Denote by $\Phi^\wedge$ the principal eigenvector associated to $\lambda_1(S^\wedge)$. Then, we have:

$$-\Delta(-\Phi^\wedge) = A(-\Phi^\wedge) + F \text{ in } \Omega, \Phi^\wedge = 0 \text{ on } \partial\Omega$$

where $F = [\lambda_1(S^\wedge) + (A^\wedge - A)](-\Phi^\wedge) \geq 0$ if $\lambda_1(S^\wedge) \leq 0$ and (S) does not satisfy the Maximum Principle.

Proof of Theorem 3. Multiplying (S) by u_i^- and integrating by parts, we get:

$$\int_\Omega |\nabla u_i^-|^2 = -\sum_{j=1}^n \int_\Omega a_{ij}(x)(u_j^+ - u_j^-)u_i^- - \int_\Omega f_i u_i^-$$

$$\leq \sum_{j=1}^n \int_\Omega a_{ij}(x)u_j^- u_i^-$$

$$\leq \sum_{j=1}^n \int_\Omega (a_{ij} \vee a_{ji})(x)u_j^- u_i^-.$$

By adding these inequalities, we obtain by (5)

$$\lambda_1(S^V).(U^-, U^-) \leq 0.$$

Since by hypothesis, $\lambda_1(S^V) > 0$, we obtain $(U^-, U^-) = 0$ and hence $U \geq 0$.

Acknowledgement

This collaboration was done when J.H. was visiting Professor in Université Toulouse 1 (GREMAQ -URA 947-) in May 1992; he also has been supported by project DGICYT $PB90/0620$. J.F. is grateful to Georgia Institute of Technology for financial support during this Conference.

Bibliography

[1] W. Allegretto, *Sturmian Theorems for Second Order Systems.* Proc. A.M.S., 94, 1985, p. 291-296.

[2] H. Amann, *Fixed Point Equations and Nonlinear Eigenvalue Problems in Ordered Banach Spaces.* S.I.A.M. Rev. 18, 1976, p. 620-709.

[3] R. S. Cantrell; K. Schmitt, *On the Eigenvalue Problem for Coupled Elliptic Systems.* SIAM J. Math. Anal. 17, 1986, p. 850-862.

[4] C. Cosner; J. Hernàndez; E. Mitidieri, *Maximum Principles and Applications to Reaction-Diffusion Systems.* Birkhauser; Boston; (in preparation).

[5] C. Cosner, *Eigenvalue Problems with Indefinite Weights and Reaction Diffusion Models in Population Dynamics.* In *Reaction-Diffusion Equations,* K. J. Brown and A. A. Lacey (eds), Oxford Sc. Publ., 1990, p. 117-137.

[6] R. Courant; D. Hilbert, *Methods of Mathematical Physics* Interscience, New York, 1953

[7] D. G. de Figueiredo; E. Mitidieri, *A Maximum Principle for an Elliptic System and Applications to Semilinear Problems.* S.I.A.M. J. Math. Anal., 17, 1986, p. 836-849.

[8] D. G. de Figueiredo; E. Mitidieri, *Maximum Principles for Cooperative Elliptic Systems,* Comptes Rendus Acad. Sc. Paris, 310, 1990, p. 49-52

[9] D. G. de Figueiredo; E. Mitidieri, *Maximum Principles for Linear Elliptic Systems,* (to appear)

[10] J. Fleckinger; J. Hernandez; F. de Thélin, *Principe du maximum pour un système elliptique non linéaire* Comptes Rendus Acad. Sc. Paris, t.314, Ser. I, p. 665-668, 1992.

[11] J. Fleckinger; J. Hernandez; F. de Thélin, *On Maximum Principles and Existence of Positive Solutions for some Cooperative Elliptic Systems,* (submitted to Differential and Integral Equations).

[12] P. Hess, *On the Eigenvalue Problem for Weakly Coupled Elliptic Systems,* Arch. Rat. Mech. Anal., 81, 1983, p. 151-159.

[13] M. A. Krasnosel'skii, *Positive Solutions of Operator Equations* Nordhoff, Grooningen, 1964

[14] M. H. Protter; H. Weinberger, *Maximum Principles in Differential Equations,* Prentice Hall, Englewood Cliffs, 1967.

[15] G. Sweers, *Strong Positivity in $C(\Omega)$ for Elliptic Systems*, Math. Z., 209, 1992, p. 251-271.

Exact Solutions to Flows in Fluid Filled Elastic Tubes

D. Fusco and N. Manganaro
Dipartimento di Matematica
Università di Messina

Abstract

By means of a similarity–like variable transformation we reduce the model governing flows in fluid filled elastic tubes to the form of a 2×2 quasilinear nonhomogeneous autonomous hyperbolic system of first order partial differential equations. By requiring the latter to be consistent with a pair of additional equations which define Riemann–like invariants along the concerned characteristic curves, we carry out a reduction approach for determining exact solutions to the model under interest.

1 Introduction and General Remarks

Several methods of approach have been proposed in order to determine exact solutions to nonlinear partial differential equations. Among others, group analysis and Bäcklund–like transformations have shown to be an useful tool for the study of a number of problems encountered in engineering and industrial applications of mathematics as well as in theoretical investigations of wave propagation. An exhaustive list of recent references on this subject can be found in [1] and [2]. Without the afore–mentioned framework a great deal of

Differential Equations with
Applications to Mathematical
Physics

attention has been paid to work out reduction techniques for quasi-linear systems of first order of the form

$$U_t + A(U)U_x = B(U) \qquad (1.1)$$

where

$$U = \begin{bmatrix} u \\ v \end{bmatrix} \qquad A = \begin{bmatrix} a_{11} & a_{12} \\ a_{21} & a_{22} \end{bmatrix} \qquad B = \begin{bmatrix} b_1 \\ b_2 \end{bmatrix}$$

x and t are space and time coordinates, respectively. Here and in the following a subscript means for derivative with respect to the indicated variable. Furthermore we asume the system (1.1) to be strictly hyperbolic [3]. That is tantamount to require the matrix A to admit two real distinct eigenvalues λ and μ (characteristic wave speeds) to which there correspond two left eigenvectors $l^{(\lambda)}, l^{(\mu)}$ as well as two right eigenvectors $d^{(\lambda)}, d^{(\mu)}$ spanning the Euclidean space E^2.

When $B = 0$ (e.g., source absence) a standard way to look for solutions to the model in point is represented by the hodograph transformation which is obtained by interchanging the role of dependent and independent variables. The integration of the resulting linear second order equation in the hodograph plane can be investigated by means of the reduction approach to canonical forms developed in [4]. That permits to characterize special classes of material response functions to governing models of physical interest which can be relevant to simple wave interactions [5], [6].

In cases where a source term like B must be taken into account in the governing system there has been proposed [7], [8] a variable transformation in order to link (1.1) to a model of a similar form. Hence a procedure to reduce nonhomogeneous 2×2 systems to canonical form allowing for a close integration or to linear form has been carried out and model constitutive laws concerning different physical contexts have been deduced [9–11].

As far as wave propagation is concerned, it is to be remarked that the term B does not allow the Riemann field variables defined by

$$r(U) = \int l^{(\lambda)} \cdot dU, \qquad\qquad s(U) = \int l^{(\mu)} \cdot dU \qquad (1.2)$$

to be invariant along the characteristic curves associated to (1.1). Such a circumstance recently motivated in [12] an "ad hoc" technique to search for exact solutions to (1.1). The leading idea of this method of approach lies in the investigation of the consistency of (1.1) with a pair of additional equations of the form

$$F_t + \lambda(r,s)F_x = 0 \qquad (1.3)$$

$$G_t + \mu(r,s)G_x = 0 \qquad (1.4)$$

where the functions $F(r,s)$ and $G(r,s)$ are to be determined and they satisfy the condition

$$\frac{\partial(F,G)}{\partial(r,s)} \neq 0. \qquad (1.5)$$

It is very easy to ascertain that the functions F and G fulfilling (1.3) and (1.4) play a role similar to that of the standard Riemann invariants r and s of the homogeneous case. However in the present case only particular solutions of (1.1) are to be expected to satisfy also the additional equations (1.3) and (1.4) since the latter act as "constraints." In other words, for admissible F and G we will determine the solution $r(x,t), s(x,t)$ (or $U(x,t)$) for which (1.3) and (1.4) hold.

As most of the reduction techniques based upon hodograph–like transformations, the approach proposed in [12] can be used for determining exact solutions to 2×2 autonomous models. The main aim of the present paper is to show, in a specific case, that the afore–mentioned method of approach in combination with a similarity reduction suggested by group analysis permits to obtain exact solutions to 2×2 nonautonomous systems as well. We illustrate the procedure for the model governing flows in fluid–filled elastic tubes [13] supplemented by constitutive laws involving response functions of suitable form.

2 The Governing Model and Similarity Reduction

Flows in fluid–filled elastic tubes can be described by the following system of equations [13]

$$p_t + vp_x + \frac{S}{S_p} v_x = -\frac{vS_x + \Psi}{S_p} \tag{2.1}$$

$$v_t + kp_x + vv_x = f - kP_x \tag{2.2}$$

where p is the transmural pressure, v is the fluid velocity, $S = S(p, x)$ is the cross–sectional area and it is assumed $S/S_p > 0$, $P = P(x, t)$ is the external pressure, $k = 1/\rho$ with ρ being (constant) density, $\Psi = \Psi(p, v, x)$ represents the outflow function and $f = f(p, v, x, t)$ is the viscous retarding force. S, Ψ and f are the concerned material response functions which have to be specified in the present case. In general they depend upon the field variables p and v as well as upon the independent variables x and/or t so that the governing model (2.1), (2.2) results to be nonautonomous.

In [14] there has been shown that the system of equations under interest is invariant with respect to infinitesimal transformation groups if the involved response functions obey the restrictions

$$S = S_0(x) \exp\left[\int \frac{d\Pi}{g(\Pi)}\right] \tag{2.3}$$

$$\Psi = -vS \left[\frac{d(\ln(S_0))}{dx} + \frac{a}{2}\right] + Q(\Pi, w) \tag{2.4}$$

$$f = k(ap + \bar{b}) + \frac{a}{2} v^2 + kP_x + H(\Pi, w) \exp\left(\int \frac{a}{2} dx\right) \tag{2.5}$$

where

$$\Pi = (p - p_0) \exp\left(-\int a(x)dx\right) - \int b(x) \exp\left(-\int a(x)dx\right) dx \tag{2.6}$$

$$w = v \exp\left(-\int \frac{a}{2} dx\right) \tag{2.7}$$

$a(x)$, $b(x)$, $S_0(x)$, $g(\Pi)$, $Q(\Pi, w)$ and $H(\Pi, w)$ are arbitrary functions with $g(\Pi) > 0$. Moreover $p_0 = \text{const.}$ and $\bar{b} = b - p_0$.

It is possible to show [15], [16] that by means of the similarity transformation

$$p = \left[\Pi(\bar{x}, t) + \int \bar{b} \exp \left(- \int a\, dx \right) dx \right] \exp \left(\int a\, dx \right) \qquad (2.8)$$

$$v = w(\bar{x}, t) \exp \left(\int \frac{a}{2}\, dx \right) \qquad (2.9)$$

$$\bar{x} = \int \exp \left(- \int \frac{a}{2}\, dx \right) dx \qquad (2.10)$$

the system (2.1), (2.2) can be reduced to the autonomous form

$$\Pi_t + w\Pi_{\bar{x}} + g(\Pi)w_{\bar{x}} = -g(\Pi)Q(\Pi, w) \qquad (2.11)$$

$$w_t + k\Pi_{\bar{x}} + ww_{\bar{x}} = H(\Pi, w) \qquad (2.12)$$

which falls into the class (1.1).

The characteristic wave speeds associated to (2.11), (2.12) are given by

$$\lambda = w + [kg(\Pi)]^{1/2} \qquad \mu = w - [kg(\Pi)]^{1/2} \qquad (2.13)$$

so that in the present case we have

$$l^{(\lambda)} = [(k/g)^{1/2}, 1], \qquad l^{(\mu)} = [-(k/g)^{1/2}, 1]$$

whereupon the Riemann variables (1.2) specialize to

$$r = w + \int [k/g]^{1/2} d\Pi \qquad s = w - \int [k/g]^{1/2} d\Pi. \qquad (2.14)$$

According to the analysis carried on in [12] for later convenience we write the system (2.11), (2.12) in terms of the variables (2.14), namely

$$r_t + \lambda r_x = \beta_1 \qquad (2.15)$$

$$s_t + \mu s_x = \beta_2 \qquad (2.16)$$

where

$$\beta_1 = l^{(\lambda)} \cdot B = H - Q[kg]^{1/2} \qquad \beta_2 = l^{(\mu)} \cdot B = H + Q[kg]^{1/2} \quad (2.17)$$

3   Existence of Riemann–Like Invariant Quantities and Exact Solutions

It is well known that if the wave speeds λ and μ satisfy the exceptionality conditions [3]

$$\nabla\lambda \cdot d^{(\lambda)} = 0 \qquad \nabla\mu \cdot d^{(\mu)} = 0 \tag{3.1}$$

where $\nabla = \left(\frac{\partial}{\partial u}, \frac{\partial}{\partial v}\right)$ then $r = \mu$ and $s = \lambda$. Consequently, for 2×2 homogeneous hyperbolic and completely exceptional (CEX) systems the Riemann invariants are given by the characteristic speeds. A classical example is given by the system of isentropic fluid–dynamics supplemented by a Vón–Karman–like $p - \rho$ law.

Bearing in mind (2.15) and (2.16) the afore–mentioned result concerning Riemann invariants is no longer true for 2×2 nonhomogeneous CEX systems (otherwise it turns out to be $B = 0$). Within the theoretical framework outlined in the introduction let us require the system (2.15), (2.16) to be consistent with two additional equations of the form (1.3) and (1.4) where $F = \mu$ and $G = \lambda$, respectively, so that (1.5) is fulfilled. Of course, taking into account the remark made above about nonhomogeneous 2×2 systems, we assume that the characteristic wave speeds do not satisfy the exceptionality conditions (3.1).

Looking for solutions of (2.11), (2.12) (or equivalently of (2.15), (2.16)) such that $\frac{\partial(r,s)}{\partial(\bar{x},t)} \neq 0$ and owing to (1.5) we can perform the following change of variables

$$\bar{x} = \bar{x}(F,G) = \bar{x}(\lambda,\mu) \qquad t = t(F,G) = t(\lambda,\mu) \tag{3.2}$$

whereupon the set of equations (2.11), (2.12), (1.3), (1.4) takes the form

$$\beta_1 = \frac{r_\lambda}{t_\lambda} \qquad \beta_2 = \frac{s_\mu}{t_\mu} \tag{3.3}$$

$$\bar{x}_\lambda = \lambda t_\lambda \qquad \bar{x}_\mu = \mu t_\mu \tag{3.4}$$

Cross differentiation in (3.4) produces the wave–like equation

$$t_{\lambda\mu} = 0 \tag{3.5}$$

so that the functions $\bar{x}(\lambda, \mu)$ and $t(\lambda, \mu)$ satisfying the pair of equations (3.4) are given by

$$\begin{aligned}
\bar{x} &= \lambda M'(\lambda) + \mu N'(\mu) - M(\lambda) - N(\mu) \\
t(\lambda, \mu) &= M'(\lambda) + N'(\mu)
\end{aligned} \tag{3.6}$$

where $M(\lambda)$ and $N(\mu)$ are arbitrary functions and upper prime means for derivative with respect to the indicated variable. The next step in our approach is to insert (3.6) into the pair of equations (3.3) and to determine appropriately the functions $M(\lambda)$ and $N(\mu)$ in order that the resulting conditions are satisfied. Thus, from (3.6) we will get the particular solution $w(\bar{x}, t), \Pi(\bar{x}, t)$ of the system (2.11), (2.12) for which (1.3) and (1.4) hold with $F = \mu$ and $G = \lambda$, respectively. The solution in point, by means of the transformation (2.8) to (2.10) will prove a particular solution $p(x, t), v(x, t)$ to the nonautonomous governing system (2.1), (2.2).

In the present case we have

$$r(\lambda, \mu) = \frac{\lambda + \mu}{2} + \Gamma(\xi) \qquad s(\lambda, \mu) = \frac{\lambda + \mu}{2} - \Gamma(\xi) \tag{3.7}$$

where

$$\xi = \frac{\lambda - \mu}{2} = [kg(\Pi)]^{1/2} \qquad \Gamma(\xi) = \int [k/g]^{1/2} d\Pi. \tag{3.8}$$

In order to show some possible solutions to the model under investigation, as far as the relations (2.3) to (2.5) defining the response functions are concerned, in the following we assume that the viscous retarding force is of the form [13]

$$f(p, v, x) = \varphi(p, x)v \tag{3.9}$$

as well as that

$$a = 0 \qquad \bar{b} + P_x = 0. \tag{3.10}$$

According to (3.9) and (3.10) in (2.5) we set

$$H(\Pi, w) = \hat{H}(\Pi)w. \tag{3.11}$$

Relations (2.17) yield $Q = \frac{\beta_2 - \beta_1}{2\xi}$ and $H = \frac{\beta_1 + \beta_2}{2}$ so that owing to (3.6), (3.7) the pair of equations (3.3) specializes to

$$Q = \frac{1}{4\xi} \left\{ 1 + \frac{d\Gamma}{d\xi} \right\} \{n(\mu) - m(\lambda)\} \tag{3.12}$$

$$\hat{H}(\xi)(\lambda + \mu) = \frac{1}{2} \left\{ 1 + \frac{d\Gamma}{d\xi} \right\} \{m(\lambda) + n(\mu)\} \tag{3.13}$$

where $m(\lambda) = \frac{1}{M''(\lambda)}$ and $n(\mu) = \frac{1}{N''(\mu)}$.

A direct inspection shows that the system of equations (3.12), (3.13) are satisfied if $M(\lambda)$ and $N(\mu)$ fulfill the relations

$$m(\lambda) = m_0 + m_1\lambda + m_2\lambda^2 \qquad n(\mu) = -m_0 + m_1\mu - m_2\mu^2 \tag{3.14}$$

where m_0, m_1, m_2 are constant and in turn the functions Q and $\hat{H}$ involved there adopt the form

$$Q(\Pi, w) = -\frac{1}{\xi} \left\{ \frac{1}{2} + \left[\frac{dg}{d\Pi}\right]^{-1} \right\} \{m_0 + m_1\xi + m_2(w^2 + \xi^2)\} \tag{3.15}$$

$$\hat{H}(\Pi) = \left\{ \frac{1}{2} + \left[\frac{dg}{d\Pi}\right]^{-1} \right\} (m_1 + 2m_2\xi) \tag{3.16}$$

with $g(\Pi)$ arbitrary. In deducing (3.15) and (3.16) use has been made of relations (3.8).

By prescribing $g(\Pi)$, i.e., through (2.3) the cross–sectional area law, the insertion of (3.15) and (3.16) into (2.4) and (2.5) will define possible model laws for the outflow functions Ψ and for the viscous force f (linearly dependent upon velocity according to (3.9)).

In particular by assuming

$$g(\Pi) = \frac{\Pi}{h} \tag{3.17}$$

with $h \neq 0$ constant, (2.3) specializes to

$$S(x, p) = S_0(x) \left\{ \frac{1}{\Pi_0} [p - p_0 + P(x)] \right\}^h \tag{3.18}$$

with $S_0(x)$ and $\Pi_0 = $ const. arbitrary, whereas (2.4) and (2.5), respectively, reduce to

$$
\Psi = -v \frac{dS_0}{dx} \left\{ \frac{p - p_0 + P}{\Pi_0} \right\}^h + \left(\frac{1}{2} + h \right) \left(\frac{h}{k} \right)^{1/2} (p - p_0 + P)^{1/2}
$$
$$
\times \left\{ m_0 + m_2 \left(\frac{k}{h} \right)^{1/2} (p - p_0 + P)^{1/2} \right.
$$
$$
\times \left[\left(\frac{k}{h} \right)^{1/2} (p - p_0 + P)^{1/2} + 1 \right] + m_2 v^2 \Big\} \tag{3.19}
$$

$$
f(p, v, x) = \left(\frac{1}{2} + h \right) \left\{ m_1 + 2m_2 (k\Pi_0/h)^{1/2} (S/S_0)^{1/2h} \right\} v. \tag{3.20}
$$

Within the present framework as far as the exact solutions to system of equations (2.11), (2.12) are concerned, from (3.6) several possibilites arise in connection with different choices of the parameters m_0, m_1 and m_2 involved in relations (3.14) as well as in (3.19) and (3.20). Here we will consider only two cases where explicit solutions to the system under investigation can be obtained.

i) $m_0 = 0$, $m_1 \neq 0$ and m_2 arbitrary. By inverting (3.6) and making use of the variable transformation (2.8) to (2.10) we gain

$$
p(x, t) = p_0 - P(x) + \frac{m_1^2[(1 - e^{m_2\chi})^2 - 4m_2^2 e^{m_2\chi + m_1\tau}]}{16m_2^2 e^{2m_2\chi}(1 + m_2^2 e^{m_1\tau})^2}
$$
$$
\times \left\{ [(1 - e^{m_2\chi})^2 - 4m_2^2 e^{m_2\chi + m_1\tau}]^{1/2} \right.
$$
$$
\pm (1 + e^{m_2\chi}) \Big\}^2 \tag{3.21}
$$

$$
v(x, t) = \frac{m_1(1 - e^{-m_2\chi})}{4m_2(1 + m_2^2 e^{m_1\tau})} \{ 1 + e^{m_2\chi} \pm [(1 - e^{m_2\chi})^2
$$
$$
- 4m_2^2 e^{m_2\chi + m_1\tau}]^{1/2} \} \tag{3.22}
$$

where

$$
\chi = x + \hat{M} \qquad \tau = t - \hat{N} \tag{3.23}
$$

and $\hat{M}$ and $\hat{N}$ are arbitrary constants coming out from integrating (3.14).

ii) $m_0 = 0, m_1 = 0, m_2$ arbitrary. Here an approach similar to that above yields

$$p(x,t) = p_0 - P(x) + \frac{h \left(e^{(m_2/2)\chi} - e^{-(m_2/2)\chi}\right)^4}{4km_2^2\tau^2} \qquad (3.24)$$

$$v(x,t) = \frac{e^{m_2\chi} - e^{-m_2\chi}}{2m_2\tau} \qquad (3.25)$$

In both cases i) and ii) considered above there are no restrictions on the function $P(x)$ simulating external pressure in the governing model (2.11), (2.12).

4   Conclusions and Final Remarks

The method of approach we developed herein in order to determine exact solutions to the nonautonomous system governing flows in fluid–filled elastic tubes was essentially based on two steps. First, by considering the general classes of material response functions (2.3) to (2.5) allowing for the existence of group symmetries to the model in point as shown in [14], we used the similarity–like variable transformation (2.8) to (2.10) in order to reduce the system of equations (2.1), (2.2) to the autonomous form (2.11), (2.12). Furthermore for the latter system we worked out a procedure for finding out the concerned solutions for which the model (2.11), (2.12) is consistent with two additional equations like (1.3) and (1.4) with a prescribed form of F and G suggested by a well established result for 2×2 quasilinear homogeneous hyperbolic systems of first order. Of course, along the same lines of the analysis worked out hitherto other forms of F and G can be considered. In these cases a leading idea to prescribe F and G is to achieve, by means of the transformation (3.2), a hodograph–like system (see (3.4)) which can be reduced to a canonical form allowing for an explicit integration [4], [5]. In the process we have been able to provide a vehicle for characterizing possible model constitutive laws to the governing system under interest. About that concern we remark that we have some freedom to choose the function $g(\Pi)$ which characterizes the cross–sectional area law and which is involved also in (3.15) and (3.16). Finally, we showed some (explicit) exact solutions to (2.1), (2.2) which can be obtained by means of the present method of approach. Nevertheless, the relations (3.6) with $M(\lambda)$ and $N(\mu)$ defined by (3.12) may provide further exact solutions to the model in point although they will be determined in general in an implicit way. Apart their own theoretical value these solutions can be used for testing numerical procedures to the system (2.1), (2.2) as well as for studying wave propagation into nonconstant states representing nonuniform tube flow regimes where dissipation is taken into account.

Acknowledgements. This work was partially supported by M.U.R.S.T. through "Fondi per la Ricerca Scientifica 40% and 60%" and by C.N.R. through G.N.F.M.

Bibliography

[1] C. Rogers and W. F. Ames, *Nonlinear Boundary Value Problems in Science and in Engineering,* Academic Press, New York (1989).

[2] C. Rogers and W. F. Ames, *Nonlinear Equations in Applied Sciences,* Academic Press, New York (1992).

[3] A. Jeffrey, *Quasilinear Hyperbolic Systems and Waves,* Pitman, London (1976).

[4] C. Currò and D. Fusco, *Reduction to linear canonical forms and generation of conservation laws for a class of quasilinear hyperbolic systems,* Int. J. Non-Linear Mech. 23, 25–35 (1988).

[5] C. Currò and D. Fusco, *On a class of quasilinear hyperbolic reducible systems allowing for special wave interactions,* Z. Angew. Math. Phys. 38, 580–594 (1987).

[6] B. Seymour and E. Varley, *Exact solutions describing soliton-like interactions in a nondispersive medium,* (SIAM) J. Appl. Math. 42, 804–821 (1982).

[7] B. Seymour and E. Varley, *Exact solutions for large amplitude waves in dispersive and dissipative systems*, Stud. Appl. Math. 72, 241–262 (1985).

[8] D. Fusco and N. Manganaro, *Linearization of a hyperbolic model for non–linear heat conduction through hodograph–like and Bäcklund transformations*, Int. J. Non–Linear Mech. 24 (2), 99–103 (1989).

[9] D. Fusco and N. Manganaro, *Prominent features of a variable transformation for a class of quasilinear hyperbolic systems of first order*, Nonlinear Wave Motion, edited by A. Jeffrey, Longman, 71–82 (1989).

[10] D. Fusco and N. Manganaro, *Recent contributions to wave propagation in nonlinear dissipative media*, Numerical and Applied Mathematics, edited by W. F. Ames, Baltzer Scientific Publishing, 101–105 (1989).

[11] D. Fusco and N. Manganaro, *A class of linearizable models and generation of material response functions to nonlinear hyperbolic heat conduction*, J. Math. Phys. 32, 3043–3046 (1991).

[12] D. Fusco and N. Manganaro, *Generation of exact solutions to a class of quasilinear hyperbolic models via reduction techniques*, to appear.

[13] T. J. Pedley, *The Fluid Mechanics of Large Blood Vessels*, Cambridge University Press, Cambridge (1980).

[14] D. Fusco, *Group analysis and constitutive laws for fluid filled elastic tubes*, Int. J. Non–Linear Mech. 19 (6), 565–574 (1984).

[15] N. Manganaro, *Linearization of non autonomous models describing fluid–filled elastic tubes and nonlinear elastic rods with variable cross section*, in print on Acta Mechanica.

[16] A. Donato and F. Oliveri, *When non autonomous equations are equivalent to autonomous ones*, to appear.

[17] G. Boillat and T. Ruggeri, *Characteristic shocks: completely and strictly exceptional systems,* Boll. U.M.I. 5 (15–A), 197–204 (1978).

Spectral Deformations and Soliton Equations

F. Gesztesy
Department of Mathematics
University of Missouri, Columbia, MO 65211, USA

R. Weikard
Department of Mathematics
University of Alabama at Birmingham, Birmingham, AL 35294, USA

1 Introduction

The main purpose of this paper is to describe the construction of new solutions V of the Korteweg–deVries (KdV) hierarchy of equations by deformations of a given finite–gap solution V_0. In order to describe the nature of these deformations we assume for a moment that the given real–valued quasi–periodic finite–gap solution V_0 is described in terms of the Its–Matveev formula [34] (see, e.g., (3.43)). The basic ingredients underlying this formula are a compact hyperelliptic curve K_n of genus n,

$$K_n : \quad y^2 = \prod_{m=0}^{2n} (E_m - z), \quad E_0 < E_1 < \cdots < E_{2n} \qquad (1.1)$$

and an associated Dirichlet divisor

$$\mathcal{D}_{\hat\mu_1(x_0)+\cdots+\hat\mu_n(x_0)}, \qquad (1.2)$$

Differential Equations with
Applications to Mathematical
Physics

$$\hat{\mu}_j(x_0) = \left(\mu_j(x_0), \left(\prod_{m=0}^{2n} (E_m - \mu_j(x_0)) \right)^{1/2} \right),$$

$$\mu_j(x_0) \in [E_{2j-1}, E_{2j}], \quad 1 \leq j \leq n, \quad x_0 \in \mathbb{R} \text{ fixed}$$

(see Section 3). Here the parameters $\{E_m\}_{m=0}^{2n}$ in (1.1) (characterizing the branch points of K_n) and the projections $\{\mu_j(x_0)\}_{j=1}^{n}$ in (1.2) are spectral parameters of the underlying one–dimensional Schrödinger differential expression

$$\tau_0 = -\frac{d^2}{dx^2} + V_0 \quad . \tag{1.3}$$

in the following sense: The spectrum $\sigma(H_0)$ of the self–adjoint operator

$$H_0 = -\frac{d^2}{dx^2} + V_0 \quad \text{on} \quad H^2(\mathbb{R}) \tag{1.4}$$

in $L^2(\mathbb{R})$ is given by

$$\sigma(H_0) = \bigcup_{j=1}^{n} [E_{2(j-1)}, E_{2j-1}] \cup [E_{2n}, \infty) \tag{1.5}$$

and the spectrum $\sigma(H_{0,x_0}^D)$ of the Dirichlet operator H_{0,x_0}^D associated with τ_0 and an additional Dirichlet boundary condition at $x_0 \in \mathbb{R}$

$$H_{0,x_0}^D = -\frac{d^2}{dx^2} + V_0, \tag{1.6}$$

$$\mathcal{D}(H_{0,x_0}^D) = \{ g \in H^1(\mathbb{R}) \cap H^2(\mathbb{R}\backslash\{x_0\}) \,|\, g(x_0) = 0 \}$$

is given by

$$\sigma(H_{0,x_0}^D) = \{\mu_j(x_0)\}_{j=1}^{n} \cup \sigma(H_0). \tag{1.7}$$

Deformations of the spectral parameters E_m, $m = 0, ..., 2n$ and $\mu_j(x_0)$, $j = 1, ..., n$ in the corresponding Its–Matveev formula then yield new solutions V of the KdV hierarchy. In particular, it follows from (1.5) that deformations of $\{E_m\}_{m=0}^{2n}$ produce non–isospectral deformations of solutions of the KdV hierarchy, whereas deformations of $\{\hat{\mu}_j(x_0)\}_{j=1}^{n}$ are isospectral with respect to H_0.

Perhaps the simplest and best known non–isospectral deformation is the one where one or several spectral bands are contracted into points, e.g.,

$$[E_{2(m_0-1)},\ E_{2m_0-1}] \longrightarrow \lambda_{m_0}. \tag{1.8}$$

In this case K_n degenerates into the singular curve $\hat{K}_n$

$$K_n \longrightarrow \hat{K}_n\ :\ y^2 = (\lambda_{m_0} - z)^2 \prod_{\substack{m=0 \\ m \neq 2m_0-1,\, 2m_0}}^{2n} (E_m - z), \tag{1.9}$$

$$V_0 \longrightarrow V_1(\lambda_{m_0}) \tag{1.10}$$

and the resulting solution $V_1(\lambda_{m_0})$ represents a one–soliton solution on the background of another finite–gap solution $\tilde{V}_0$ corresponding to the hyperelliptic curve

$$\tilde{K}_{n-1}:\ y^2 = \prod_{\substack{m=0 \\ m \neq 2m_0-1,\, 2m_0}}^{2n} (E_m - z) \tag{1.11}$$

of genus $n - 1$. Applying this procedure n–times finally yields the celebrated n–soliton solutions $V_n(\lambda_1, \ldots, \lambda_n)$ of the KdV hierarchy (see [48], [49]).

On the other hand, varying $\hat{\mu}_j(x_0)$, $1 \leq j \leq n$ independently from each other traces out the isospectral manifold of solutions associated with the base solution V_0.

In Section 2 we give a brief account of the KdV hierarchy using a recursive approach. Section 3 describes real–valued quasi–periodic finite–gap solutions and the underlying Its–Matveev formula in some detail. (It also describes the mathematical terminology in connection with hyperelliptic curves needed in our main Section 5.) Section 4 introduces isospectral and non–isospectral deformations in a systematic way by alluding to single and double commutation techniques. In Section 5 we present our main new result on the isospectral set $I_{IR}(V_0)$ of smooth real–valued quasi–periodic finite–gap solutions of a given base solution V_0. (To be precise, we only represent the stationary, i.e., time–independent case since the insertion of the proper time–dependence poses no difficulties.) Finally, in Section 6 we sketch some generalizations and open problems in connection with infinite–gap solutions and consider the limit of N–soliton solutions as $N \to \infty$ in some detail.

Throughout this paper we confine ourselves to the KdV hierarchy. However, our methods extend to other $1 + 1$–dimensional completely integrable nonlinear evolution equations and to higher–dimensional systems such as the KP hierarchy. Work on these extensions is in progress and will appear elsewhere.

2   The KdV Hierarchy

In order to describe the hierarchy of KdV equations we first recall the recursive approach to the underlying Lax pairs (see, e.g., [3], [44], [46] for details). Consider the differential expressions

$$L(t) \quad = \quad -\frac{d^2}{dx^2} + V(x,t), \tag{2.1}$$

$$\hat{P}_{2n+1}(t) \quad = \quad \sum_{j=0}^{n} [-\frac{1}{2}\hat{f}_{j,x}(x,t) + \hat{f}_j(x,t)\frac{d}{dx}]L(t)^{n-j},$$

where the $\{\hat{f}_j\}_{j=0}^{n}$ satisfy the recursion relation

$$\hat{f}_0 \quad = \quad 1, \tag{2.2}$$

$$2\hat{f}_{j,x} \quad = \quad -\frac{1}{2}\hat{f}_{j-1,xxx} + 2V\hat{f}_{j-1,x} + V_x\hat{f}_{j-1}, \quad 1 \le j \le n.$$

Define also $\hat{f}_{n+1}$ by

$$2\hat{f}_{n+1,x} = -\frac{1}{2}\hat{f}_{n-1,xxx} + 2V\hat{f}_{n-1,x} + V_x\hat{f}_{n-1}. \tag{2.3}$$

Then one can show that

$$[\hat{P}_{2n+1}, L] = 2\hat{f}_{n+1,x}, \tag{2.4}$$

where $[.,.]$ denotes the commutator. Explicitly one computes from (2.2) for the first few $\hat{f}_n$

$$\hat{f}_0 = 1, \tag{2.5}$$

$$\hat{f}_1 = \frac{1}{2}V + c_1, \tag{2.6}$$

$$\hat{f}_2 = -\frac{1}{8}V_{xx} + \frac{3}{8}V^2 + \frac{c_1}{2}V + c_2, \tag{2.7}$$

$$\hat{f}_3 = \frac{1}{32}V_{xxxx} - \frac{5}{16}VV_{xx} - \frac{5}{32}V_x^2 + \frac{5}{16}V^3$$
$$+ \frac{c_1}{2}\left[-\frac{1}{4}V_{xx} + \frac{3}{4}V^2\right] + \frac{c_2}{2}V + c_3, \tag{2.8}$$

where $\{c_j\}_{j \in \mathbb{N}}$ are integration constants. We shall use the convention that all homogeneous quantities, defined by $c_l \equiv 0$, $l \in \mathbb{N}$, are denoted by $f_j := \hat{f}_j(c_l \equiv 0)$, $P_{2n+1} := \hat{P}_{2n+1}(c_l \equiv 0)$, $l \in \mathbb{N}$, i.e.,

$$f_0 = 1, \tag{2.9}$$

$$f_1 = \frac{1}{2}V, \tag{2.10}$$

$$f_2 = -\frac{1}{8}V_{xx} + \frac{3}{8}V^2, \tag{2.11}$$

$$f_3 = \frac{1}{32}V_{xxxx} - \frac{5}{16}VV_{xx} - \frac{5}{32}V_x^2 + \frac{5}{16}V^3. \tag{2.12}$$

The KdV hierarchy is then defined as the sequence of evolution equations

$$KdV_n(V) : = V_t - [P_{2n-1}, L] = V_t - 2f_{n+1,x}(V) = 0,$$
$$n \in \mathbb{N} \cup \{0\}. \tag{2.13}$$

(Since the $\hat{f}_{n+1}$ are differential polynomials in V we somewhat abuse notation by writing $\hat{f}_{n+1}(V)$ for $\hat{f}_{n+1}(x,t)$.) The first few equations of the KdV hierarchy (2.13) then read

$$KdV_0(V) = V_t - V_x = 0, \tag{2.14}$$

$$KdV_1(V) = V_t + \frac{1}{4}V_{xxx} - \frac{3}{2}VV_x = 0, \tag{2.15}$$

$$KdV_2(V) = V_t - \frac{1}{16}V_{xxxxx} + \frac{5}{8}VV_{xxx}$$
$$+ \frac{5}{4}V_xV_{xx} - \frac{15}{8}V^2V_x = 0, \tag{2.16}$$

with $KdV_1(.)$ the usual KdV equation. The inhomogeneous version associated with (2.13) is

$$
\begin{aligned}
V_t - [\hat{P}_{2n+1}, L] &= V_t - 2\hat{f}_{n+1,x}(V) \qquad\qquad (2.17)\\
&= V_t - 2\sum_{j=0}^{n} c_{n-j}\, f_{j+1,x}(V) = 0, \quad c_0 = 1.
\end{aligned}
$$

The special case of the stationary KdV hierarchy characterized by $V_t = 0$ then reads

$$
f_{n+1,x}(V) = 0, \quad \text{resp.} \quad \sum_{j=0}^{n} c_{n-j} f_{j+1,x}(V) = 0. \qquad (2.18)
$$

Particularly simple solutions of (2.18) for $n = 1, 2$ are

$$
V(x) = 2\mathcal{P}(x + w'; g_2, g_3), \qquad\qquad (2.19)
$$

$$
KdV_1(2\mathcal{P}) = 0, \qquad\qquad (2.20)
$$

$$
V(x) = 6\mathcal{P}(x + w'; g_2, g_3), \qquad\qquad (2.21)
$$

$$
KdV_2(6\mathcal{P}) - \frac{21}{8} g_2 KdV_0(6\mathcal{P}) = 0, \qquad\qquad (2.22)
$$

where $\mathcal{P}(z; g_2, g_3)$ denotes the Weierstrass elliptic function with invariants g_2, g_3 and half–periods ω, ω', $\omega > 0$, $-i\omega' > 0$ [2].

Next define the polynomial $\hat{F}_n$ in z

$$
\hat{F}_n(z, x, t) = \sum_{j=0}^{n} z^j\, \hat{f}_{n-j}(V(x,t)) = \prod_{j=1}^{n} [z - \mu_j(x,t)], \quad n \in \mathbb{N} \cup \{0\},
$$
$$(2.23)$$

whose zeros we denote by $\{\mu_j(x,t)\}_{j=1}^{n}$. Then (2.17) becomes

$$
V_t = -\frac{1}{2}\hat{F}_{n,xxx} + 2(V - z)\hat{F}_{n,x} + V_x\hat{F}_n. \qquad (2.24)
$$

In the following we specialize to the stationary case $V_t = 0$. However, as will become clear from the paragraph following (3.42) (see also the end of Sections 4 and 6), corresponding solutions for any time–dependent element of the KdV hierarchy can easily be obtained.

Assuming $V_t = 0$ we get

$$-\frac{1}{2}\hat{F}_{n,xxx} + 2(V - z)\hat{F}_{n,x} + V_x\hat{F}_n = 0. \tag{2.25}$$

Integrating (2.25) once results in

$$\hat{F}_{n,xx}\hat{F}_n - \frac{1}{2}\hat{F}_{n,x}^2 - 2(V - z)\hat{F}_n^2 = -2\hat{R}_{2n+1}(z), \tag{2.26}$$

where the integration constant $-2\hat{R}_{2n+1}(z)$ is easily seen to be a polynomial in z of degree $2n + 1$. Thus we may write

$$\hat{R}_{2n+1}(z) = \prod_{m=0}^{2n}(E_m - z) \tag{2.27}$$

denoting by $\{E_m\}_{m=0}^{2n}$ the zeros of $\hat{R}_{2n+1}$. A comparison of powers of z in (2.26) then yields the trace relation

$$V(x) = \sum_{m=0}^{2n} E_m - 2\sum_{j=1}^{n} \mu_j(x) \tag{2.28}$$

and the first–order system of differential equations

$$\mu_j'(x) = 2\hat{R}_{2n+1}(\mu_j(x))^{1/2}\prod_{\substack{l=1 \\ l \neq j}}^{n}[\mu_l(x) - \mu_j(x)]^{-1}, \quad 1 \leq j \leq n. \tag{2.29}$$

Since $V_t = 0$ implies

$$[\hat{P}_{2n+1}, L] = 0 \tag{2.30}$$

the (inhomogeneous) stationary KdV hierarchy is defined in terms of commuting ordinary differential operators. By a result of Burchnall and Chaundy [7], [8], (2.30) implies that $\hat{P}_{2n+1}$ and L fulfill an algebraic equation. One readily verifies that the polynomial $\hat{R}_{2n+1}$ enters this algebraic equation in the form

$$\hat{P}_{2n+1}^2 = \hat{R}_{2n+1}(L) = \prod_{m=1}^{2n}(E_m - L). \tag{2.31}$$

Hence one is led to hyperelliptic curves

$$y^2 = \hat{R}_{2n+1}(z) = \prod_{m=0}^{2n} (E_m - z) \qquad (2.32)$$

in a natural way. Returning to our simple examples (??)–(??), one computes for n=1:

$$V(x) = 2\mathcal{P}(x + \omega'; g_2, g_3), \qquad (2.33)$$

$$P_3^2 = -L^3 + \frac{g_2}{4} L - \frac{g_3}{4} \qquad (2.34)$$

(an elliptic curve), and for n=2:

$$V(x) = 6\mathcal{P}(x + \omega'; g_2, g_3), \qquad (2.35)$$

$$\hat{P}_5 = P_5 - \frac{21}{8} g_2 P_1, \qquad (2.36)$$

$$(P_5 - \frac{21}{8} g_2 P_1)^2 = (L^2 - 3g_3)(-L^3 + \frac{9}{4} g_2 L + \frac{27}{4} g_3). (2.37)$$

3 Finite–Gap Potentials, Its–Matveev Formula

Any V satisfying a stationary higher order KdV equation of the type

$$\hat{f}_{n+1,x}(V) = \sum_{j=0}^{n} c_{n-j} f_{j+1,x}(V) = 0 \qquad (3.1)$$

will be called a (stationary) finite–gap potential. In order to explain this terminology we make the following two hypotheses:

(H.3.1) $V \in C^\infty(\mathbb{R})$ is real–valued.

(H.3.2) $E_0 < E_1 < \cdots < E_{2n}$.

In particular, (H.3.2) implies simple zeros of $\hat{R}_{2n+1}$ and hence yields a nonsingular hyperelliptic curve (??). In addition one can show that (3.1) together with (H.3.1) and (H.3.2) imply quasi–periodicity and hence boundedness of V (see (3.36)). Hypotheses (H.3.1) and (H.3.2) will be assumed throughout the end of Section 5. Moreover, the one–dimensional Schrödinger operator H in $L^2(\mathbb{R})$ defined by

$$H = -\frac{d^2}{dx^2} + V \quad \text{on} \quad H^2(\mathbb{R}) \tag{3.2}$$

($H^p(\Omega)$, $\Omega \subseteq \mathbb{R}$, $p \in \mathbb{N}$ the usual Sobolev spaces) is self–adjoint with spectrum $\sigma(H)$ given by

$$\sigma(H) = \bigcup_{j=1}^{n} [E_{2(j-1)}, E_{2j-1}] \cup [E_{2n}, \infty). \tag{3.3}$$

Thus H has finitely many spectral gaps ρ_n,

$$\rho_0 = (-\infty, E_0), \quad \rho_j = (E_{2j-1}, E_{2j}), \quad 1 \le j \le n. \tag{3.4}$$

Moreover, $\mu_j(y)$ defined in (2.23) are the eigenvalues of the Dirichlet operator H_y^D in $L^2(\mathbb{R})$

$$\begin{aligned} H_y^D &= -\frac{d^2}{dx^2} + V, \tag{3.5} \\ \mathcal{D}(H_y^D) &= \{ g \in H^1(\mathbb{R}) \cap H^2(\mathbb{R}\backslash\{y\}) \,|\, g(y) = 0 \} \end{aligned}$$

with a Dirichlet boundary condition at $y \in \mathbb{R}$. In addition,

$$\mu_j(y) \in \bar{\rho}_j, \quad y \in \mathbb{R}, \quad 1 \le j \le n. \tag{3.6}$$

(See, e.g., [57] for proofs of (3.3)–(3.6).)

In order to describe the Its–Matveev formula [34] for potentials satisfying (3.1) and Hypotheses (H.3.1) and (H.3.2) we need to discuss the hyperelliptic curve

$$y^2 = \hat{R}_{2n+1}(z) = \prod_{m=0}^{2n} (E_m - z), \quad E_0 < E_1 < \cdots < E_{2n} \tag{3.7}$$

in more detail. (See [15]–[17], [24], [26], [30], [44], [46], [48], [50], [57] for reviews on the remaining material of Section 3. Our terminology will follow the one in [24] and [26].)

We employ the usual topological model associated with (3.7) by considering two copies of the cut plane

$$\Pi_0 = \mathbb{C} \backslash \bigcup_{j=0}^{n} \rho_j \tag{3.8}$$

and joining the upper and lower rims of the cuts $\bar{\rho}_j$ crosswise. This leads to the compact hyperelliptic curve K_n consisting of points

$$P = (z, \hat{R}_{2n+1}(z)^{1/2}), \quad z \in \mathbb{C} \text{ and } P_\infty \tag{3.9}$$

(P_∞ the point at infinity obtained by one–point compactification) with branch points

$$(E_m, 0), \quad 0 \le m \le 2n, \quad P_\infty. \tag{3.10}$$

We also need the projection

$$\Pi : \begin{cases} K_n & \longrightarrow \mathbb{C} \cup \{\infty\} \\ P = (z, \hat{R}_{2n+1}(z)^{1/2}) & \longrightarrow z \\ P_\infty & \longrightarrow \infty \end{cases} \tag{3.11}$$

and the involution (sheet exchange map)

$$* : \begin{cases} K_n & \longrightarrow K_n \\ P = (z, \hat{R}_{2n+1}(z)^{1/2}) & \longrightarrow P^* = (z, -\hat{R}_{2n+1}(z)^{1/2}). \end{cases} \tag{3.12}$$

The upper sheet Π_+ of K_n is then declared as follows. Define

$$\lim_{\epsilon \downarrow 0} \hat{R}_{2n+1}(\lambda + i\epsilon)^{1/2} = -|\hat{R}_{2n+1}(\lambda + i0)^{1/2}|, \quad \lambda < E_0 \tag{3.13}$$

on Π_+ and analytically continue with respect to λ. Local coordinates ζ near $P_0 = (z_0, \hat{R}_{2n+1}(z_0)^{1/2})$, P^∞ then read

$$\zeta = \begin{cases} (z - z_0), & z_0 \in \mathbb{C} \backslash \{E_m\}_{m=0}^{2n} \\ (z - E_m)^{1/2}, & z_0 = E_m, \; 0 \le m \le 2n \\ z^{-1/2}, & z_0 = \infty. \end{cases} \tag{3.14}$$

A convenient homology basis $\{a_j, b_j\}_{j=1}^n$ on K_n, $n \in \mathbb{N}$ is then chosen as follows: the cycle a_j surrounds the cut $\bar{\rho}_j$ clockwise on Π_+ while b_j starts at the lower rim of $\bar{\rho}_j$ on Π_+, intersects a_j, then encircles E_0 clockwise thereby changing into the lower sheet Π_-, and returns on Π_- to its initial point. The cycles are chosen in such a way that their intersection matrix reads

$$a_j \circ b_l = \delta_{j,l}, \quad 1 \le j, l \le n. \tag{3.15}$$

A basis for the holomorphic differentials (Abelian differentials of the first kind, DFK) on K_n is given by

$$\eta_j = \hat{R}_{2n+1}(z)^{-1/2} z^{j-1} dz, \quad 1 \le j \le n. \tag{3.16}$$

We choose the standard normalization

$$\omega_j = \sum_{l=1}^{n} c_{j,l} \, \eta_l \quad \text{with} \quad \int_{a_j} \omega_l = \delta_{j,l}, \quad 1 \le j, l \le n \tag{3.17}$$

and define the b–periods of ω_l by

$$\tau_{j,l} = \int_{b_j} \omega_l, \quad 1 \le j, l \le n. \tag{3.18}$$

Riemann's period relations and (H.3.2) then imply

$$\tau_{j,l} = \tau_{l,j}, \quad \tau = iT, \quad T = (T_{j,l}) > 0. \tag{3.19}$$

Abelian differentials of the second kind (DSK) $\omega^{(2)}$ are characterized by vanishing residues and conveniently normalized by

$$\int_{a_j} \omega^{(2)} = 0, \quad 1 \le j \le n. \tag{3.20}$$

The Riemann theta–function θ and Jacobi variety $J(K_n)$ associated with K_n are then defined as

$$\theta(\underline{z}) = \sum_{\underline{m} \in \mathbb{Z}^n} e^{2\pi i (\underline{m}, \underline{z}) + \pi i (\underline{m}, \tau \underline{m})}, \quad \underline{z} \in \mathbb{C}^n \tag{3.21}$$

and

$$J(K_n) = \mathbb{C}^n / L_n, \tag{3.22}$$

where L_n denotes the period lattice

$$L_n = \{ \underline{z} = (\underline{N} + \tau \underline{M}) \in \mathbb{C}^n \mid \underline{M}, \underline{N} \in \mathbb{Z}^n \}. \tag{3.23}$$

Divisors $\mathcal{D}$ on K_n are defined as integer–valued maps

$$\mathcal{D} : K_n \longrightarrow \mathbb{Z} \tag{3.24}$$

where only finitely many $\mathcal{D}(P) \neq 0$. The degree $\deg(\mathcal{D})$ of $\mathcal{D}$ is defined by

$$\deg(\mathcal{D}) = \sum_{P \in K_n} \mathcal{D}(P). \tag{3.25}$$

The set of all divisors on K_n is denoted by $\text{Div}(K_n)$ and forms an Abelian group under addition. The set of positive divisors will be denoted by $\text{Div}_+(K_n)$,

$$\text{Div}_+(K_n) = \{\mathcal{D} \in \text{Div}(K_n) \,|\, \mathcal{D} : K_n \longrightarrow \mathbb{N} \cup \{0\}\} \tag{3.26}$$

(one writes $\mathcal{D} \geq 0$ for $\mathcal{D} \in \text{Div}_+(K_n)$) and the set of positive divisors of degree $r \in \mathbb{N}$ is as usual identified with the r-th symmetric product $\sigma^r K_n$ of K_n. We also use the notation

$$\mathcal{D}_{P_1 + \cdots + P_r} : \begin{cases} K_n & \longrightarrow \mathbb{N} \cup \{0\} \\ P & \longrightarrow \begin{cases} m & \text{if } P \text{ occurs } m\text{-times in } \{P_1, \ldots, P_r\} \\ 0 & \text{if } P \notin \{P_1, \ldots, P_r\} \end{cases} \end{cases} \tag{3.27}$$

for divisors in $\sigma^r K_n$. The Abel (Jacobi) map with base point $P_0 \in K_n$ is then defined by

$$\underline{A}_{P_0} : \begin{cases} K_n & \longrightarrow J(K_n) \\ P & \longrightarrow \left\{ \int_{P_0}^{P} \omega_j \right\}_{j=1}^{n} \quad (\text{mod} \, L_n) \end{cases} \tag{3.28}$$

respectively by

$$\underline{\alpha}_{P_0} : \begin{cases} \text{Div}(K_n) & \longrightarrow J(K_n) \\ \mathcal{D} & \longrightarrow \sum_{P \in K_n} \mathcal{D}(P) \underline{A}_{P_0}(P). \end{cases} \tag{3.29}$$

If $f \not\equiv 0$ is a meromorphic function on K_n, the divisor (f) of f is defined by

$$(f) : \begin{cases} K_n & \longrightarrow \mathbb{Z} \\ P & \longrightarrow \nu_f(P), \end{cases} \tag{3.30}$$

where $\nu_f(P)$ denotes the order of f at P. Divisors of the type (3.30) are called principal. Two divisors $\mathcal{D}, \mathcal{E} \in \text{Div}(K_n)$ are called linearly equivalent, $\mathcal{D} \sim \mathcal{E}$ iff they differ by a principal divisor, i.e., iff

$$\mathcal{D} = \mathcal{E} + (f) \tag{3.31}$$

for some meromorphic $f \not\equiv 0$ on K_n. The equivalence class of $\mathcal{D}$ is denoted by $[\mathcal{D}]$ (if $\mathcal{D} \geq 0$, $|\mathcal{D}|$ usually denotes the set of positive divisors linearly equivalent to $\mathcal{D}$). By Abel's theorem,

$$\mathcal{D} \sim \mathcal{E} \text{ iff } \begin{cases} \deg(\mathcal{D}) = \deg(\mathcal{E}) \\ \underline{A}_{P_0}(\mathcal{D}) = \underline{A}_{P_0}(\mathcal{E}). \end{cases} \tag{3.32}$$

The Jacobi inversion theorem states

$$\underline{\alpha}_{P_0}(\sigma^n K_n) = J(K_n). \tag{3.33}$$

Finally, a positive divisor $\mathcal{D} \in \sigma^n K_n$ is called nonspecial iff the equivalence class $|\mathcal{D}|$ of positive divisors of $\mathcal{D}$ only consists of $\mathcal{D}$ itself, i.e., iff

$$|\mathcal{D}| = \{\mathcal{D}\}. \tag{3.34}$$

Otherwise $\mathcal{D} \geq 0$ is called special. One can show that $\mathcal{D}_{P_1+\cdots+P_n} \in \sigma^n K_n$ is special iff there exists at least one pair (P, P^*) such that

$$(P, P^*) \in \{P_1, \ldots, P_n\}. \tag{3.35}$$

After these preliminaries we can describe in detail the Its–Matveev formula [34] for real–valued finite–gap potentials V satisfying (3.1). It reads

$$V(x) = \sum_{m=0}^{2n} E_m - 2 \sum_{j=1}^{n} \lambda_j \tag{3.36}$$

$$-2 \frac{d^2}{dx^2} \ln \theta \left(\underline{\zeta}_{P_\infty} + \underline{\alpha}_{P_\infty} (\mathcal{D}_{\hat{\mu}_1(x_0)+\cdots+\hat{\mu}_n(x_0)}) + \frac{(x - x_0)}{2\pi} \underline{U}_0 \right).$$

Here

$$\underline{\zeta}_{P_\infty} = \frac{1}{2} \left\{ j + \sum_{l=1}^{n} \tau_{j,l} \right\}_{j=1}^{n} \in \mathbb{C}^n \tag{3.37}$$

denotes the vector of Riemann constants, $\underline{U}_0$ given by

$$U_{0,j} = \int_{b_j} \omega_0^{(2)}, \quad \int_{a_j} \omega_0^{(2)} = 0, \quad 1 \leq j \leq n \tag{3.38}$$

denotes the vector of b–periods of the normalized DSK

$$\omega_0^{(2)} = -2^{-1} i \, \hat{R}_{2n+1} (z)^{-1/2} \prod_{j=1}^{n} (\lambda_j - z) \, dz \qquad (3.39)$$

$$= [\zeta^{-2} + 0(1)] \, d\zeta \quad \text{near} \quad P_\infty$$

with a single pole at P_∞. (3.39) also identifies the numbers $\{\lambda_j\}_{j=1}^{n}$ in (3.36). (One infers $\lambda_j \in \rho_j$, $1 \le j \le n$.) Moreover, the Dirichlet divisor $\mathcal{D}_{\hat{\mu}_1(x)+\cdots+\hat{\mu}_n(x)}$ is obtained as follows.

$$\hat{\mu}_j(x) = (\mu_j(x), \hat{R}_{2n+1} (\mu_j(x))^{1/2}), \quad 1 \le j \le n, \qquad (3.40)$$

where $\{\mu_j(x)\}_{j=1}^{n}$ satisfy the system (2.29) with prescribed initial conditions

$$\hat{\mu}_j(x_0) = (\mu_j(x_0), \hat{R}_{2n+1} (\mu_j(x_0))^{1/2}), \quad 1 \le j \le n \qquad (3.41)$$

at x_0. In particular, the Abel map linearizes the system (2.29) since (modulo L_n)

$$\underline{\alpha}_{P_\infty} \left(\mathcal{D}_{\hat{\mu}_1(x)+\cdots+\hat{\mu}_n(x)}\right) = \underline{\alpha}_{P_\infty} \left(\mathcal{D}_{\hat{\mu}_1(x_0)+\cdots+\hat{\mu}_n(x_0)}\right)$$

$$+ \frac{(x - x_0)}{2\pi} \underline{U}_0, \quad x \in \mathbb{R}. \quad (3.42)$$

So far we have only discussed the stationary case. However, (3.36) easily extends to the time–dependent situation [34]. E.g.,

$$V(x,t) = \sum_{m=0}^{2n} E_m - 2 \sum_{j=1}^{n} \lambda_j \qquad (3.43)$$

$$-2\partial_x^2 \ln \theta \left(\underline{\zeta}_{P_\infty} + \underline{\alpha}_{P_\infty} \left(\mathcal{D}_{\hat{\mu}_1(x_0,t_0)+\cdots+\hat{\mu}_n(x_0,t_0)}\right)\right.$$

$$\left. + \frac{(x - x_0)}{2\pi} \underline{U}_0 + \frac{3(t - t_0)}{2\pi} \underline{U}_2\right)$$

satisfies the KdV$_1$ equation (see (2.16)), i.e.,

$$KdV_1(V) = V_t + \frac{1}{4} V_{xxx} - \frac{3}{2} V V_x = 0, \qquad (3.44)$$

where $\underline{U}_2$ is the vector of b–periods of the normalized DSK $\omega_2^{(2)}$ with a single pole at P_∞ of the type

$$\omega_2^{(2)} = [\zeta^{-4} + 0(1)]\,d\zeta \quad \text{near} \quad P_\infty, \tag{3.45}$$

$$U_{2,j} = \int_{b_j} \omega_2^{(2)}, \quad \int_{a_j} \omega_2^{(2)} = 0, \quad 1 \le j \le n. \tag{3.46}$$

In this case the Dirichlet divisor $\mathcal{D}_{\hat\mu_1(x,t)+\cdots+\hat\mu_n(x,t)}$ is obtained as follows.

$$\hat\mu_j(x,t) = (\mu_j(x,t),\ \hat{R}_{2n+1}\,(\mu_j(x,t))^{1/2}), \quad 1 \le j \le n, \tag{3.47}$$

where $\{\mu_j(x,t)\}_{j=1}^n$ satisfy the system

$$\partial_x\,\mu_j(x,t) = 2\hat{R}_{2n+1}\,(\mu_j(x,t))^{1/2} \prod_{\substack{l=1 \\ l\ne j}}^{n} [\mu_l(x,t) - \mu_j(x,t)]^{-1},$$

$$\partial_t\,\mu_j(x,t) = 2\left[\sum_{m=0}^{2n} E_m - 2\sum_{\substack{l=1 \\ l\ne j}}^{n} \mu_l(x,t)\right] \partial_x\,\mu_j(x,t),$$

$$1 \le j \le n \tag{3.48}$$

with prescribed initial conditions

$$\hat\mu_j(x_0,t_0) = (\mu_j(x_0,t_0),\ \hat{R}_{2n+1}\,(\mu_j(x_0,t_0))^{1/2}), \quad 1 \le j \le n \tag{3.49}$$

at (x_0,t_0). Again the Abel map linearizes the system (3.48) since

$$\underline{\alpha}_{P_\infty}\left(\mathcal{D}_{\hat\mu_1(x,t)+\cdots+\hat\mu_n(x,t)}\right) = \underline{\alpha}_{P_\infty}\left(\mathcal{D}_{\hat\mu_1(x_0,t_0)+\cdots+\hat\mu_n(x_0,t_0)}\right)$$

$$+ \frac{(x - x_0)}{2\pi}\underline{U}_0 + \frac{3(t - t_0)}{2\pi}\underline{U}_2, \quad (x,t) \in \mathbb{R}^2 \tag{3.50}$$

(modulo L_n).

4 Spectral Deformations, Commutation Techniques

Since virtually all explicitly known solutions of the KdV hierarchy, such as soliton solutions, rational solutions, and solitons on the background of quasi–periodic finite–gap solutions, can be obtained from

the Its–Matveev formula upon suitable deformations (singulariza-
tions) of the underlying hyperelliptic curve K_n (see e.g. [17]–[20],
[26], [48], [49], [64] and the references therein), we propose a systematic study of such deformations in this section. Our main strategy
will be to exploit single and double commutation techniques to be
explained below.

We illustrate the main idea by the following simple example. Consider again the potential (2.19)

$$V(x) = 2\mathcal{P}(x + \omega'; g_2, g_3) + \mathcal{P}(\omega'; g_2, g_3) \qquad (4.1)$$

associated with the nonsingular elliptic curve (see (2.34))

$$
\begin{aligned}
y^2 &= (-e_1 + e_3 - z)(-e_2 + e_3 - z)(-z), & (4.2)\\
e_1 &= \mathcal{P}(\omega; g_2, g_3),\ e_2 = \mathcal{P}(\omega + \omega'; g_2, g_3),\ e_3 = \mathcal{P}(\omega'; g_2, g_3).
\end{aligned}
$$

(For convenience we added $\mathcal{P}(\omega')$ in (4.1) in order to guarantee $E_2 = 0$.) Then $H = -\dfrac{d^2}{dx^2} + V$ has spectrum (see (3.3))

$$\sigma(H) = [-e_1 + e_3, -e_2 + e_3] \cup [0, \infty). \qquad (4.3)$$

Fix $\kappa > 0$ and deform

$$[-e_1 + e_3, -e_2 + e_3] \longrightarrow -\kappa^2 \qquad (4.4)$$

by taking $\omega \to \infty$, $\omega' = (i\pi/2\kappa)$. Then V in (4.1) converges to the
one–soliton potential V_1

$$
\begin{aligned}
V(x) &= 2\mathcal{P}(x + \omega'; g_2, g_3) + \mathcal{P}(\omega'; g_2, g_3) & (4.5)\\
&\longrightarrow V_1(x) = -2\kappa^2 [\cosh(\kappa x)]^{-2}
\end{aligned}
$$

and the associated elliptic curve (4.2) degenerates into a singular
curve

$$y^2 = (-e_1 + e_3 - z)(-e_2 + e_3 - z)(-z) \longrightarrow y^2 = (-\kappa^2 - z)^2(-z). \quad (4.6)$$

The corresponding operator $H_1 = -\dfrac{d^2}{dx^2} + V_1$ then has the spectrum

$$\sigma(H_1) = \{-\kappa^2\} \cup [0, \infty). \qquad (4.7)$$

A further degeneration $\kappa \to 0$ finally yields

$$V(x) = 0 \quad \text{and} \quad y^2 = (-z)^3. \tag{4.8}$$

This point of view has been adopted in [48] and [49] and the general n–soliton potentials have been derived from the Its–Matveev formula by a singularization of K_n where all compact spectral bands degenerate into a single point

$$[E_{2(j-1)}, E_{2j-1}] \longrightarrow -\kappa_j^2, \ 1 \le j \le n, \ \kappa_1 > \kappa_2 > \cdots > \kappa_n, \ E_{2n} = 0 \tag{4.9}$$

(see (3.3)).

Here we shall in a sense reverse the above point of view. Instead of starting with a finite–gap potential such as (4.1) and degenerating compact spectral bands into single points (such as in (4.4) with the result (4.5)–(4.7)), we shall start with a finite–gap potential V_0 and insert eigenvalues into its spectral gaps. In the context of the above example this amounts to starting with

$$V_0(x) = 0, \quad y^2 = -z \tag{4.10}$$

and inserting the eigenvalue $-\kappa^2$ into the spectral gap $\rho_0 = (-\infty, 0)$ of V_0 to arrive at

$$V_1(x) = -2\kappa^2[\cosh(\kappa x)]^{-2}, \ y^2 = (-\kappa^2 - z)^2(-z). \tag{4.11}$$

The spectral deformations described so far were clearly non–isospectral. In addition we will also discuss various isospectral deformations of potentials below. In short, these isospectral deformations either "insert eigenvalues" at points where there were already eigenvalues or they formally insert eigenvalues with certain "defects" such as zero or infinite norming constants. In either case no new eigenvalue is actually inserted and the deformation is isospectral. A systematic and detailed approach to these ideas can be found in [25]–[27].

We start with the single commutation method or Crum–Darboux method [11]–[14], [18], [19], [36], [61]. Assume that $V_0 \in L^1_{\text{loc}}(\mathbb{R})$ is real–valued and that the differential expression

$$\tau_0 = -\frac{d^2}{dx^2} + V_0(x) \tag{4.12}$$

is nonoscillatory and in the limit point case at $\pm\infty$. Consider the self–adjoint realization H_0 of τ_0 in $L^2(\mathbb{R})$

$$H_0 = -\frac{d^2}{dx^2} + V_0, \tag{4.13}$$

$$\mathcal{D}(H_0) = \{g \in L^2(\mathbb{R}) \,\big|\, g,\, g' \in AC_{\mathrm{loc}}(\mathbb{R}),\, \tau_0\, g \in L^2(\mathbb{R})\}$$

(here $AC_{\mathrm{loc}}(\cdot)$ denotes the set of locally absolutely continuous functions) with

$$E_0 = \inf[\sigma(H_0)] > -\infty. \tag{4.14}$$

The basic idea behind the single commutation method is the following: choose

$$\lambda_1 \in \rho_0 = (-\infty, E_0) \tag{4.15}$$

and factor

$$H_0 = AA^* + \lambda_1 = -\frac{d^2}{dx^2} + V_0 \tag{4.16}$$

with

$$A = \frac{d}{dx} + \phi,\ \phi(x) = \psi'_0(\lambda_1, x)/\psi_0(\lambda_1, x),\ H_0\, \psi_0(\lambda_1) = \lambda_1\, \psi_0(\lambda_1) \tag{4.17}$$

for some real–valued distributional solution $\psi_0(\lambda_1, x)$. Commuting A and A^* yields

$$H_1 = A^*A + \lambda_1 = -\frac{d^2}{dx^2} + V_1, \tag{4.18}$$

$$V_1(x) = V_0(x) - 2\frac{d^2}{dx^2} \ln \psi_0(\lambda_1, x). \tag{4.19}$$

We note that $\tau_1 = -\dfrac{d^2}{dx^2} + V_1(x)$ is in the limit point case at $\pm\infty$ and that

$$\sigma(H_1)\backslash\{\lambda_1\} = \sigma(H_0). \tag{4.20}$$

Depending on the choice of $\psi_0(\lambda_1, x)$, λ_1 either belongs to $\sigma(H_1)$ and one has inserted an eigenvalue λ_1 into $\rho_0 = (-\infty, E_0)$ which represents the non–isospectral case, or $\lambda_1 \notin \sigma(H_1)$, i.e., $\sigma(H_1) =$

$\sigma(H_0)$ which is the isospectral case. The above procedure can easily be iterated and we only summarize the final results.

Consider weak solutions $\psi_{0,\pm}(\lambda_1, x)$ such that

$$0 < \psi_{0,\pm}(\lambda, .) \in L^2((R, \pm\infty)), \ R \in \mathbb{R}, \ \lambda < E_0,$$
$$H_0 \psi_{0,\pm}(\lambda) = \lambda\psi_{0,\pm}(\lambda), \ \lambda < E_0. \tag{4.21}$$

Pick

$$\lambda_1 < \lambda_2 < \cdots < \lambda_N < E_0 \tag{4.22}$$

and define in $L^2(\mathbb{R})$

$$H(\lambda_1, \epsilon_1, \ldots, \lambda_N, \epsilon_N) = -\frac{d^2}{dx^2} + V(\lambda_1, \epsilon_1, \ldots, \lambda_N, \epsilon_N), \tag{4.23}$$

$$V(\lambda_1, \epsilon_1, \ldots, \lambda_N, \epsilon_N, x) = V_0(x)$$
$$-2\frac{d^2}{dx^2} \ln W(\psi_{0,\epsilon_1}(\lambda_1), \ldots, \psi_{0,\epsilon_N}(\lambda_N))(x),$$
$$\epsilon_l \in \{+, -\}, \quad 1 \le l \le N. \tag{4.24}$$

Then $\tau_N = -\dfrac{d^2}{dx^2} + V(\lambda_1, \epsilon_1, \ldots, \lambda_N, \epsilon_N, x)$ is in the limit point case at $\pm\infty$ and $H(\lambda_1, \epsilon_1, \ldots, \lambda_N, \epsilon_N)$ and H_0 are isospectral, i.e.,

$$\sigma(H(\lambda_1, \epsilon_1, \ldots, \lambda_N, \epsilon_N)) = \sigma(H_0) \tag{4.25}$$

(in fact, one can show that they are unitarily equivalent [13]). If on the other hand one replaces $\psi_{0,\epsilon_l}(\lambda_l, x)$ in (4.24) by a genuine linear combination of $\psi_{0,+}(\lambda_l, x)$ and $\psi_{0,-}(\lambda_l, x)$

$$\psi_{0,\epsilon_l}(\lambda_l, x) \longrightarrow \alpha\psi_{0,+}(\lambda_l, x) + \beta\psi_{0,-}(\lambda_l, x), \ \alpha > 0, \ \beta > 0 \tag{4.26}$$

then $\lambda_l \in \rho_0 = (-\infty, E_0)$ becomes actually an eigenvalue of the resulting operator. Since we are going to use the single commutation method only in the isospectral context in Section 5 we shall not give any further details on the non–isospectral case.

In the special case of V_0 in (4.13) being a finite–gap potential of the type (3.36),

$$V_0(x) = \sum_{m=0}^{2n} E_m - 2\sum_{j=1}^{n} \lambda_j \tag{4.27}$$

$$-2\frac{d^2}{dx^2} \ln \theta \left(\underline{\varsigma}_{P_\infty} + \underline{\alpha}_{P_\infty}(\mathcal{D}_{\hat{\mu}_1^0(x_0)+\cdots+\hat{\mu}_n^0(x_0)}) + \frac{(x-x_0)}{2\pi}\underline{U}_0\right),$$

(4.24) becomes

$$V(\lambda_1, \epsilon_1, \ldots, \lambda_N, \epsilon_N, x) = \sum_{m=0}^{2n} E_m - 2\sum_{j=1}^{n} \lambda_j \tag{4.28}$$

$$-2\frac{d^2}{dx^2} \ln \theta(\underline{\zeta}_{P_\infty} - \underline{\alpha}_{P_\infty}(\mathcal{D}_{Q_1 + \cdots + Q_N})$$

$$+\underline{\alpha}_{P_\infty}\left(\mathcal{D}_{\hat{\mu}_1^0(x_0) + \cdots + \hat{\mu}_n^0(x_0)}\right) + \frac{(x - x_0)}{2\pi}\underline{U}_0),$$

$$Q_l = \left(\lambda_l, -\epsilon_l\left|\hat{R}_{2n+1}(\lambda_l + i0)^{1/2}\right|\right), \quad \epsilon_l \in \{+, -\}, \quad 1 \le l \le N.$$

In this particular context it can be shown that (4.23)–(4.25) extend to the case $\lambda_N \le E_0$ (in addition to (4.22)).

The single commutation method has the obvious drawback that λ_1 in (4.15) is confined to being below $E_0 = \inf[\sigma(H_0)]$ since for $\lambda_1 > \inf[\sigma(H_0)]$, ψ_0 in (4.17), (4.19) would have at least one zero by Sturm's oscillation theory and hence V_1 in (4.19) would necessarily be singular. In order to overcome this drawback and insert an eigenvalue λ_1 into any spectral gap of H_0 one is led to the double commutation method (going back at least to [23] and described in detail in [13], [14], [22], [25–27], [38]), a refinement of two single commutations at the same spectral point λ_1.

Assuming

$$\lambda_1 \in \mathbb{R}\backslash\sigma(H_0) \tag{4.29}$$

one factors again

$$H_0 = A_\pm A_\pm^* + \lambda_1 = -\frac{d^2}{dx^2} + V_0, \tag{4.30}$$

$$H_{1,\pm} = A_\pm^* A_\pm + \lambda_1 = -\frac{d^2}{dx^2} + V_{1,\pm}, \tag{4.31}$$

$$V_{1,\pm}(x) = V_0(x) - 2\frac{d^2}{dx^2} \ln \psi_{0,\pm}(\lambda_1, x), \tag{4.32}$$

where

$$A_\pm = \frac{d}{dx} + \phi_\pm, \quad \phi_\pm(x) = \psi'_{0,\pm}(\lambda_{1,x})/\psi_{0,\pm}(\lambda_1, x), \tag{4.33}$$

$$\psi_{0,\pm}(\lambda_{1,}.) \in L^2((R,\pm\infty)), \ R \in \mathbb{R}, \ H_0\,\psi_{0,\pm,}(\lambda_1) = \lambda_1\,\psi_{0,\pm}(\lambda_1)$$

and $V_{1,\pm}$ are now singular in general. Introducing

$$\Psi_{\gamma_1,\pm}(x) = \psi_{0,\pm}(\lambda_1,x)^{-1}\,[1 \mp \gamma_{1,\pm}\int_{\pm\infty}^{x} dx'\,\psi_{0,\pm}(\lambda_1,x')^2],$$
$$\gamma_{1,\pm} \geq 0, \tag{4.34}$$

$$\Phi_{\pm}(x) = \Psi'_{\gamma_1,\pm}(x)/\Psi_{\gamma_1,\pm}(x), \tag{4.35}$$

$$B_{\pm} = \frac{d}{dx} + \Phi_{\pm}, \quad B_{\pm}^{+} = -\frac{d}{dx} + \Phi_{\pm}, \tag{4.36}$$

one infers by inspection that

$$H_{1,\pm} = A_{\pm}^{*}A_{\pm} + \lambda_1 = B_{\pm}\,B_{\pm}^{+} + \lambda_1. \tag{4.37}$$

A further commutation of $B_{\pm}$ and $B_{\pm}^{+}$ then leads to

$$H_{\gamma_1,\pm} = B_{\pm}^{+}\,B_{\pm} + \lambda_1 = -\frac{d^2}{dx^2} + V_{\gamma_1,\pm}, \tag{4.38}$$

$$V_{\gamma_1,\pm}(x) = V_0(x) - 2\frac{d^2}{dx^2}\ln[1 \mp \gamma_{1,\pm}\int_{\pm\infty}^{x} dx'\,\psi_{0,\pm}(\lambda_1,x')^2]. \tag{4.39}$$

One can prove that $\tau_{\gamma_1,\pm} = -\frac{d^2}{dx^2} + V_{\gamma_1,\pm}$ is in the limit point case at $\pm\infty$ and that

$$\sigma(H_{\gamma_1,\pm}) = \sigma(H_0) \cup \{\lambda_1\} \quad \text{iff} \quad 0 < \gamma_{1,\pm} < \infty. \tag{4.40}$$

Hence $\gamma_{1,\pm} \in (0,\infty)$ represents the non–isospectral case. The two cases $\gamma_{1,\pm} = 0,\infty$ on the other hand represent the isospectral case, i.e.,

$$\sigma(H_{\infty,\pm}) = \sigma(H_0), \tag{4.41}$$

where

$$H_{\infty,\pm} = -\frac{d^2}{dx^2} + V_{\infty,\pm}, \tag{4.42}$$

$$V_{\infty,\pm}(x) = V_0(x) - 2\frac{d^2}{dx^2}\ln[\mp\int_{\pm\infty}^{x} dx'\,\psi_{0,\pm}(\lambda_1,x')^2]. \tag{4.43}$$

This procedure can easily be iterated and we summarize again the final results.

Consider weak solutions $\psi_{0,\pm}(\lambda,x)$ such that

$\psi_{0,\pm}(\lambda,.) \in L^2((R,\pm\infty))$ is real–valued, $R \in \mathbb{R}$,

$$H_0\,\psi_{0,\pm}(\lambda) = \lambda\psi_{0,\pm}(\lambda), \quad \lambda \in \mathbb{R}\backslash\sigma(H_0). \qquad (4.44)$$

Pick

$$\lambda_j \in \mathbb{R}\backslash\sigma(H_0), \quad 1 \leq j \leq N, \quad \lambda_j \neq \lambda_l \text{ for } j \neq l \qquad (4.45)$$

and define in $L^2(\mathbb{R})$

$$H_{\gamma_1,\dots,\gamma_N,\pm}(\lambda_1,\dots,\lambda_N) = -\frac{d^2}{dx^2} + V_{\gamma_1,\dots,\gamma_N,\pm}(\lambda_1,\dots,\lambda_N), \qquad (4.46)$$

$$V_{\gamma_1,\dots,\gamma_N,\pm}(\lambda_1,\dots,\lambda_N,x) = V_0(x)$$
$$-2\frac{d^2}{dx^2}\ln\det\left\{\left[\delta_{l,l'} \mp \gamma_{l,\pm}\int_{\pm\infty}^{x}dx'\,\psi_{0,\pm}(\lambda_l,x')\,\psi_{0,\pm}(\lambda'_l,x')\right]_{l,l'=1}^{N}\right\},$$

$$\gamma_{l,\pm} \geq 0, \quad 1 \leq l \leq N. \qquad (4.47)$$

Then $\tau_{N,\pm} = -\dfrac{d^2}{dx^2} + V_{\gamma_1,\dots,\gamma_N,\pm}(\lambda_1,\dots,\lambda_N,x)$ is in the limit point case at $\pm\infty$ and

$$\sigma(H_{\gamma_1,\dots,\gamma_N,\pm}(\lambda_1,\dots,\lambda_N)) = \sigma(H_0) \cup \{\lambda_l\}_{l=1}^{N} \text{ iff } \gamma_{l,\pm} \in (0,\infty),$$
$$1 \leq l \leq N, \qquad (4.48)$$

illustrating the nonisospectral case. Similarly, defining

$$H_{\infty,\pm}(\lambda_1,\dots,\lambda_N) = -\frac{d^2}{dx^2} + V_{\infty,\pm}(\lambda_1,\dots,\lambda_N), \qquad (4.49)$$

$$V_{\infty,\pm}(\lambda_1,\dots,\lambda_N,x) = V_0(x) \qquad (4.50)$$
$$-2\frac{d^2}{dx^2}\ln\det\left\{\left[\mp\int_{\pm\infty}^{x}dx'\psi_{0,\pm}(\lambda_l,x')\psi_{0,\pm}(\lambda_{l'},x')\right]_{1\leq l,\,l'\leq N}\right\}$$

yields the isospectral counterpart, i.e.,

$$\sigma(H_{\infty,\pm}(\lambda_1,\ldots,\lambda_N)) = \sigma(H_0) \tag{4.51}$$

(actually, one can show that $H_{\infty,\pm}(\lambda_1,\ldots,\lambda_N)$ and H_0 are unitarily equivalent [25]).

In the particular case where V_0 is the finite–gap potential (4.27), equation (4.50) becomes

$$V_{\infty,\pm}(\lambda_1,\ldots,\lambda_N,x) = \sum_{m=0}^{2n} E_m - 2\sum_{j=1}^{n} \lambda_j \tag{4.52}$$

$$-2\frac{d^2}{dx^2} \ln \theta \left(\underline{\zeta}_{P_\infty} \mp 2\underline{\alpha}_{P_\infty}(\mathcal{D}_{Q_1+\cdots+Q_N}) \right.$$

$$\left. +\underline{\alpha}_{P_\infty}(\mathcal{D}_{\hat\mu_1^0(x)+\cdots+\hat\mu_n^0(x_0)}) + \frac{(x-x_0)}{2\pi}\underline{U}_0 \right),$$

$$Q_l = \left(\lambda_l, -\left| \hat{R}_{2n+1}(\lambda_l+i0)^{1/2} \right| \right), \quad 1 \le l \le N.$$

A comparison of (4.52) and (4.28) reveals that in the finite–gap context one double commutation at λ_1 corresponds to two single commutations at λ_1 and λ_2 in the limit $\lambda_2 \to \lambda_1$. Actually this fact is independent of the finite–gap context and holds in general. Indeed, taking into account the identity

$$\int_{\pm\infty}^{x} dx'\, \psi_{0,\pm}(\lambda_1,x')\psi_{0,\pm}(\lambda_2,x') \tag{4.53}$$

$$= (\lambda_1 - \lambda_2)^{-1} W(\psi_{0,\pm}(\lambda_1),\ \psi_{0,\pm}(\lambda_2))(x),$$

$$\lambda_1, \lambda_2 \in \mathbb{R}\backslash\sigma(H_0), \quad \lambda_1 \ne \lambda_2$$

and the fact that $W(\psi_{0,+}(\lambda_1), \psi_{0,-}(\lambda_1))$ is a nonzero constant, one infers, e.g.,

$$V(\lambda_1,\epsilon_1,\lambda_2,\epsilon_2,x) = V_0(x) - 2\frac{d^2}{dx^2} \ln W(\psi_{0,\epsilon_1}(\lambda_1),\psi_{0,\epsilon_2}(\lambda_2))(x)$$

$$\xrightarrow[\lambda_2 \to \lambda_1]{} \begin{cases} V_0(x), & \epsilon_1 = -\epsilon_2 \\ V_{\infty,\epsilon_1}(\lambda_1,x), & \epsilon_1 = \epsilon_2. \end{cases} \tag{4.54}$$

Finally, with a slight adjustment only, one can also use directly formulas (4.39) resp. (4.47) to produce potentials isospectral to V_0.

E.g., if λ_1 is already an eigenvalue of H_0,

$$\lambda_1 \in \sigma_p(H_0) \tag{4.55}$$

then $H_{\gamma_1,\pm}$ in (??) and (??), with $\psi_{0,+}(\lambda_1,x) = c\,\psi_{0,-}(\lambda_1,x)$ the corresponding eigenfunction of H_0, are well defined. In this case one only changes the corresponding norming constant of the eigenfunction of $H_{\gamma_1,\pm}$ associated with λ_1 and hence $H_{\gamma_1,\pm}$ and H_0 are isospectral

$$\sigma(H_{\gamma_1,\pm}) = \sigma(H_0). \tag{4.56}$$

(A further extension, allowing $\gamma_{1,\pm} = -\|\psi_{0,\pm}(\lambda_1)\|_2^2$, removes the eigenvalue λ_1 from H_0, i.e., $\sigma(H_{\gamma_1,\pm}) = \sigma(H_0)\backslash\{\lambda\}$ in this case.) These facts are illustrated, e.g., in [1], [58].

It should perhaps be pointed out again at this occasion that the substitution

$$\psi_{0,\pm}(\lambda_j,x) \longrightarrow \psi_{0,\pm}(\lambda_j,x,t) \tag{4.57}$$

in (??), (??), (??), where $\psi_{0,\pm}(\lambda_j,x,t)$ satisfies

$$H_0\,\psi_{0,\pm}(\lambda_j) = \lambda_j\,\psi_{0,\pm}(\lambda_j),\ \partial_t\,\psi_{0,\pm}(\lambda_j) = P_{2n+1}\,\psi_{0,\pm}(\lambda_j),\ 1 \le j \le N \tag{4.58}$$

and V_0 satisfies the n-th KdV equation

$$KdV_n(V_0) = 0, \tag{4.59}$$

produces again solutions $V(\lambda_1,\epsilon_1,\ldots,\lambda_N,\epsilon_N,x,t)$ and $V_{\gamma_1,\ldots,\gamma_N,\pm}(\lambda_1,\ldots,\lambda_N,x,t)$, $V_{\infty\pm}(\lambda_1,\ldots,\lambda_N,x,t)$ of the n-th KdV equation.

5 Isospectral Sets of Quasi–Periodic Finite–Gap Potentials

In this section we fix a real–valued quasi–periodic finite–gap potential $V_0(x)$ satisfying Hypotheses (H.3.1) and (H.3.2) and

$$\hat{f}_{n+1,x}(V_0) = \sum_{j=0}^{n} c_{n-j}\,f_{j+1,x}(V_0) = 0 \tag{5.1}$$

for some fixed $\{c_j\}_{j=0}^n \subset \mathbb{R},\quad c_0 = 1$

with the associated nonsingular compact hyperelliptic curve $K_n = K_n(V_0)$

$$K_n : y^2 = \hat{R}_{2n+1}(z) = \prod_{m=0}^{2n} (E_m - z), \quad E_0 < E_1 < \cdots < E_{2n} \quad (5.2)$$

(cf. (2.23), (2.26), and (2.27)). Thus V_0 can be represented by the Its–Matveev formula (4.27)

$$V_0(x) = \sum_{m=0}^{2n} E_m - 2 \sum_{j=1}^{n} \lambda_j \tag{5.3}$$

$$-2 \frac{d^2}{dx^2} \ln \theta \left(\underline{\zeta}_{P_\infty} + \underline{\alpha}_{P_\infty} \left(\mathcal{D}_{\hat{\mu}_1^0(x_0) + \cdots + \hat{\mu}_n^0(x_0)} \right) + \frac{(x - x_0)}{2\pi} \underline{U}_0 \right).$$

The isospectral set $I_{\mathbb{R}}(V_0)$ of real–valued quasi–periodic finite–gap potentials of V_0 is then defined by

$$I_{\mathbb{R}}(V_0) = \{ V \in C^\infty(\mathbb{R}), \text{ real–valued } \big| \hat{f}_{n+1,x}(V) = 0 ,$$
$$K_n(V) = K_n(V_0) \}, \tag{5.4}$$

where $\hat{f}_{n+1,x}$ is given in terms of the sequence $\{c_j\}_{j=0}^{n}$, $c_0 = 1$ in (5.1) and $K_n(V) = K_n(V_0)$ denotes the fixed hyperelliptic curve (5.2).

In order to give an explicit realization of $I_{\mathbb{R}}(V_0)$ we need to introduce the following sets $\mathcal{D}_{\mathbb{R}_\pm} \subset \sigma^n K_n$ of positive divisors in "real position" (see Section 3 for the terminology employed)

$$\mathcal{D}_{\mathbb{R}_-} = \{ \mathcal{D}_{P_1 + \cdots + P_n} \in \sigma^n K_n \mid \Pi(P_j) \in \bar{\rho}_0 = [-\infty, E_0], \, 1 \leq j \leq n \}, \tag{5.5}$$

$$\mathcal{D}_{\mathbb{R}_+} = \{ \mathcal{D}_{P_1 + \cdots + P_n} \in \sigma^n K_n \mid \Pi(P_j) \in \bar{\rho}_{\pi(j)} = [E_{2\pi(j)-1}, E_{2\pi(j)}],$$
$$1 \leq j \leq n \}, \tag{5.6}$$

where π denotes some permutation of $\{1, \ldots, n\}$.

The Its–Matveev formula (3.36) and the fact that Dirichlet divisors $\mathcal{D}_{\hat{\mu}_1(x) + \cdots + \hat{\mu}_n(x)}$ are nonspecial then yields the following theorem (see, e.g., [4], [5], [17], [21], [35], [44], [48], [50], [57]).

Theorem 5.1 *The map*

$$
i_+ : \begin{cases} I_{\mathbb{R}}(V_0) & \longrightarrow \mathcal{D}_{\mathbb{R}_+} \\ V_{\hat{\mu}_1,\dots,\hat{\mu}_n} & \longrightarrow \mathcal{D}_{\hat{\mu}_1(x_0)+\cdots+\hat{\mu}_n(x_0)} \end{cases} \tag{5.7}
$$

is bijective, where

$$
V_{\hat{\mu}_1,\dots,\hat{\mu}_n}(x) = \sum_{m=0}^{2n} E_m - 2 \sum_{j=1}^{n} \lambda_j \tag{5.8}
$$

$$
-2\frac{d^2}{dx^2} \ln \theta(\underline{\zeta}_{P_\infty} + \underline{\alpha}_{P_\infty}(\mathcal{D}_{\hat{\mu}_1(x_0)+\cdots+\hat{\mu}_n(x_0)}) + \frac{(x-x_0)}{2\pi} \underline{U}_0)
$$

and the associated Dirichlet divisor $\mathcal{D}_{\hat{\mu}_1(x)+\cdots+\hat{\mu}_n(x)}$ *is obtained from (3.40) by solving the system (2.29) with initial conditions (3.41).*

Next we state the following "real" version of the Jacobi inversion theorem (3.33).

Lemma 5.2 *Denote by* $[\underline{z}]$ *the equivalence class of* $\underline{z} \in \mathbb{C}^n$ *in* $J(K_n) = \mathbb{C}^n/L_n$. *Then*

$$
\underline{\alpha}_{P_\infty}(\mathcal{D}_{\mathbb{R}_-}) = \{[\underline{x}] \in J(K_n) \mid \underline{x} \in \mathbb{R}^n\}. \tag{5.9}
$$

Sketch of proof. Due to the fact that $\hat{R}_{2n+1}(z)^{1/2}$ is real–valued iff $z \in \bigcup_{j=0}^{n} \bar{\rho}_j$ and

$$
\underline{A}_{P_\infty}((E_{2j}, 0)) = \frac{1}{2} \left[(0,\dots,\underbrace{0}_{j},1,\dots,1) + (\tau_{j,1},\dots,\tau_{j,n}) \right],
$$

$$
\underline{A}_{P_\infty}((E_{2j-1}, 0)) = \frac{1}{2} \left[(0,\dots,\underbrace{0}_{j-1},1,\dots,1) + (\tau_{j,1},\dots,\tau_{j,n}) \right] \tag{5.10}
$$

one can show that

$$
\underline{\alpha}_{P_\infty}(\mathcal{D}_{Q_1+\cdots+Q_n}) \subseteq \{[\underline{x}] \in J(K_n) \mid \underline{x} \in \mathbb{R}^n\} \text{ iff } \mathcal{D}_{Q_1+\cdots+Q_n} \in \mathcal{D}_{\mathbb{R}_-}. \tag{5.11}
$$

(5.9) then follows from (3.33) by restricting $\underline{\alpha}_{P_\infty}$ to $\mathcal{D}_{\mathbb{R}_-}$.

Next we introduce the notion of admissibility of divisors: a positive divisor $\mathcal{D}_{P_1} + \cdots + P_n \in \sigma^n K_n$ is called admissible iff there is no pair $(P, P^*) \in \{P_1, \ldots, P_n\}$ with $P \in K_n \backslash \{P_\infty\}$. The set of all admissible divisors is denoted by $\mathcal{A}$.

We note that admissible divisors $\mathcal{D}_{P_1 + \ldots + P_n} \in \mathcal{A}$ are either nonspecial or their speciality stems from one or more points P_∞ contained in $\{P_1, \ldots, P_n\}$.

Lemma 5.3 *Given* $\mathcal{D}_{\hat{\mu}_1^0 + \cdots + \hat{\mu}_n^0} \in \mathcal{D}_{\mathbb{R}_+}$ *and* $\mathcal{D}_{\hat{\mu}_1 + \cdots + \hat{\mu}_n} \in \mathcal{D}_{\mathbb{R}_+}$ *there exists a unique divisor* $\mathcal{D}_{Q_1 + \cdots + Q_n} \in \mathcal{D}_{\mathbb{R}_-} \cap \mathcal{A}$ *such that*

$$\underline{\alpha}_{P_\infty}(\mathcal{D}_{\hat{\mu}_1 + \cdots + \hat{\mu}_n}) = \underline{\alpha}_{P_\infty}(\mathcal{D}_{\hat{\mu}_1^0 + \cdots + \hat{\mu}_n^0}) - \underline{\alpha}_{P_\infty}(\mathcal{D}_{Q_1 + \cdots + Q_n}). \quad (5.12)$$

Sketch of proof. Since $\hat{R}_{2n+1}(z)^{1/2}$ is real–valued if $z \in \bigcup_{j=1}^{n} \overline{\rho}_j$, (5.12) is equivalent to

$$\underline{\alpha}_{P_\infty}(\mathcal{D}_{Q_1 + \cdots + Q_n}) = -\sum_{j=1}^{n} \underline{A}_{\hat{\mu}_j^0}(\hat{\mu}_{\pi(j)}) \in \{[\underline{x}] \in J(K_n) \mid \underline{x} \in \mathbb{R}^n\}$$

$$(5.13)$$

for some permutation π of $\{1, \ldots, n\}$. Thus the existence of some $\mathcal{D}_{Q_1 + \cdots + Q_n} \in \mathcal{D}_{\mathbb{R}_-}$ satisfying (5.12) follows from Lemma 5.2. If $\mathcal{D}_{Q_1 + \cdots + Q_n}$ is nonspecial then $\mathcal{D}_{Q_1 + \cdots + Q_n} \in \mathcal{A}$ is clearly the unique solution of (5.12). If on the other hand $n \geq 2$ and $\{Q_1, \ldots, Q_n\}$ contains a pair (P, P^*) with $\Pi(P) \in (-\infty, E_0]$, say $Q_1 = P, Q_2 = P^*$, then simply replace Q_1 and Q_2 by P_∞ since

$$\mathcal{D}_{Q_1 + Q_2 + Q_3 + \cdots + Q_n} \sim \mathcal{D}_{P_\infty + P_\infty + Q_3 + \cdots + Q_n} \quad (5.14)$$

by Abel's theorem (3.32). By continuing this process of replacing pairs (P, P^*), $P \neq P_\infty$ by (P_∞, P_∞) one finally ends up with a unique admissible divisor linearly equivalent to the original $\mathcal{D}_{Q_1 + \cdots + Q_n}$.

Our new main result on $I_{\mathbb{R}}(V_0)$ then reads

Theorem 5.4 [27] *The map*

$$i_- : \begin{cases} I_{\mathbb{R}}(V_0) & \longrightarrow \mathcal{D}_{\mathbb{R}_-} \cap \mathcal{A} \\ V_{\hat{\mu}_1, \ldots, \hat{\mu}_n} & \longrightarrow \mathcal{D}_{Q_1 + \cdots + Q_n} \end{cases} \quad (5.15)$$

is bijective, where $\mathcal{D}_{Q_1+\cdots+Q_n} \in \mathcal{D}_{\mathbb{R}_-} \cap \mathcal{A}$ *is the unique solution of*

$$\underline{\alpha}_{P_\infty}\left(\mathcal{D}_{Q_1+\cdots+Q_n}\right) \;=\; \underline{\alpha}_{P_\infty}\left(\mathcal{D}_{\hat{\mu}_1^0(x_0)+\cdots+\hat{\mu}_n^0(x_0)}\right) \tag{5.16}$$
$$-\underline{\alpha}_{P_\infty}\left(\mathcal{D}_{\hat{\mu}_1(x_0)+\cdots+\hat{\mu}_n(x_0)}\right).$$

Moreover,

$$V_{\hat{\mu}_1,\ldots,\hat{\mu}_n}(x) = \sum_{m=0}^{2n} E_m - 2\sum_{j=1}^{n} \lambda_j$$
$$-2\frac{d^2}{dx^2}\ln\theta\left(\underline{\zeta}_{P_\infty} + \underline{\alpha}_{P_\infty}\left(\mathcal{D}_{\hat{\mu}_1(x_0)+\cdots+\hat{\mu}_n(x_0)}\right) + \frac{(x-x_0)}{2\pi}\underline{U}_0\right)$$
$$= \sum_{m=0}^{2n} E_m - 2\sum_{j=1}^{n} \lambda_j$$
$$-2\frac{d^2}{dx^2}\ln\theta\left(\underline{\zeta}_{P_\infty} - \underline{\alpha}_{P_\infty}\left(\mathcal{D}_{Q_1+\cdots+Q_n}\right)\right.$$
$$\left.+\underline{\alpha}_{P_\infty}\left(\mathcal{D}_{\hat{\mu}_1^0(x_0)+\cdots+\hat{\mu}_n^0(x_0)}\right) + \frac{(x-x_0)}{2\pi}\underline{U}_0\right)$$
$$= V(\lambda_{j_1}, \epsilon_{j_1}, \ldots, \lambda_{j_m}, \epsilon_{j_m}, x)$$
$$= V_0(x) - 2\frac{d^2}{dx^2}\ln W(\psi_{0,\epsilon_{j_1}}(\lambda_{j_1}), \ldots, \psi_{0,\epsilon_{j_m}}(\lambda_{j_m}))(x), \tag{5.17}$$

where

$$\{Q_1, \ldots, Q_n\} = \{P_\infty, \ldots, P_\infty, Q_{j_1}, \ldots, Q_{j_m}\}, \tag{5.18}$$
$$Q_{j_l} = \left(\lambda_{j_l}, -\epsilon_{j_l}\left|\hat{R}_{2n+1}(\lambda_{j_l} + i0)^{1/2}\right|\right),$$
$$\lambda_{j_l} \in (-\infty, E_0], \; 1 \le l \le m \le n.$$

Sketch of proof. Existence and uniqueness of $\mathcal{D}_{Q_1+\cdots+Q_n} \in \mathcal{D}_{\mathbb{R}_-} \cap$ $\mathcal{A}$ in (5.15) associated with $V_{\hat{\mu}_1,\ldots,\hat{\mu}_n}$ by (5.16) follows from Lemma 5.3. (5.17) and (5.18) are a consequence of (4.24) and (4.28).

Remark 5.5 An explicit realization of $I_{\mathbb{R}}(V_0)$ in the case where V_0 is a real–valued periodic finite–gap potential has first been derived by Finkel, Isaacson, and Trubowitz [21]. We also refer to [9], [35], [37], [51]–[53], [59], and [62] for further investigations in this direction.

Our realization (5.17) of $I_{\mathbb{R}}(V_0)$ differs from the one in [21] in two respects. First of all, for fixed genus n, (5.17) involves at most an $n \times n$ Wronskian as opposed to a $2n \times 2n$ Wronskian in [21] (involving n additional Dirichlet eigenfunctions) and secondly, (5.17) does not assume periodicity but applies to the quasi–periodic finite–gap case. The upshot of (5.17) is the following: the entire isospectral torus $I_{\mathbb{R}}(V_0)$ of the given base potential V_0 is generated by at most n–single commutations associated with $(\lambda_1, \epsilon_1, \ldots, \lambda_n, \epsilon_n)$, where the points $Q_j = (\lambda_j, -\epsilon_j | \hat{R}_{2n+1} (\lambda_j + i0)^{1/2} |)$, $1 \leq j \leq n$ vary independently of each other on both rims of the cut $\overline{\rho}_0 = [-\infty, E_0]$ (avoiding pairs of the type (Q, Q^*), $Q \neq P_\infty$ in $\{Q_1, \ldots, Q_n\}$).

One can prove an analogous representation for $I_{\mathbb{R}}(V_0)$ by using the isospectral double commutation approach (4.49)–(4.52) [27].

6   Some Generalizations

In our final section we comment on some natural generalizations of the approach in Sections 4 and 5 and mention some open problems.

a)   Infinitely Many Spectral Gaps in $\sigma(H_o)$:

The case where $V_0 \in \mathbb{C}^\infty(\mathbb{R})$ is real–valued and periodic of period $a > 0$ with infinitely many spectral gaps in $\sigma(H_0)$ is well understood [21], [35], [37], [46], [47], [54], [55], [59], [62]. If

$$\sigma(H_0) = \bigcup_{j \in \mathbb{N}} [E_{2(j-1)}, E_{2j-1}], \qquad (6.1)$$

then V_0 can be approximated uniformly on $\mathbb{R}$ by a sequence of real–valued finite–gap potentials $V_{0,n}$ (of the same period a) associated with K_n in (5.2) as $n \to \infty$. In this context determinants of the type (4.24) and (4.50) converge to Fredholm determinants as $n \to \infty$ (we shall illustrate this in some detail in a similar context at the end of this section).

These results have been, extended to particular classes of real–valued almost periodic potentials $V_0 \in C^\infty(\mathbb{R})$ with suitable conditions on the asymptotic behavior of E_j as $j \to \infty$ in [10], [39]–[44].

It should perhaps be pointed out that with the exceptions of [4]–[6], [31], [32], [60], the corresponding complex–valued analog received much less attention in the literature. In particular, the Jacobi inversion problem on the noncompact Riemann surface K_∞ associated with V_0 in the complex–valued periodic or almost–periodic infinite–gap case (a crucial step in the corresponding generalization of the Its–Matveev formula) appears to be open.

b)   Harmonic Oscillators etc:

The double commutation approach in connection with (4.55) and (4.56) can be used to produce families of isospectral unbounded potentials with purely discrete spectra. In order to see the connection with spectral deformations in Section 4 consider the harmonic oscillator example

$$V_0(x) = x^2 - 1 \tag{6.2}$$

and the (suitably scaled) Mathieu potential

$$V_\epsilon(x) = 2\epsilon^{-2}[1 - \cos(\epsilon x)] - 1, \quad \epsilon > 0. \tag{6.3}$$

As is well known [57], all periodic and anti–periodic eigenvalues of $-\dfrac{d^2}{dx^2} + V_\epsilon$ restricted to $[x_0, x_0 + (2\pi/\epsilon)]$, $\epsilon > 0$ are simple and hence

$$H_\epsilon = -\frac{d^2}{dx^2} + V_\epsilon \quad \text{on} \quad H^2(\mathbb{R}), \quad \epsilon > 0 \tag{6.4}$$

has infinitely many spectral gaps for all $\epsilon > 0$

$$\sigma(H_\epsilon) = \bigcup_{j \in \mathbb{N}} [E_{2(j-1)}(\epsilon), \, E_{2j-1}(\epsilon)]. \tag{6.5}$$

As $\epsilon \downarrow 0$,

$$V_\epsilon(x) \xrightarrow[\epsilon \downarrow 0]{} V_0(x) = x^2 - 1 \tag{6.6}$$

and, since

$$E_{2(j-1)}(\epsilon), \, E_{2j-1}(\epsilon) \xrightarrow[\epsilon \downarrow 0]{} 2(j-1), \quad j \in \mathbb{N}, \tag{6.7}$$

one infers

$$\sigma(H_\epsilon) \xrightarrow[\epsilon \downarrow 0]{} \sigma(H_0) = \{2(j-1)\}_{j \in \mathbb{N}} \tag{6.8}$$

(see, e.g., [33], [63]). In this scaling limit $\epsilon \downarrow 0$, the noncompact Riemann surface $K_\infty(\epsilon)$ associated with V_ϵ, $\epsilon > 0$ degenerates into a highly singular curve consisting of infinitely many double points $\{2(j-1)\}_{j \in \mathbb{N}}$. A careful study of this limit on the level of degenerating hyperelliptic curves and their θ–functions, to the best of our knowledge, has not been undertaken yet. Isospectral families of the limit potential $V_0(x) = x^2 - 1$ have been constructed in [45] and [56] but apart from the harmonic oscillator case we are not aware of any other detailed study of isospectral families for unbounded potentials with purely discrete spectra.

Finally, we mention another possible generalization in a bit more detail:

c) N–Soliton Solutions as $N \to \infty$:

Here we choose

$$H_0 = -\frac{d^2}{dx^2} \quad \text{on} \quad H^2(\mathbb{R}), \quad V_0(x) = 0 \tag{6.9}$$

and choose double commutation to insert N eigenvalues

$$\{\lambda_j = -\kappa_j^2\}_{j=1}^N, \ \kappa_j > 0, \ 1 \le j \le N, \ \kappa_j \ne \kappa_{j'} \ \text{for} \ j \ne j' \tag{6.10}$$

into the spectral gap $\rho_0 = (-\infty, 0)$ of H_0. The result is the N–soliton potential [22], [38]

$$V_N(x) = -2\frac{d^2}{dx^2} \ln \det[1_N + C_N(x)], \tag{6.11}$$

$$C_N(x) = \left[\frac{c_l\, c_{l'}}{\kappa_l + \kappa_{l'}} e^{-(\kappa_l + \kappa_{l'})x} \right]_{1 \le l, l' \le N}, \tag{6.12}$$

where

$$c_l > 0, \quad 1 \le l \le N \tag{6.13}$$

are (norming) constants (related to $\gamma_{l,+}$ in (4.47) by $c_l^2 = \gamma_{l,+}$, $1 \le l \le N$, i.e., $V_N(x) = V_{c_1^2,\dots,c_N^2,+}(\lambda_1,\dots,\lambda_N,x)$). Introducing

$$H_N = -\frac{d^2}{dx^2} + V_N \text{ on } H^2(\mathbb{R}), \tag{6.14}$$

one verifies that

$$\sigma(H_N) = \{-\kappa_j^2\}_{j=1}^N \cup [0,\infty) \tag{6.15}$$

with purely absolutely continuous essential spectrum of multiplicity two

$$\sigma_{ess}(H_N) = \sigma_{ac}(H_N) = [0,\infty), \tag{6.16}$$

$$\sigma_p(H_N) \cap [0,\infty) = \sigma_{sc}(H_N) = \emptyset \tag{6.17}$$

and simple discrete eigenvalues $\{-\kappa_j^2\}_{j=1}^N$. (Here $\sigma_{ess}(.)$, $\sigma_{ac}(.)$, $\sigma_{sc}(.)$, and $\sigma_p(.)$ denote the essential, absolutely continuous, singularly continuous, and point spectrum (the set of eigenvalues) respectively.) The unitary scattering matrix $S_N(k)$ in $\mathbb{C}^2$ associated with the pair (H_N, H_0) is reflectionless and reads

$$S_N(k) = \begin{pmatrix} T_N(k) & 0 \\ 0 & T_N(k) \end{pmatrix}, \tag{6.18}$$

$$T_N(k) = \prod_{j=1}^N \left(\frac{k + i\kappa_j}{k - i\kappa_j}\right), \ k \in \mathbb{C} \setminus \{i\kappa_j\}_{j=1}^N$$

($\lambda = k^2$ the spectral parameter of H_0). As briefly mentioned in Section 4, the singular curve associated with H_N is of the type

$$K_{0,N}: \quad y^2 = \left[\prod_{j=1}^N (-\kappa_j^2 - z)^2\right](-z) \tag{6.19}$$

which can be obtained from the nonsingular curve

$$K_N: \quad y^2 = \prod_{m=0}^{2n}(E_m - z), \quad E_0 < E_1 < \cdots < E_{2N} = 0 \tag{6.20}$$

by degenerating the compact spectral bands $[E_{2(j-1)}, E_{2j-1}]$ into the eigenvalues $-\kappa_j^2$

$$[E_{2(j-1)}, E_{2j-1}] \longrightarrow -\kappa_j^2, \quad 1 \leq j \leq N. \tag{6.21}$$

At this point it seems natural to ask what happens if $N \to \infty$. This can be answered as follows.

Theorem 6.1 [28], [29] *Assume* $\{\kappa_j > 0\}_{j \in \mathbb{N}} \in l^\infty(\mathbb{N})$, $\kappa_j \neq \kappa_{j'}$ *for* $j \neq j'$ *and choose* $\{c_j > 0\}_{j \in \mathbb{N}}$ *such that* $\{c_j^2/\kappa_j\}_{j \in \mathbb{N}} \in l^1(\mathbb{N})$. *Then* V_N *converges pointwise to some* $V_\infty \in C^\infty(\mathbb{R}) \cap L^\infty(\mathbb{R})$ *as* $N \to \infty$ *and*

(i) $\lim\limits_{x \to +\infty} V_\infty(x) = 0$ *and*

$$\lim\limits_{n \to \infty} \sup\limits_{x \in K} \left| V_N^{(m)}(x) - V_\infty^{(m)}(x) \right| = 0, \quad m \in \mathbb{N} \cup \{0\} \tag{6.22}$$

for any compact $K \subset \mathbb{R}$.

(ii) Denoting

$$H_\infty = -\frac{d^2}{dx^2} + V_\infty \ \ on \ H^2(\mathbb{R}) \tag{6.23}$$

we have

$$\sigma_{ess}(H_\infty) = \{-\kappa_j^2\}'_{j \in \mathbb{N}} \cup [0, \infty), \tag{6.24}$$

$$\sigma_{ac}(H_\infty) = [0, \infty), \tag{6.25}$$

$$[\sigma_p(H_\infty) \cup \sigma_{sc}(H_\infty)] \cap (0, \infty) = \emptyset, \tag{6.26}$$

$$\{-\kappa_j^2\}_{j \in \mathbb{N}} \subseteq \sigma_P(H_\infty) \subseteq \overline{\{-\kappa_j^2\}_{j \in \mathbb{N}}}. \tag{6.27}$$

The spectral multiplicity of H_∞ on $(0, \infty)$ equals two while $\sigma_p(H_\infty)$ is simple. In addition, if $\{\kappa_j\}_{j \in \mathbb{N}}$ is a discrete subset of $(0, \infty)$ (i.e., 0 is its only limit point) then

$$\sigma_{sc}(H_\infty) = \emptyset, \tag{6.28}$$

$$\sigma(H_\infty) \cap (-\infty, 0) = \sigma_d(H_\infty) = \{-\kappa_j^2\}_{j \in \mathbb{N}}. \tag{6.29}$$

More generally, if $\{\kappa_j\}'_{j\in\mathbb{N}}$ is countable then (6.28) holds.

Here A' denotes the derived set of $A \subset \mathbb{R}$ (i.e., the set of accumulation points of A) and $\sigma_d(.)$ denotes the discrete spectrum (cf. also the paragraph following (6.17)).

We refer to [29] for a complete proof of this result. Here we only mention that the condition $\{c_j^2/\kappa_j\}_{j\in\mathbb{N}} \in l^1(\mathbb{N})$ implies convergence in trace norm topology of the $N \times N$ matrix $C_N(x)$ (see (6.12)) embedded into $l^2(\mathbb{N})$ to the trace class operator $C_\infty(x)$ in $l^2(\mathbb{N})$ given by

$$C_\infty(x) = \left[\frac{c_l\,c_{l'}}{\kappa_l + \kappa_{l'}}\, e^{-(\kappa_l+\kappa_{l'})x}\right]_{l,l'\in\mathbb{N}}. \qquad (6.30)$$

Moreover, one has in analogy to (6.11),

$$V_\infty(x) = -2\frac{d^2}{dx^2} \ln \det{}_1[1 + C_\infty(x)], \qquad (6.31)$$

where $\det_1(.)$ denotes the Fredholm determinant associated with $l^2(\mathbb{N})$.

We emphasize that Theorem 6.1 solves the following inverse spectral problem: *Given any bounded and countable subset* $\{-\kappa_j^2\}_{j\in\mathbb{N}}$ *of* $(-\infty,0)$, *construct a (smooth and real-valued) potential V such that* $H = -\dfrac{d^2}{dx^2} + V$ *has a purely absolutely continuous spectrum equal to* $[0,\infty)$ *and the set of eigenvalues of H includes the prescribed set* $\{-\kappa_j^2\}_{j\in\mathbb{N}}$. (*In particular,* $\{-\kappa_j^2\}_{j\in\mathbb{N}}$ *can be dense in a bounded subset of* $(-\infty,0)$.)

Under the stronger hypothesis $\{\kappa_j\}_{j\in\mathbb{N}} \in l^1(\mathbb{N})$ one obtains

Theorem 6.2 [28], [29] *Assume* $\{\kappa_j > 0\}_{j\in\mathbb{N}} \in l^1(\mathbb{N})$, $\kappa_j \neq \kappa_{j'}$ *for* $j \neq j'$ *and choose* $\{c_j > 0\}_{j\in\mathbb{N}}$ *such that* $\{c_j^2/\kappa_j\}_{j\in\mathbb{N}} \in l^1(\mathbb{N})$. *Then in addition to the conclusions of Theorem 6.1 we have*

(i)

$$\lim_{n\to\infty} \|V_N^{(m)} - V_\infty^{(m)}\|_p = 0, \quad 1 \leq p \leq \infty, \quad m \in \mathbb{N} \cup \{0\}. \qquad (6.32)$$

(ii)
$$\sigma_{ess}(H_\infty) = \sigma_{ac}(H_\infty) = [0, \infty), \qquad (6.33)$$
$$\sigma_p(H_\infty) \cap (0, \infty) = \sigma_{sc}(H_\infty) = \emptyset, \qquad (6.34)$$
$$\sigma_d(H_\infty) = \{-\kappa_j^2\}_{j \in \mathbb{N}}. \qquad (6.35)$$

The unitary scattering matrix $S_\infty(k)$ in $\mathbb{C}^2$ associated with the pair (H_∞, H_0) is reflectionless and given by

$$S_\infty(k) = \begin{pmatrix} T_\infty(k) & 0 \\ 0 & T_\infty(k) \end{pmatrix}, \qquad (6.36)$$

$$T_\infty(k) = \prod_{j=1}^{\infty} \left(\frac{k + i\kappa_j}{k - i\kappa_j} \right), \ k \in \mathbb{C} \backslash \{\{i\kappa_j\}_{j \in \mathbb{N}} \cup \{0\}\}.$$

Note that Theorem 6.2 constructs a new class of reflectionless potentials involving an infinite negative point spectrum of H_∞ accumulating at zero.

For a detailed proof of Theorem 6.2 see [29]. We remark that the condition $\{\kappa_j\}_{j \in \mathbb{N}} \in l^1(\mathbb{N})$ implies that $V_\infty \in L^1(\mathbb{R})$ (but $V_\infty \notin L^1(\mathbb{R}; (1 + |x|)\, dx)$) and that the product $T_N(k)$ converges absolutely to $T_\infty(k)$ as $N \to \infty$.

We conclude with the observation that the simple substitution

$$c_j \longrightarrow c_j\, e^{\kappa_j^3 t}, \quad j \in \mathbb{N} \qquad (6.37)$$

in (6.30) and (6.31), denoting the result in (6.31) by $V_\infty(x, t)$, produces solutions of the KdV$_1$ equation (see (2.8))

$$KdV_1(V_\infty) = V_{\infty,t} + \frac{1}{4} V_{\infty,xxx} - \frac{3}{2} V_\infty V_{\infty,x} = 0. \qquad (6.38)$$

In particular, substitutions of the type (6.37) together with Theorem 6.2 provide new soliton solutions of the KdV hierarchy [28],[29].

Acknowledgements

We would like to thank W. Karwowski, K. Unterkofler, and Z. Zhao for numerous discussions on this subject. It is a great pleasure to thank W. F. Ames, E. M. Harrell, and J. V. Herod for their kind invitation to a highly stimulating conference.

Bibliography

[1] P. B. Abraham and H. E. Moses, Phys. Rev. **A22** (1980), 1333–1340.

[2] M. Abramowitz and I. A. Stegun, *Handbook of Mathematical Functions*, Dover, New York, 1972.

[3] S. I. Al'ber, Commun. Pure Appl. Math. **34** (1981), 259–272.

[4] B. Birnir, Commun. Pure Appl. Math. **39** (1986), 1–49.

[5] B. Birnir, Commun. Pure Appl. Math. **39** (1986), 283–305.

[6] B. Birnir, SIAM J. Appl. Math. **47** (1987), 710–725.

[7] J. L. Burchnall and T. W. Chaundy, Proc. London Math. Soc. Ser. 2, **21** (1923), 420–440.

[8] J. L. Burchnall and T. W. Chaundy, Proc, Roy. Soc. London **A118** (1928), 557–583.

[9] M. Buys and A. Finkel, J. Diff. Eqs. **55** (1984), 257–275.

[10] W. Craig, Commun. Math. Phys **126** (1989), 379–407.

[11] M. M. Crum, Quart. J. Math. Oxford (2), **6** (1955), 121–127.

[12] G. Darboux, C. R. Acad. Sci. (Paris) **94** (1882), 1456–1459.

[13] P. A. Deift, Duke Math. J. **45** (1978), 267–310.

[14] P. Deift and E. Trubowitz, Commun. Pure Appl. Math. **32** (1979), 121–251.

[15] B. A. Dubrovin, Russ. Math. Surv. **36:2** (1981), 11–92.

[16] B. A. Dubrovin, I. M. Krichever, and S. P. Novikov, in *Dynamical Systems IV*, V. I. Arnold, S. P. Novikov (eds.), Springer, Berlin, 1990, pp. 173–280.

[17] B. A. Dubrovin, V. B. Matveev, and S. P. Novikov, Russ. Math. Surv. **31:1** (1976), 59–146.

[18] F. Ehlers and H. Knörrer, Comment. Math. Helv. **57** (1982), 1–10.

[19] N. M. Ercolani and H. Flaschka, Phil. Trans. Roy. Soc. London **A315** (1985), 405–422.

[20] N. Ercolani and H. P. McKean, Invent. math. **99** (1990), 483–544.

[21] A. Finkel, E. Isaacson, and E. Trubowitz, SIAM J. Math. Anal. **18** (1987), 46–53.

[22] C. S. Gardner, J. M. Greene, M. D. Kruskal, and R. M. Miura, Commun. Pure Appl. Math. **27** (1974), 97–133.

[23] I. M. Gel'fand and B. M. Levitan, Amer. Math. Soc. Transl. Ser. 2, **1** (1955), 253–304.

[24] F. Gesztesy, in *Ideas and Methods in Mathematical Analysis, Stochastics, and Applications*, Vol. 1, S. Albeverio, J. E. Fenstad, H. Holden, and T. Lindstrøm (eds.), Cambridge Univ. Press, 1992, pp. 428-471.

[25] F. Gesztesy, A complete spectral characterization of the double commutation method, preprint, 1992.

[26] F. Gesztesy and R. Svirsky, *(m) KdV –solitons on the background of quasi–periodic finite–gap solutions*, preprint, 1991.

[27] F. Gesztesy and R. Weikard, in preparation.

[28] F. Gesztesy, W. Karwowski, and Z. Zhao, Bull. Amer. Math. Soc., to appear.

[29] F. Gesztesy, W. Karwowski, and Z. Zhao, Duke Math. J., to appear.

[30] P. G. Grinevich and I. M. Krichever, in *Soliton theory: a survey of results*, A. P. Fordy (ed.), Manchester Univ. Press, Manchester, 1990, pp. 354–400.

[31] V. Guillemin and A. Uribe, Trans. Amer. Math. Soc. **279** (1983), 759-771.

[32] V. Guillemin and A. Uribe, Commun. Part. Diff. Eqs. **8** (1983), 1455-1474.

[33] E. M. Harrell, Ann. Phys. **119** (1979), 351-369.

[34] A. R. Its and V. B. Matveev, Theoret. Math. Phys. **23** (1975), 343-355.

[35] K. Iwasaki, Ann. Mat. Pure Appl. Ser.4, **149** (1987), 185-206.

[36] C. G. J. Jacobi, J Reine angew. Math. **17** (1837), 68-82.

[37] T. Kappeler, Ann. Inst. Fourier, Grenoble **41** (1991), 539-575.

[38] I. Kay and H. E. Moses, J. Appl. Phys. **27** (1956), 1503-1508.

[39] S. Kotani and M. Krishna, J. Funct. Anal. **78** (1988), 390-405.

[40] B. M. Levitan, Math. USSR Izvestija **18** (1982), 249-273.

[41] B. M. Levitan, Math USSR Izvestija **20** (1983), 55-87.

[42] B. M. Levitan, Trans. Moscow Math. Soc. **45:1** (1984), 1-34.

[43] B. M. Levitan, Math. USSR Sbornik **51** (1985), 67-89.

[44] B. M. Levitan, *Inverse Sturm-Liouville Problems*, VNU Science Press, Utrecht, 1987.

[45] B. M. Levitan, Math USSR Sbornik **60** (1988), 77-106.

[46] V. A. Marchenko, *Sturm-Liouville Operators and Applications*, Birkhäuser, Basel, 1986.

[47] V. A. Marchenko, I. V. Ostrovskii, Math. USSR Sbornik **26** (1975), 493-554.

[48] V. B. Matveev, *Abelian Functions and Solitons*, 1976, preprint.

[49] H. P. McKean, in *Partial Differential Equations and Geometry*, C. I. Byrnes (ed.), Marcel Dekker, New York, 1979, pp. 237–254.

[50] H. P. McKean, in *Global Analysis*, M. Grmela and J. E. Marsden (eds.), Lecture Notes in Math., Vol. **755**, 1979, pp. 83–200.

[51] H. P. McKean, Commun. Pure Appl. Math. **38** (1985), 669–678.

[52] H. P. McKean, Rev. Mat. Iberoamericana **2** (1986), 235–261.

[53] H. P. McKean, J. Stat. Phys. **46** (1987), 1115–1143.

[54] H. P. McKean and E. Trubowitz, Commun. Pure Appl. Math. **29** (1976), 143–226.

[55] H. P. McKean and E. Trubowitz, Bull. Amer. Math. Soc. **84** (1978), 1042–1085.

[56] H. P. McKean and E. Trubowitz, Commun. Math. Phys. **82** (1982), 471–495.

[57] N. W. McLachlan, *Theory and Application of Mathieu Functions*, Clarendon Press, Oxford, 1947.

[58] S. Novikov, S. V. Manakov, L. P. Pitaevskii, and V. E. Zakharov, *Theory of Solitons*, Consultants Bureau, New York, 1984.

[59] D. L. Pursey, Phys. Rev. **D33** (1986), 1048–1055.

[60] J. Ralston and E. Trubowitz, Ergod. Th. Dyn. Syst. **8** (1988), 301–358.

[61] P. Sarnak, Commun. Math. Phys. **84** (1982), 377–401.

[62] U.-W. Schmincke, Proc. Roy. Soc. Edinburgh **80A** (1978), 67–84.

[63] T. Sunada, Duke Math. J. **47** (1980), 529–546.

[64] M. I. Weinstein and J. B. Keller, SIAM J. Appl. Math. **45** (1985), 200–214.

[65] J. Zagrodziński, J. Phys. **A17** (1984), 3315–3320.

Nuclear Cusps, Magnetic Fields and the Lavrentiev Phenomenon In Thomas-Fermi Theory

Gisèle Ruiz Goldstein and Jerome A. Goldstein
Department of Mathematics
Louisiana State University
Baton Rouge, LA 70803

Chien-an Lung
Department of Mathematics
Tulane University
New Orleans, LA 70118

1 Introduction

Conventional Thomas-Fermi theory is concerned with minimizing the functional

$$E(\rho) = C_0 \int_{\mathbb{R}^3} \rho(x)^{5/3} dx + \int_{\mathbb{R}^3} V(x)\rho(x) dx \qquad (1)$$
$$+ \frac{1}{2} \int_{\mathbb{R}^3} \int_{\mathbb{R}^3} \frac{\rho(x)\rho(y)}{|x - y|} dx dy$$

subject to the constraints $\rho \geq 0$, $\int_{\mathbb{R}^3} \rho(x) dx = N$ (where $N > 0$ is given), and each of the three integrals in (1) is finite. The function

Differential Equations with
Applications to Mathematical
Physics

ρ which minimizes E is the *ground state electron density* in Thomas-Fermi theory corresponding to the potential V. More precisely, if H is the Hamiltonian of a quantum mechanical system of N electrons under the influence of a potential V, then if ψ is a normalized wave function and ρ is its corresponding density, then $E(\rho)$ is an approximation to the energy expectation value $\langle H\psi, \psi \rangle$ (cf. [13], [6]). Thus minimizing $E(\rho)$ gives an approximation to the ground state energy and the density corresponding to the ground state wave function.

Ever since the original rigorous treatment of the minimization problem for E by E. Lieb and B. Simon [12], [13], much attention has focussed on various extensions. Of particular concern here is the nuclear cusp condition, which we now prepare to describe. The Euler-Lagrange equation for the convex functional E given by (1) is

$$Q := \frac{5}{3}C_0\rho^{2/3} + G\rho + V + \lambda = 0. \tag{2}$$

on the set where $\{\rho > 0\}$ and $Q \geq 0$ on $\{\rho = 0\}$. Here $-\lambda$ is the chemical potential, which is a Lagrange multiplier corresponding to the constraint $\int_{\mathbb{R}^3} \rho(x)dx = N$, and

$$G\rho(x) := (\frac{1}{|\cdot|} * \rho)(x) = \int_{\mathbb{R}^3} \frac{\rho(y)}{|x-y|}\, dy.$$

Consider an atom, so that $V(x) = -Z/|x|$ where Z is the positive charge of the nucleus, which is located at the origin. Since

$$\int_{|y|<1} \frac{\rho(y)}{|x-y|}\, dy \approx \int_{|y|<1} \frac{\rho(y)}{|y|}\, dy$$

for x close to zero and (by Hölder's inequality)

$$\int_{|y|<1} \frac{\rho(y)}{|y|}\, dy \leq \left(\int_{|y|<1} \rho(y)^{5/3}\, dy\right)^{3/5} \left(\int_{|y|<1} |y|^{-5/2}dy\right)^{2/5}$$

$$= \|\rho\|_{5/3}\, 4\pi \int_0^1 r^{-5/2}\, r^2\, dr = 8\pi\|\rho\|_{5/3} < \infty,$$

it follows that $G\rho + \lambda$ is bounded near $x = 0$, whence (see (2))

$$\rho(x) \approx \text{const}|x|^{-3/2} \tag{3}$$

near $x = 0$. Thus ρ is unbounded near the nucleus, which is physically incorrect. The behavior of the true quantum mechanical density was pointed out by T. Kato in 1957 [11]; namely

$$\rho(x) \approx \text{const} \cdot \exp\{-2Z|x|\}$$

as $|x| \to 0$. (Cf. also Thirring [16, 240] and the Hoffmann-Ostenhofs, et al [9], [10].)

An explanation for this is that the true ground state density is continuous at the origin but its gradient $\nabla\rho$ has a jump discontinuity there; thus $\Delta\rho$ should exist (near the origin) as a finite signed measure. R. Parr and S. Ghosh [16] formally suggested how to incorporate the nuclear cusp condition (3) into Thomas-Fermi theory, and J. Goldstein and G. Rieder* [4] established this rigorously. See the monograph of R. Parr and W. Yang [17] for more details.

Now consider the case of an atom but let a magnetic field be present. The magnetic field will spin polarize the system, so the density becomes $\vec{\rho} = (\rho_1, \rho_2)$ where ρ_1 [resp. ρ_2] is the density of the spin up [resp. spin down] electrons. If $\rho = \rho_1 + \rho_2$ is the total electron density, then the Thomas-Fermi energy is

$$E(\vec{\rho}) = \sum_{i=1}^{2} C_1 \int_{\mathbb{R}^3} \rho_i(x)^{5/3} dx + \int_{\mathbb{R}^3} V(x)\rho(x) dx \qquad (4)$$
$$+ \frac{1}{2} \int_{\mathbb{R}^3} \int_{\mathbb{R}^3} \frac{\rho(x)\rho(y)}{|x-y|} dx\, dy + \int_{\mathbb{R}^3} B(x)(\rho_1(x) - \rho_2(x)) dx$$

where the function B describes the magnetic field. This problem was treated in detail recently by Goldstein and Rieder [7]. The purpose of the present paper is to incorporate the nuclear cusp condition into the context of (4).

Section 2 is devoted to an explanation of the solution to this problem. In Section 3 we discuss the Lavrentiev phenomenon aspect of our results and make further remarks.

---

*G. R. Rieder is now G. R. Goldstein.

2   The Nuclear Cusp Condition

Of concern is (4), where C_1 is a positive constant and $V(x) = -Z/|x|$ with $Z > 0$. We want to consider only those $\overrightarrow{\rho}$ for which $\rho_1, \rho_2 \geq 0$ and each of the integrals in (4) is finite. We have $\int_{\mathbb{R}^3} \rho_i(x)dx = N_i$, and $N = N_1 + N_2$ is the total number of electrons. We may specify only N or we may specify both N_1 and N_2. This defines the domains of E, denoted by $D_1[N]$ and $D_1[N_1, N_2]$ respectively, and we consider the problem of minimizing E over each of them. These problems were solved in [7]. Near the origin, the Euler-Lagrange equations (i.e., $\partial E/\partial \rho_1 = 0 = \partial E/\partial \rho_2$) are

$$\frac{5}{3}C_1\rho_j^{2/3} - \frac{Z}{|x|} + G(\rho_1 + \rho_2) - (-1)^j B + \lambda_j = 0$$

for $j = 1, 2$. Here $\lambda_1 = \lambda_2 = \lambda$ is the Lagrange multiplier corresponding to the constraint $\int_{\mathbb{R}^3} \rho(x)dx = N$ when N is given or else λ_j corresponds to $\int_{\mathbb{R}^3} \rho_j(x)dx = N_j$ when both N_1 and N_2 are specified. Here is the key idea which originated with Parr and Ghosh.

Assume that $\Delta\rho$ is a tempered distribution on $\mathbb{R}^3$. Then for each $k > 0$ it is not difficult to show that $\int_{\mathbb{R}^3} e^{-2k|x|}\Delta\rho(x)dx$ exists; call it $M \in \mathbb{R}$. (For a proof of the existence of M see [14].) Integration by parts gives

$$\int_{\mathbb{R}^3} \Delta(e^{-2k|x|})\rho(x)dx = M. \tag{5}$$

Now let $D_2[N; M]$, $D_2[N_1, N_2; M]$ be the domains $D_1[N]$, $D_2[N_1, N_2]$, further restricted by requiring that (5) holds. (These domains depend on $k > 0$ which is fixed.) Of concern is E acting on the domain $\cup\{D_2[N; M] : M \in \mathbb{R}\}$ and $\cup\{D_2[N_1, N_2; M] : M \in \mathbb{R}\}$. The Lagrange multiplier μ corresponding to the constraint (5) has the effect in the Euler-Lagrange equations of replacing the potential $V(x) = -Z/|x|$ by

$$\tilde{V}(x) = -\frac{Z}{|x|} + \mu\Delta(e^{-2k|x|}).$$

If $\mu = Z/4k^2$, this becomes

$$\tilde{V}(x) = \frac{-Z}{|x|}(1 - e^{2k|x|}) - kZe^{-2k|x|};$$

thus $\lim_{x \to 0} \tilde{V}(x) = -3kZ$ and the singularity at the origin has disappeared. Thus (recall the argument involving (2)) we expect ρ to be bounded near the origin. The arguments of [4] can now be extended to handle the present case.

We now stop being informal and state some precise results. Consider

$$E(\vec{\rho}) = C_p \int_{\mathbb{R}^3} \sum_{i=1}^{2} \rho_i(x)^p dx + \frac{1}{2} \int_{\mathbb{R}^3} \int_{\mathbb{R}^3} \frac{\rho(x)\rho(y)}{|x-y|} dx dy \quad (6)$$

$$- Z \int_{\mathbb{R}^3} \frac{\rho(x)}{|x|} dx + \int_{\mathbb{R}^3} B(x)(\rho_1(x) - \rho_2(x)) dx$$

where $3/2 < p < \infty, C_p > 0, Z > 0, B(x) = b_1 + b_2(x)$ with $b_1 \in \mathbb{R}, b_2(x) \to 0$ as $|x| \to \infty, \Delta b_2 \in L^1(\mathbb{R}^3)$ and $\int_{\mathbb{R}^3} \Delta b_2(x) dx = 0, b_2 \in L^\infty(\mathbb{R}^3) \cap L^{3/2}(\mathbb{R}^3)$, and finally $\tilde{V}(x) + |b_2(x)|$ is negative on a set of positive measure. (Note that $\tilde{V}(0) = -3kZ < 0$, so that this last condition holds if b_2 is small near the origin.) The earlier definitions of $D_2[N_1, M]$ etc. involved the choice of $p = 5/3$; these definitions should be modified in the obvious way to accommodate the power p appearing in the kinetic energy integral in the definition of $E(\vec{\rho})$.

Theorem 1 *Let the conditions in the above paragraph hold. Let $k > 0$ and let $0 < N \leq Z$. Then E given by (6) has a unique minimum $\vec{\rho}$ on the domain $\cup\{D_2[N; M] : M \in \mathbb{R}\}$. Moreover, $\vec{\rho}$ has compact support if $N < Z$. Furthermore, $\rho = \rho_1 + \rho_2$ is radially symmetric and is nonincreasing on $[0, \infty)$ if the magnetic field B is constant. If B is a C^1 function of $|x|$ only in a neighborhood of the origin, then one may choose*

$$k = \left[\frac{C_p p(p-1)}{4} \left\{ \rho_1(0)^{p-1} + \rho_2(0)^{p-1} \right\} \right]^{1/2}$$

and conclude that

$$\rho(x) \approx const \, e^{-2Z|x|} \quad (7)$$

(to first order) near $x = 0$.

In the above theorem, E fails to have a minimum on $\cup\{D_2[N; M] : M \in \mathbb{R}\}$ when $N > Z$.

Theorem 2 *Let the conditions of the paragraph preceding Theorem 1 hold. Let $k > 0$ and let $N_1, N_2 > 0$ be given and satisfy $N_1 + N_2 \leq Z$. Then E given by (6) has a unique minimum on the domain $\cup\{D_2[N_1, N_2; M] : M \in \mathbb{R}\}$. Moreover, if $N = N_1 + N_2 < Z$, then both ρ_1, ρ_2 have compact support while if $N = Z$ and $N_1 < N_2$ [or $N_2 < N_1$], then ρ_1 [or ρ_2] has compact support. If B is a C^1 function of $|x|$ only in a neighborhood of the origin, then k may be chosen so that $\rho = \rho_1 + \rho_2$ satisfies the nuclear cusp condition (7) near $x = 0$, to first order.*

Here E fails to have a minimum on $\cup\{D_2[N_1, N_2; M] : M \in \mathbb{R}\}$ if

$$N > \frac{1}{4\pi} \int_{\mathbb{R}^3} [\Delta(\widetilde{V} + B)]_+,$$

where the subscript denotes "positive part". When B is a constant (and thus $b_2 \equiv 0$), this condition can be replaced by

$$N > Z = \frac{1}{4\pi} \int_{\mathbb{R}^3} \Delta\widetilde{V} = \frac{1}{4\pi} \int_{\mathbb{R}^3} \Delta V.$$

By making different choices of k, we can make ρ_1 or ρ_2 (rather than ρ) satisfy the nuclear cusp condition. But it is not clear if we can make both ρ_1 and ρ_2 (and hence ρ) satisfy it simultaneously. We conjecture that this can be done.

3   The Lavrentiev Phenomenon

Of concern is the classical calculus of variations. Consider the functional

$$\widetilde{E}[u] = \int_a^b L(r, u(r), u'(r)) dr \tag{8}$$

with two domains

$$D_1(\widetilde{E}) = \{u \in \text{Lip } [a,b] : u(a) = A, u(b) = B\}$$
$$D_2(\widetilde{E}) = \{u \in AC[a,b] : u(a) = A, u(b) = B\}.$$

Here $-\infty < a < b < \infty$, A and B are given, L is a given function, and "Lip", "AC" denote Lipschitz continuous and absolutely continuous functions, respectively. The *Lavrentiev phenomenon* is said to occur when

$$\inf\{\tilde{E}(u) : u \in D_1(\tilde{E})\} > \inf\{\tilde{E}(u) : u \in D_2(\tilde{E})\}. \tag{9}$$

See, for example [15] or [8] for a nice discussion of this notion. It only occurs for very special integrands L.

The nuclear cusp condition in Thomas-Fermi theory gives rise to a similar phenomenon, which may also be termed a Lavrentiev phenomenon. For simplicity we work with the functional E defined by (1) rather than (4). Consider an atom, and define $E[\rho]$ by (1) with $V(x) = -Z/|x|$. Thus E can be written as

$$
\begin{aligned}
E[\rho] &= \int_0^\infty (J(\rho)(r) + V(r)\rho(r))r^2 dr \\
&\quad + \int_0^\infty \int_0^\infty F(r_1, r_2)\rho(r_1)\rho(r_2)r_1^2 r_2^2 dr_1 dr_2 \\
&= \int_0^\infty L_1(r, \rho(r))dr + \int_0^\infty \int_0^\infty L_2(r_1, r_2)\rho(r_1)\rho(r_2)dr_1 dr_2.
\end{aligned}
$$

Here $J(s)$ is $C_0 s^{5/3}$ or $C_p s^p$ $(p > 3/2)$, and F is obtained as follows. The ground state density ρ is radially symmetric; for such radial functions ρ, a spherical coordinate representation gives

$$
\int_{\mathbb{R}^3} \int_{\mathbb{R}^3} \frac{\rho(x_1)\rho(x_2)}{|x_1 - x_2|} dx_1 dx_2 = \\
\int_0^\infty \int_0^\infty F(r_1, r_2)\rho(r_1)\rho(r_2)r_1^2 \, r_2^2 \, dr_1 \, dr_2
$$

where $x_j = (r_j, \theta_j, \varphi_j)$ in spherical coordinates and

$$
F(r_1\, r_2) = \int_0^{2\pi} \int_0^{2\pi} \int_0^\pi \int_0^\pi Q \sin\varphi_1 \sin\varphi_2 d\varphi_1 d\varphi_2 d\theta_1 d\theta_2
$$

with

$$
\begin{aligned}
Q &= [r_1^2 + r_2^2 - 2r_1 r_2\{\sin\varphi_1 \sin\varphi_2[\cos\theta_1 \cos\theta_2 + \sin\theta_1 \sin\theta_2] \\
&\quad + \cos\varphi_1 \cos\varphi_2\}]^{1/2}.
\end{aligned}
$$

Let $u(r) = \int_0^r \rho(s)ds$ and consider u as a basic variable rather than ρ. Thus $u' = \rho$. Define $\tilde{E}[u]$ to be $E[\rho]$. Then

$$\tilde{E}[u] = \int_0^\infty L_1(r, u'(r))dr$$
$$+ \int_0^\infty \int_0^\infty L_2(r_1, r_2)u'(r_1)u'(r_2)dr_1 dr_2. \quad (10)$$

This has two domains (at least), namely (given $N > 0$)

$$D_1(\tilde{E}) = \{u \in Lip[0, \infty] : u(0) = 0, u(\infty) = N, u \text{ is nondecreasing,}$$
$$u' \in L^1(0, \infty), \text{ and each integral in (10) exists}\}.$$
$$D_2(\tilde{E}) = \{u \in AC[0, \infty] : u(0) = 0, u(\infty) = N, u \text{ is nondecreasing,}$$
$$u' \in L^1(0, \infty), \text{ and each integral in (10) exists}\}.$$

Minimizing $\tilde{E}$ over $D_2(\tilde{E})$ [resp. $D_1(\tilde{E})$] with $N \leq Z$ gives the usual Thomas-Fermi ground state (resp. the one satisfying the nuclear cusp condition). We get a different ground state (namely $u' = \rho$ is unbounded as $r \to 0$ in the $D_1(\tilde{E})$ case but is bounded as $r \to 0$ in the $D_2(\tilde{E})$ case). Thus (taking into account uniqueness) (9) holds.

In minimizing (8), when $L(r, u, u') = L(u, u')$ is independent of the r variable, the Lavrentiev phenomenon normally does not hold [3]. This is not the case with (10).

4 Remarks, Open Problems, and Acknowledgements

In the case when one specifies both N_1 and N_2 it would be of interest to show that both ρ_1 and ρ_2 satisfy the nuclear cusp condition. Also, in the case of a constant magnetic field, Bénilan, Goldstein and Rieder [1], [2] found a critical point of the energy functional E given by a modification of (4) incorporating the Fermi-Amaldi correction. This allows one to find $\vec{\rho}$ whenever $N_1 + N_2 \leq Z + 1$, that is, singly negative ions are allowed. It would be of interest to incorporate the nuclear cusp condition into this context.

The results of this paper can be easily extended from atoms to molecules. In this case $V(x) = -Z/|x|$ is replaced by $V(x) =$

$-\sum_{j=1}^{M} Z_j/|x - R_j|$. The nuclear cusp condition says that

$$\rho(x) \approx \text{const} \cdot \exp\left\{-2Z_j/|x - R_j|\right\}$$

near R_j for $j = 1, 2, \ldots, M$.

It would be of interest to study the Lavrentiev phenomenon for (10) simply as a problem in the calculus of variations.

We gratefully acknowledge that all three authors were partially supported by two NSF grants. We also thank Peter Wolenski for some stimulating and helpful discussions concerning the Lavrentiev phenomenon.

Bibliography

[1] Bénilan, Ph., J. A. Goldstein and G. R. Rieder, *The Fermi-Amaldi correction in spin polarized Thomas-Fermi theory,* in Differential Equations and Mathematical Physics (ed. by C. Bennewitz), Academic Press (1991), 25-37.

[2] Bénilan, Ph., J. A. Goldstein and G. R. Rieder, *A nonlinear elliptic system arising in electron density theory,* Comm. PDE, to appear.

[3] Clarke, F. H. and R. B. Vinter, *Regularity properties of solutions to the basic problem in the calculus of variations,* Trans. Amer., Math. Soc. 289 (1985), 73-98.

[4] Goldstein, J.A. and G. R. Rieder, *A rigorous modified Thomas-Fermi theory for atomic systems,* J. Math. Phys. 28 (1987), 1198-1202.

[5] Goldstein, J. A. and G. R. Rieder, *Spin polarized Thomas-Fermi theory,* J. Math. Phys. 29 (1988), 709-716.

[6] Goldstein, J. A. and G. R. Rieder, *Recent rigorous results in Thomas-Fermi theory,* in Nonlinear Semigroups, Partial Differential Equations, and Attractors (ed. by T. L. Gill and W. W.

Zachary), Lecture Notes in Math. No 1294, Springer, (1989) 68-82.

[7] Goldstein, J. A. and G. R. Rieder, *Thomas-Fermi theory with an external magnetic field,* J. Math Phys. 32 (1991), 2907-2919.

[8] Heinricher, A. C. and V. J. Mizel, *The Lavrentiev phenomenon for invariant variational problems,* Arch. Rat. Mech. Anal. 102 (1988), 57-93.

[9] Hoffmann-Ostenhof, M., T. Hoffmann-Ostenhof, and H. Stremnitzer, *Electronic wave functions near coalescence points,* to appear.

[10] Hoffmann-Ostenhof, M. and R. Seiler, *Cusp conditions for eigenfunctions of N-electron systems,* Phys. Rev. A23 (1981), 21-23.

[11] Kato, T., *On the eigenfunctions of many particle systems in quantum mechanics,* Comm. Pure Appl. Math. 10 (1957), 151-171.

[12] Lieb, E. H. and B. Simon, *Thomas-Fermi theory revisted,* Phys. Rev. Lett. 33 (1973), 681-683.

[13] Lieb, E. H. and B. Simon, *The Thomas-Fermi theory of atoms, molecules, and solids,* Adv. Math. 23 (1977), 22-116.

[14] Lung. C. A., *The Nuclear Cusp Condition in Spin Polarized Thomas-Fermi Theory,* Ph.D. Thesis, Tulane University, 1992.

[15] Mizel, V. J., *Developments in one-dimensional calculus of variations affecting continuum mechanics,* in Proceedings of KIT Mathematics Workshop, Vol. 4, Taejon, Korea (1989), 83-103.

[16] Parr, R. G. and S. K. Ghosh, *Thomas-Fermi theory for atomic systems,* Proc. Nat. Acad. Sci. USA 83 (1983), 3577-3579.

[17] Parr, R. G. and W. Yang, *Density-Functional Theory of Atoms and Molecules,* Oxford U. Press,1989.

[18] Thirring, W., *Quantum Theory of Atoms and Molecules*, Vol. 3, Springer, 1979.

On Schrödinger Equation in Large Dimension and Connected Problems in Statistical Mechanics

Bernard Helffer
Ecole Normale Supérieure de Paris
DMI-ENS, 45 rue d'Ulm
F75230 Paris Cédex

Abstract

In this article we announce three results concerning semi-classical techniques in statistical mechanics. The two first results concern the Schrödinger equation and are obtained in collaboration with J. Sjöstrand. The last one is a stationary phase theorem and can be considered as an adaptation of a result of J. Sjöstrand in a different context.

1   Introduction

If $V^{(m)}$ is a suitable family of C^∞ potentials on $I\!\!R^m$ parametrized by m, there appears to be three connected problems related to the properties of the thermodynamic limit in different contexts of statistical mechanics.

(I) Study the asymptotic behavior of the quantity:

$$\left[\ln \left((1/h\pi)^{(m/2)} \int \exp(-V^{(m)}(x)/h)dx \right) \right] /m$$

Differential Equations with
Applications to Mathematical
Physics

as m tends to ∞ and control this limit with respect to h as h tends to $+0$.

(II) If $\mu_1(m,h)$ is the largest eigenvalue of the operator:

$$K^m(h) = \exp(-V^{(m)}(x)/2) \cdot \exp(h^2 \Delta^{(m)}) \cdot \exp(-V^{(m)}(x)/2), \tag{1}$$

study the asymptotic behavior of the quantity $-\ln \mu_1(m,h)/m$ as m tends to ∞ and control this limit with respect to h as h tends to $+0$.

(III) If $\lambda_1(m,h)$ is the smallest eigenvalue of the Schrödinger operator:

$$S^m(h) = -h^2 \Delta^{(m)} + V^{(m)}(x), \tag{2}$$

what is the asymptotic behavior of the quantity $\lambda_1(m,h)/m$ as m tends to ∞ and control this limit with respect to h as h tends to $+0$?

These three questions are of course strongly related. If you think of a potential which is invariant by circular permutation of the variables and "near" in a suitable sense of the harmonic oscillator, all these questions are well analyzed for fixed m as h tends to zero. (I) can be treated by application of the stationary phase theorem, (III) corresponds to a semiclassical analysis of the Schrödinger operator at the bottom (see [8] and [20]) and the study of (II) can be considered as a pseudo-differential extension of (II) (see [2] or [5]). In particular this study gives for example that

$$-\ln(\mu_1(m,h)) = \lambda_1(m,h) + O_m(h^2). \tag{3}$$

One can get better by proving first (using Segal's lemma) (cf [18]) the universal inequality:

$$-\ln(\mu_1(m,h)) \leq \lambda_1(m,h). \tag{4}$$

By monotonicity, one observes also (in the strictly convex case) that if $V_0^{(m)}$ is a quadratic potential s.t. $V_0^{(m)} \leq V^{(m)}$, then we have also:

$$-\ln(\mu_1^0(m,h)) \leq -\ln(\mu_1(m,h)) \tag{5}$$

where $\mu_0^1(m,h)$ (the largest eigenvalue of $K^m(h; V_0^{(m)})$) is explicitly computable. Equations (4) and (5) give for example (in the case when the limits exist):

$$- \lim_{m\to\infty} \ln(\mu_1^0(m,h))/m \leq - \lim_{m\to\infty} \ln(\mu_1(m,h))/m \tag{6}$$

$$- \lim_{m\to\infty} \ln(\mu_1(m,h))/m \leq \lim_{m\to\infty} (\lambda_1(m,h))/m \tag{7}$$

in the strictly convex case.

Another link is that, by Golden Thompson inequality (cf [16]), we have:

$$Tr(\exp-(-\Delta + W)) \leq Tr(\exp(-W/2) \cdot \exp\Delta \cdot \exp(-W/2))$$
$$= C_m \int \exp(-W)dx^m.$$

The difficult problem is of course a good control with respect to m as m is large. Another interesting (and more difficult) problem appears in the same context, in the cases (II) and (III):

(IV) Study the lim inf and the lim sup of $m \to \mu_2(m,h)/\mu_1(m,h)$ as $m \to \infty$ where $\mu_2(m,h)$ is the second eigenvalue of $K^m(h)$.

 (V) Study the lim inf and the lim sup of $m \to (\lambda_2(m,h) - \lambda_1(m,h))$ as $m \to \infty$ where $\lambda_2(m,h)$ is the second eigenvalue of $S^m(h)$.

The problem (V) corresponds to the well known problem of the study of the splitting between the two first eigenvalues. Our motivation comes from the reading of a course of M. Kac ([13]) in which he develops partially heuristical ideas in order to prove the existence of phase transition by semi-classical techniques. The results we shall present here correspond to a class containing the model potential:

$$V^{(m)}(x;\nu) = (1/4) \sum_{k=1}^{m} x_k^2 - \sum_{k=1}^{m} \ln\cosh((\nu/2)^{1/2}(x_k + x_{k+1})),$$

with the convention $(x_{m+1} = x_1)$.

Let us briefly recall how M. Kac arrived at this potential. He studied the following model (called Model A in section 7 in [13]) whose hamiltonian is given by:

$$E_{V(N,M)}(\sigma) = - \sum_{(P,Q) \in V(N,M) \times V(N,M)} v_{(P,Q)} \sigma_P \cdot \sigma_Q$$

with $V(N,M) = [1, \ldots, N] \times (Z/\pi Z)$ in Z^2, $\sigma_P \in \{-1, +1\}$, $J \in I\!R_+^*$, $h \in I\!R_+^*$, $v_{P,P} = 0$ and

$$v_{P,Q} = Jh \exp(-h|k - k'|)\{\delta_{l',l} + (1/2)(\delta_{l',l+1} + \delta_{l',l-1})\}$$

if $P = (k,l) \neq Q = (k',l')$.

He observes that the free energy per spin in the thermodynamic limit $-\psi/kT$ can be computed as:

$$-\psi/kT = \ln 2 - h/2 + \lim_{m \to \infty} (\ln \mu_1(m,h)/m)$$

where $\mu_1(m,h)$ is the largest eigenvalue of the m-dimensional integral operator K given by:

$$K = \exp(-Q^{(m)}/2) \cdot \exp(-h(-\Delta^{(m)})) \cdot \exp(-Q^{(m)}/2)$$

with

$$Q^{(m)}(y) = (\tanh(h/2)/2) \sum_{k=1}^{m} y_k^2 - \sum_{k=1}^{m} \ln \cosh((\nu h/2)^{1/2}(y_k + y_{k+1})),$$

$\nu = J/kT$. A scaling argument $x_k = h^{1/2} y_k$ permits one to arrive essentially to the problem posed in (II).

The detailed proofs are or will be given elsewhere ([2], [5], [6], [7], [9], [10], [23], [24], [25]).

2   Schrödinger Equation in Large Dimension

Let us consider

$$S^{(m)}(x, hD_x; \nu) = -h^2 \Delta^{(m)} + V^{(m)}(x; \nu) \tag{8}$$

with

$$V^{(m)}(x;\nu) = (1/4)\sum_{k=1}^{m} x_k^2 - \sum_{k=1}^{m} \ln \cosh((\nu h/2)^{1/2}(x_k + x_{k+1})), \quad (9)$$

with the convention: $(x_{m+1} = x_1)$. If $\nu < 1/4$, the potential is convex (single well) but, if $\nu > 1/4$, we are in the situation of a double well. Let $\lambda_j(m, h; \nu)$ be the sequence of the eigenvalues of $S^{(m)}$; we are interested by the problems (III) and (V). Let us present the results which were obtained in this case.

Theorem 2.1 (Cf [9], [25]) *For every ν in $\mathbb{R}_+$, the limit $\Lambda(h,\nu) = \lim_{m\to\infty}(\lambda_1(m; h, \nu)/m)$ exists.*

This is not surprising and it is proved following the ideas of statistical mechanics (see [18]). Let us observe that

$$|\Lambda(h,\nu) - (\lambda_1(m; h, \nu)/m)| = hO(1/m)$$

by easy arguments and that J. Sjöstrand [25] proves recently an exponentially rapid convergence to the thermodynamic limit.

Theorem 2.2 (Cf [9],[25]) *If $\nu \neq 1/4$,*

$$\Lambda(h,\nu) = \lim_{m\to\infty}(\lambda_1(m; h, \nu)/m)$$

admits a complete asymptotic expansion: $\Lambda(h,\nu) \approx h\sum_{j30}\Lambda_j(\nu).h^j$ as h tends to 0. Moreover, if we denote the corresponding semiclassical expansions for $\lambda_1(m; h, \nu)/m$ by:

$$\lambda_1(m; h, \nu)/m \approx h \cdot \sum_{j\geq 0}\lambda_j(m, \nu).h^j$$

there exists $k_0(\nu) > 0$ s.t. for each j, there exists a constant $C_j(\nu)$, s.t. $|\Lambda_j(\nu) - \lambda_j(m,\nu)| \leq C_j(\nu).\exp(-k_0 m)$. $k_0(\nu)$ and $C_j(\nu)$ can be chosen locally independent of ν in $\mathbb{R}_+ \setminus \{1/4\}$.

The study around $\nu = 1/4$ is not complete (see however [13], [6] for partial result for fixed m).

Theorem 2.3 (Cf [24],[10]) *If $\nu < 1/4$ then the splitting between the two first eigenvalues λ_2 and λ_1 is controlled by:*

$$h(1 - 4\nu)^{1/2} \leq \lambda_2(m,h,\nu) - \lambda_1(m,h,\nu) \leq 4\lambda_1(m,h,\nu)/m. \quad (10)$$

The majorization is easily obtained by estimates of the type used in the proof of [17] (see [9]) and is true for any ν in $I\!R_+$. The minorization ([24]) can be obtained by using the maximum Principle (see [22]) or the Brascamp-Lieb inequalities [1] as explained in [10] and the strict convexity of the potential is the decisive and unique assumption.

Theorem 2.4 (Cf [9]) *Let $\nu > 1/4$ and let us consider $\mathcal{N}$ the set in $I\!N \times I\!R_+$ defined by*

$$m \leq C \cdot h^{-N_0}, \quad (11)$$

(we write shortly $m = O(h^{-N_0})$) for some C and N_0; then there exists C_ν, h_ν and $\varepsilon_\nu > 0$ such that for all the (m,h) in $\mathcal{N}$ satisfying $0 < h \leq h_\nu$:

$$\lambda_2(m,h,\nu) - \lambda_1(m,h,\nu) \leq C_\nu \cdot \exp -(\varepsilon_\nu \cdot m/h). \quad (12)$$

Remark 2.5 *Here we observe a very different behavior in comparison with the case $\nu < 1/4$ (cf Theorem 2.3) but we have unfortunately a restriction on m. This is probably a technical difficulty. We were hoping to prove simply that (conjecture given by M. Kac):*

$$\lim_{m \to \infty} (\lambda_2(m,h,\nu) - \lambda_1(m,h,\nu)) = 0.$$

This property would have been a sign of a "transition of phase".

The proof of Theorem 2.3 and Theorem 2.4 is based on the following strategy initiated in [23] and [24]. We can distinguish four steps.

Step 1: Control in the WKB approximation (just look for approximate eigenfunctions of the type $\exp(-f(x,h)/h)$) the dependence on the dimension m as initiated in [23] and [24]. Of course it is a construction which depends only of the germ of the potential at the bottom, but in order to have reasonable estimates we have to assume holomorphy in a complex open l^∞-ball. This will give us the formal expansion of the first eigenvalue.

Step 2: One compares the WKB approximation of the one well problem and the first eigenvalue of the Dirichlet problem in a sufficiently small l^∞-ball around the point where the minimum of the potential was attained.

Step 3: One compares the first eigenvalue of the Dirichlet problem in this small l^∞-ball with the first eigenvalue of the global problem in $I\!R^m$.

In these three steps, one works modulo $m.O_N(h^N)$ (for any N) but the dimension is possibly limited by $m = O(h^{-N_0})$.

Step 4: One eliminates the restriction on the dimension, because one controls the rate of convergence in the thermodynamic limit. In order to analyze the splitting between the two first eigenvalues, let us recall the following classical formula for the splitting (see for example [8], [15], [20] and [21]):

$$\lambda_2 - \lambda_1 = \inf_{\varphi \in \mathcal{H}} \left(\left(\int |h\nabla\varphi|^2 (u_{1,m})^2(x)dx \right) / \left(\int |\varphi|^2 (u_{1,m})^2(x)dx \right) \right),$$
(13)

where

$$\mathcal{H} = \left\{ \varphi \in C_0^\infty; \int \varphi(u_{1,m})^2(x)dx = 0 \right\}$$

and $u_{1,m}$ is the first positive normalized eigenfunction.

The estimates about the splitting are then deduced from a judicious choice of φ and of the information on the decay of $u_{1,m}$ in suitable domains. We observe that, under the assumption $\nu > 1/4$, the potential admits two minima and that there exists δ s.t. the region $\Omega(\delta)$ defined by:

$$\Omega(\delta) = \left\{ x \in I\!R^m, -\delta \leq \sum_i x_i \leq \delta \right\}$$
(14)

does not contain these two wells.

$$\lambda_2(m,h) - \lambda_1(m,h) \leq C_\nu.m.h^2.(\alpha(m,h,\delta)^2)/(1 - \alpha(m,h,\delta)^2) \quad (15)$$

with $\alpha(m,h,\delta) = \|u_{1,m}\|_{L^2(\Omega(\delta))}$.

Theorem 2.4 will be a consequence of the following theorem:

Theorem 2.6 (cf [9]) *There exists C, h_0 and $\delta > 0$ s.t. $\alpha(m, h, \delta) \leq C \exp(-m/Ch)$ for (m, h) in $\mathcal{N}$ and $0 < h < h_0$.*

This theorem is obtained by Agmon's type estimates with a very careful control with respect to the dimension.

3   Thermodynamic Limit in Small Temperature: A Stationary Phase Theorem in Large Dimension

In this section, we shall explain briefly how similar techniques can be used for connected problems. Actually, these theorems are frequently implicitly proved in [23], [24] or [9], [10].

We just consider the "classical" problem introduced as Problem (I). Let us consider

$$J(\beta, m, V) = (\beta/\pi)^{m/2} \int \exp(-\beta V^{(m)}(x)) dx. \qquad (16)$$

The normalization is chosen in order to get $J(\beta, m, V^{(m)}) = 1$ in the case where $V^{(m)}(x) = \sum_{i=1}^{m} x_i^2$. Let us very briefly state why we meet in this context the stationary phase theorem. We assume that

$V^{(m)}$ is convex and admits a unique non-degenerate minimum at 0 with $V^{(m)}(0) = 0$. $\qquad (17)$

It is well known that:

$$(\beta/\pi)^{m/2} \cdot \int \exp(-\beta V^{(m)}(x)) a^{(m)}(x) dx \approx \left(\sum_{j=0}^{\infty} \alpha_j(m) \beta^{-j} \right) \qquad (18)$$

as $\beta \to \infty$ but the problem is to control the behavior of the different coefficients and of the remainder. Actually, we can have a very bad behavior with respect to m as j increases (also in the "physical cases"); however, under suitable assumptions,

$$\left(\ln \left((\beta/\pi)^{m/2} \int \exp(-\beta V^{(m)}(x)) dx \right) \right) / m$$

has an expansion in powers of β^{-1} with coefficients which are bounded independently of m!

Let us write down possible assumptions in order to obtain such a result (see [23]). Let us introduce a set $\mathcal{V}$ as the disjoint union over $I\!N$ of sets $\mathcal{V}_m$: $\mathcal{V} = \bigcup_m \mathcal{V}_m$ where $\mathcal{V}_m$ is a subset of C^∞ potentials on $I\!R^m$. Let us assume that for all V in $\mathcal{V}$:

(H1) V is holomorphic in $B(0,1)$ with $|\nabla V(x)|_\infty = O(1)$ uniformly in $\mathcal{V}$ and $B(0,1)$. (Here $B(0,1)$ is the open unit ball in $\mathbb{C}^m$ with respect to the norm $|x|_\infty = \sup |x_j|$.)

(H2) $V(0) = 0$, $V'(0) = 0$, $V''(0) = D + A$, where D is diagonal (positive definite).

(H3) There exists r_1 and r_0 (independent of V in $\mathcal{V}$) such that: $\|A\|_{\mathcal{L}(\ell^p)} \leq r_1 < r_0 \leq \lambda_{\min}(D)$ for all p s.t. $1 \leq p \leq \infty$. We also assume:

(H4) $\|\nabla^2 V\|_{\mathcal{L}(\ell^p)} = O(1)$ uniformly in $\mathcal{V}$ and p. Here we write: $|x|_p = (\sum |x_j|^p)^{1/p}$ for $1 \leq p < \infty$ and $|x|_\infty = \sup_j |x_j|$.

Then we see that:

$$(V''(0))^{1/2} = \tilde{D} + \tilde{A} \tag{19}$$

with $\tilde{D}$ diagonal and

$$\|\tilde{A}\|_{\mathcal{L}(\ell^p)} \leq \tilde{r}_1 < \tilde{r}_0 \leq \lambda_{\min}(\tilde{D}) \tag{20}$$

for all p s.t. $1 \leq p \leq \infty$ and uniformly in $\mathcal{V}$.

Theorem 3.1 (cf Sjöstrand ([23]) *Under assumptions* (H1)-(H4), *then there exists*

$$f(x,m,h) \approx \sum_j f_j(x,m)h^j \tag{21}$$

and an expansion

$$E(h;m) \approx h \cdot \sum_{j \geq 0} E_j(m)h^j \tag{22}$$

s.t. in the sense of the formal series in h but in a fixed sufficiently small ℓ^∞-neighborhood of $B(0,1)$, the following equation is satisfied:

$$|\nabla f|^2 - V - h \ln \det(\nabla^2 f) \approx E(h; m). \tag{23}$$

Moreover, the functions f_j satisfy:

$$f_j(0) = 0, \quad |\nabla f_j(x)| \leq C_j \text{ in } B \tag{24}$$

and the E_j (which of course depend on m through V in $\mathcal{V}_m$) satisfy:

$$|E_j(m)| \leq C_j \cdot m \tag{25}$$

This problem is quite analogous to the problem of solving the equation:

$$|\nabla f|^2 - V - h\Delta f \approx E(h) \tag{26}$$

in order to construct a WKB solution of the type $\exp(-f(x,h)/h)$ for $S^{(m)} = -h^2\Delta + V^{(m)}$.

Of course this statement could appear mysterious and it is probably better to give the following "formal" corollary:

Corollary 3.2 *If $J(\beta, m)$ is defined by (4), then, formally,*

$$(\ln J(\beta, m))/m \approx \sum_j (E_j(m)/m)\beta^{-j} \tag{27}$$

as β tends to ∞.

"Proof of the Corollary" This is just a "formal" proof of the stationary phase theorem with a uniform control with respect to m (i.e. $V(m)$ in $\mathcal{V}_m$). In the formal integral giving $J(m, \beta)$:

$$(\beta/\pi)^{m/2} \cdot \left(\int \exp(-\beta V^{(m)}(x)) dx \right),$$

we reduce the integral to a small ℓ^∞-path in $I\!R^m$. We are then looking for a change of variable $y = f(x, \beta^{-1})$. Then the integral becomes:

$$(\beta/\pi)^{m/2} \cdot \left(\int \exp(-\beta V^{(m)}(x) - \ln \det \nabla^2 f(x, \beta^{-1})) dy \right)$$

and, at least formally, it is then clear that the Theorem 3.1 gives:

$$(\beta/\pi)^{m/2} \cdot \left(\int \exp\left(-\beta \sum y_j^2 + \beta \cdot E(\beta^{-1}) \right) dy \right).$$

So finally:

$$\ln J^f(m,\beta) \approx \beta E(m,\beta^{-1}).$$

Of course everything is for the moment formal but the control of the coefficients is what is basic for the future.

By adding assumption (5), invariance by permutation and other assumptions needed for the proof of Theorem 2.3, one can prove (for a class containing the model $V^{(m)}(x,\nu)$ (with $\nu < 1/4$)) the exponentially rapid convergence of the coefficients $(E_j(m)/m)$ and control the remainder terms. In fact the proof is parallel (and easier !) to the proof for the Schrödinger equation (steps 2, 3, 4), and we can prove the Corollary:

Corollary 3.3 (Cf [7]) *If $J(\beta,m)$ is defined by (4) as $\beta \to \infty$, then,*

$$\lim_{m \to \infty} (\ln J(\beta,m))/m \approx \sum_j (\lim_{m \to \infty} (E_j(m)/m))\beta^{-j} \qquad (28)$$

as β tends to ∞.

Bibliography

[1] H. J. Brascamp, E. Lieb: *On extensions of the Brunn-Minkovski and Prékopa-Leindler Theorems, including inequalities for Log concave functions, and with an application to diffusion equation.* Journal of Functional Analysis 22 (1976).

[2] M. Brunaud, B. Helffer: *Un problème de double puits provenant de la théorie statistico- mécanique des changements de phase (ou relecture d'un cours de M. Kac).* Preprint de l'ENS (1991).

[3] R. S. Ellis: *Entropy, large deviations, and statistical mechanics..* Grundlehren der mathematischen Wissenschaften 271 Springer, New York, (1985).

[4] E. M. Harrell: *On the rate of asymptotic eigenvalue degeneracy.* Comm. in Math. Phys., 75, 1980, 239-261.

[5] B. Helffer: *Décroissance exponentielle pour les fonctions propres d'un modèle de Kac en dimension > 1.* Preprint de l'ENS (November 1991), to appear in the Proceeding of the Lambrecht's Conference (1991).

[6] B. Helffer: *Problèmes de double puits provenant de la théorie statistico-mécanique des changements de phase: II Modèles de Kac avec champ magnétique, étude de modèles près de la température critique.* Preprint (1992), talk in Toulon (Feb 1992).

[7] B. Helffer: *Around a stationary phase theorem in large dimension,* (in preparation).

[8] B. Helffer, J. Sjöstrand: *Multiple wells in the semi-classical limit,* I. Comm. in PDE, 9(4), 337-408 (1984) II, Annales de l'IHP (section Physique théorique) Vol. 42, #2, 1985, 127-212 .

[9] B. Helffer, J. Sjöstrand: *Semiclassical expansions of the thermodynamic limit for a Schrödinger equation.* Preprint Actes du colloque Méthodes semi-classiques à l'université de Nantes, 24 Juin-30 Juin 1991 (submitted to Astérisque).

[10] B. Helffer, J. Sjöstrand: *Semiclassical expansions of the thermodynamic limit for a Schrödinger equation II.* Preprint (1991) (to appear in Helvetica Physica Acta).

[11] C. Itzykson, J. M. Drouffe: *Théorie statistique des champs 1 Savoirs actuels.* InterEditions/Editions du CNRS (1989).

[12] M. Kac: *Statistical mechanics of some one-dimensional systems, Studies in mathematical analysis and related topics: essays in honor of Georges Polya.* Stanford University Press, Stanford, California (1962), 165-169.

[13] M. Kac: *Mathematical mechanisms of phase transitions, Brandeis lectures.* (1966) Gordon and Breach, New York.

[14] M. Kac, C. J. Thompson: *Phase transition and eigenvalue degeneracy of a one dimensional anharmonic oscillator.* Studies in Applied Mathematics 48 (1969) 257-264.

[15] W. Kirsch, B. Simon: *Comparison theorems for the gap of Schrödinger operators.* Journal of Functional Analysis, Vol. 75, #2, Dec. 1987.

[16] A. Lenard: *Generalizations·of the Golden Thompson inequality.* Ind. Math. J. 21 (1971) 457-467.

[17] L. E. Payne, G. Polya, H. F. Weinberger: *On the ratio of consecutive eigenvalues.* Journal of Math. and Physics, 35 (3) (Oct. 56).

[18] D. Ruelle: *Statistical Mechanics. Math. Physics Monograph Series.* W. A. Benjamin, Inc. 1969.

[19] B. Simon: *Functional Integration and Quantum Physics.* Pure and Applied Mathematics 86, Academic Press, New York, 1979.

[20] B. Simon: *Instantons, double wells and large deviations.* Bull. AMS., 8, (1983), 323-326.

[21] B. Simon: *Semi-classical analysis of low lying eigenvalues:* I Ann. Inst. H. Poincaré 38, (1983), 295-307; II. Annals of Mathematics, 120, (1984), 89-118.

[22] I. M. Singer, B. Wong, Sh-Tung Yau, S. T. Yau: *An estimate of the gap of the first two eigenvalues of the Schrödinger operator.* Ann. d. Scuola Norm. Sup. di Pisa, Vol.XII, 2.

[23] J. Sjöstrand: *Potential wells in high dimensions I.* Ann. Inst. H. Poincaré, (to appear).

[24] J. Sjöstrand: *Potential wells in high dimensions II, more about the one well case.* Preprint March 1991, Ann. Inst. Poincaré, section physique théorique, to appear.

[25] J. Sjöstrand: *Exponential convergence of the first eigenvalue divided by the dimension, for certain sequences of Schrödinger operator.* Preprint Juin 1991. Submitted to Astérisque.

[26] W. Thirring: *Volume 4, Quantum Mechanics of Large Systems.* (translated by E. M. Harrell), Springer Verlag (1983).

[27] Colin J. Thompson: *Mathematical Statistical Mechanics.* The Macmillan Company, New York (1972).

Regularity of Solutions for Singular Schrödinger Equations

Andreas M. Hinz

University of Munich, Germany

Abstract

Applying Kato's inequality to locally integrable solutions of $-(\nabla - ib)^2 u + qu = 0$ leads to $(\triangle + q_-)|u| \geq 0$, which allows for a mean value inequality for $|u|$, as in the case of subharmonic functions. The local Kato condition on q_- enters naturally as one tries to provide local bounds on u. This in turn is the base for other regularity properties of u, such as the existence of square integrable first derivatives. But also quantitative results can be obtained from the mean value inequality. Here we were led to introduce non-local Kato classes K_ρ, where ρ is some positive, Lipschitz continuous function on $\mathbb{R}^n$ which reflects the behavior of q_- at infinity, possibly depending on directions. Self-adjointness of $T := -(\nabla - ib)^2 + q$ is another easy consequence of this approach. The main result is that T is essentially self-adjoint on C_0^∞, if it is bounded from below and q_- fulfills the local Kato condition. The famous result of Simon, Kato and Jensen, based on the assumption $q_- \in K$ (our K_1), follows immediately; but we also get self-adjointness of T if $q_- \in K_\rho$ with $\rho(x) = (1+|x|)^{-1}$, which contains the case $q_- \in K + O(|x|^2)$. Finally, we can specify the connections between the position of λ in the spectrum of T and the behavior at infinity of corresponding eigensolutions.

Differential Equations with
Applications to Mathematical
Physics

0 Introduction

Twenty years ago, Tosio Kato presented his famous inequality which
opened a new way to deal with the positive part of potentials of
Schrödinger operators in questions of regularity of weak eigensolutions. In the same paper [5] a condition on the negative part was
introduced to establish self-adjointness of the operator. In the se-
quel, however, the global aspect of this Kato condition, employed for
instance to prove mean value inequalities, has been overemphasized.
We therefore consider less restrictive global conditions on the potentials to point out which properties of the operator and its eigensolutions depend on local assumptions only and to get more quantitative
results globally. The material comes from [2], where supplementary
and more detailed information can be found, and from a collaboration with Günter Stolz [4].

We consider the Schrödinger operator $T = -(\nabla - ib)^2 + q$, where
q is a real-valued, measurable function on $\mathbb{R}^n$ and $b : \mathbb{R}^n \longrightarrow \mathbb{R}^n$
will be continuously differentiable. (In [2] there is no magnetic po-
tential b at all, while in [4] we have weaker, in fact weakest, assumptions on b; this latter approach requires some different techniques,
however.) A solution for the corresponding (generalized) eigenvalue
equation for $\lambda \in \mathbb{R}$ is a $u \in L_{1,loc}$ with $qu \in L_{1,loc}$ and

$$\forall \varphi \in C_0^\infty : \int \overline{u} T \varphi = \lambda \int \overline{u} \varphi \; ;$$

we write $Tu = \lambda u$. By putting λ into q, we may assume $\lambda = 0$.

Now Kato's inequality ([5], Lemma A) yields:

$$\triangle |u| \geq \text{re} \left(\text{sign}(\overline{u}) \cdot (\nabla - ib)^2 u \right) = q|u| \geq -q_-|u|$$

in the distributional sense, $q_- := \max\{0, -q\}$ denoting the negative
part of q. Writing v for $|u|$ and p for q_-, we are left with the
differential inequality $\triangle v + pv \geq 0$, with non-negative v and p. We
will show that the mean value inequality for subharmonic functions
(i.e. the case $p = 0$) extends to our situation and can serve as a base
for establishing local boundedness of u, self-adjointness of T, and
connections between the spectrum of T and the behavior of eigensolutions at infinity. We will, of course, need some extra assumptions

on q, but as shown only on q_-. These conditons, both local and global ones, will emerge quite naturally from our discussion of mean value inequalities.

1   Mean Value Inequalities

The following lemma is the basic tool in this report.

Lemma 1.1 *Let $v \in L_{1,loc}$ be real-valued; $f \in L_{1,loc}$ be non-negative and such that $\triangle v + f \geq 0$, i.e.*

$$\forall \varphi \in C_0^\infty, \varphi \geq 0 : \int (v \triangle \varphi + f\varphi) \geq 0 .$$

Then for almost every $x \in \mathbb{R}^n$ and for any $r > 0$:

$$v(x) \leq \frac{n}{\sigma_n r^n} \int\limits_{B(x;r)} v(y)dy + \frac{1}{(n-2)\sigma_n} \int\limits_{B(x;r)} \frac{f(y)}{|x-y|^{n-2}}dy , \qquad (1)$$

where σ_n is the area of the unit sphere in $\mathbb{R}^n$.

The proof can be found in ([2], p. 117f). The price we have to pay for the help of f in the case of negative $\triangle v$ is the second term on the right-hand side. Since our goal is local boundedness of $v = |u|$ for an eigensolution u, we somehow have to get rid of this term, for which there is no a priori bound, when $f = q_-|u|$. This can be achieved by replacing $|u|$ here by inequality (1) once again. Then the integral

$$\int\limits_{B(x;r)\cap\omega} \frac{|p(y)|}{|x-y|^{n-2}}dy$$

has to vanish for $r \to 0$, uniformly in $x \in \mathbb{R}^n$, for any compact $\omega \subset \mathbb{R}^n$. A p with this property is said to belong to the local Kato class K_{loc}. By a method developed in Hinz and Kalf [3] one can then show that for almost every $x \in \omega$ and small r:

$$v(x) \leq \frac{(1+2^{1-n})n}{\sigma_n r^n} \int\limits_{\omega_{3r}} v(y)dy + \frac{1}{(n-2)\sigma_n r^{n-2}} \int\limits_{\omega_{8r}} p(y)v(y)dy , \qquad (2)$$

where ω_ε denotes the set obtained from ω by adding an ε-rim arround. So we arrive at:

Theorem 1.2 *Let $p \in K_{loc}$ be real-valued, $v \in L_{1,loc}$ non-negative with $pv \in L_{1,loc}$ and $(\Delta + p)v \geq 0$. Then $v \in L_{\infty,loc}$.*

As an immediate consequence we get the most fundamental regularity properties of weak solutions for the Schrödinger equation:

Corollary 1.3 *Let q be real-valued and measurable with $q_- \in K_{loc}$; $b \in C^1$. Let $u \in L_{1,loc}$ with $qu \in L_{1,loc}$ be a solution of*

$$-(\nabla - ib)^2 u + qu = 0.$$

Then $u \in L_{\infty,loc} \cap W^1_{2,loc}$.

Proof. As shown in the Introduction, $\Delta|u| + q_-|u| \geq 0$ by Kato's inequality. Since $0 \leq q_-|u| \leq |qu| \in L_{1,loc}$, Theorem 1.2 applies, whence $u \in L_{\infty,loc}$.

Furthermore $(\nabla - ib)^2 u = qu \in L_{1,loc}$ and an interpolation argument ([4], Lemma 2.2) yields $\nabla u \in L_{2,loc}$. □

Another look at inequality (2) reveals that apart from this qualitative result, the right-hand side provides quantitative upper bounds for v, as soon as one can estimate $\int p(y)v(y)dy$. The same approach which led from (1) to (2), carried out with some more sophistication, shows that in fact the second term in the right-hand side of (2) is completely subordinate to the first term, such that we can reach a mean value inequality

$$v(x) \leq \frac{2n}{\sigma_n r^n} \int\limits_{B(x;r)} v(y)dy \ .$$

Since the method depends on some estimates of Caccioppoli type (see [3], Lemma 4), we have to assume $v \in W^1_{2,loc}$, which in view of applications to eigensolutions u and Corollary 1.3 is no restriction at all. As for p, in order to allow for an r as large as possible, we have to controll the decay rate of

$$\int\limits_{B(x;r)\cap\omega} \frac{|p(y)|}{|x-y|^{n-2}} dy$$

when r goes to 0. This can be done through the following definition of a global Kato class:

Definition 1.4 *Let* $\rho : \mathbb{R}^n \longrightarrow]0, \infty[$ *be globally Lipschitz continuous. Then*

$$K_\rho := \left\{ p \text{ measurable on } \mathbb{R}^n; \lim_{k \to \infty} \sup_{x \in \mathbb{R}^n} \int_{B(x; \frac{\rho(x)}{k})} \frac{|p(y)|}{|x-y|^{n-2}} dy = 0 \right\}.$$

Note that this coincides with the definition of the classical Kato class K if ρ is constant and that $K_\rho \subset K_{loc} \subset L_{1,loc}$.The mean value inequality then reads:

Theorem 1.5 *Let* $p \in K_\rho$ *be real-valued. Then there is a* $K \in \mathbb{N}$ *such that for any non-negative solution* $v \in W_{2,loc}^1$ *with* $pv \in L_{1,loc}$ *of* $(\triangle + p)v \geq 0$:

$$\forall x \in \mathbb{R}^n \ \forall 0 < r \leq \frac{\rho(x)}{K} : \quad v(x) \leq \frac{2n}{\sigma_n r^n} \int_{B(x;r)} v(y) dy .$$

As pointed out, the proof depends on Theorem 1 in [3], where a mean value inequality for v^2 has been obtained. The estimate on v then follows by a kind of reverse Hölder inequality. We refer to ([2], p. 123-127) for details.

Typical applications of Theorem 1.5 are Harnack's inequality (see [3]) and pointwise decay of eigenfunctions.

Corollary 1.6 *Let* q *be real-valued and measurable on* $\mathbb{R}^n$ *with* $q_- \in K_\rho$; $b \in C^1$. *Then for every* $u \in L_2$ *with* $qu \in L_{1,loc}$ *and which is a solution of* $-(\nabla - ib)^2 u + qu = 0$:

$$u = o(\rho^{-n/2}) \text{ at } \infty, \text{ i.e. } \rho^{n/2}(x)|u(x)| \to 0 \text{ , as } |x| \to \infty .$$

Proof. Since $q_- \in K_{loc}$, Corollary 1.3 yields $u \in W_{2,loc}^1$ and so is $|u|$ because $\partial_j |u| = \mathrm{re}(\mathrm{sign}(\overline{u}) \cdot \partial_j u)$. Again by Kato's inequality we know that $\triangle |u| + q_- |u| \geq 0$, whence Theorem 1.5 applies:

$$\forall x \in \mathbb{R}^n : |u(x)| \leq \frac{2n K^n}{\sigma_n \rho^n(x)} \int_{B(x;\rho(x))} |u(y)| dy .$$

Hölder's inequality yields

$$\forall\, x \in \mathbb{R}^n : \; |u(x)|^2 \leq \frac{4nK^{2n}}{\sigma_n \rho^n(x)} \int\limits_{B(x;\rho(x))} |u(y)|^2 dy \,.$$

As $K_\rho = K_{a\rho}$ for any constant $a > 0$, we may assume the Lipschitz constant of ρ to be $\frac{1}{2}$, such that $|x| - \rho(x) \geq \frac{|x|}{2} - \rho(0)$, and the last integral goes to 0 as $|x| \to \infty$, since $u \in L_2$. □

Genuine examples are obtained from $\rho(x) = (1 + |x|)^\delta$ with a $\delta \leq 1$, including the classical case $(\delta = 0)$ of $q_- \in K$, where $|u(x)| \to 0$, but giving faster decay for $\delta > 0$ and weaker bounds if $\delta < 0$ (these are potentials q which might go to $-\infty$ as $|x| \to \infty$). If ρ is not spherically symmetric, we get direction depending bounds on eigenfunctions.

2   Self-Adjointness

Based on the results of the last section, the following general criterion for essential self-adjointness of T on C_0^∞ is easy to derive. To get a well-defined symmetric operator in L_2, we have to assume $q \in L_{2,loc}$ real-valued from now on.

Theorem 2.1 *Let* $q \in L_{2,loc}$ *with* $q_- \in K_{loc}$, $b \in C^1$, *and let*

$$T := -(\nabla - ib)^2 + q \,|\, C_0^\infty$$

be bounded from below. Then T *is essentially self-adjoint in* L_2.

Proof. Without loss $T \geq 1$. We show $\overline{TC_0^\infty} = L_2$.

Consider $u \in \overline{TC_0^\infty}^\perp$, whence $u \in L_2$ and $Tu = 0$. By Corollary 1.3, $u \in L_{\infty,loc} \cap W^1_{2,loc}$. For $\varepsilon > 0$ and $k \in \mathbb{N}$ consider $\psi := u_\varepsilon \eta_k^2$, where u_ε denotes the classical regularization of u, and η_k is obtained from a smooth cut-off function η (i.e. $\eta(t) = 1$ for $t \leq \frac{1}{2}$, 0 for $t \geq 1$ and otherwise in $[0,1]$) by putting $\eta_k(x) = \eta(\frac{|x|}{k})$. Then a thorough calculation shows that

$$0 = (u, T(\psi_\varepsilon))$$

$$= (u_\varepsilon \eta_k, T(u_\varepsilon \eta_k)) - \| \, |\nabla \eta_k| u_\varepsilon \|^2 + 2i \cdot \mathrm{im}(\eta_k \nabla u_\varepsilon, u_\varepsilon \nabla \eta_k)$$

$$+ \left((\{-i\nabla \cdot b + |b|^2 + q\}u)_\varepsilon - \{-i\nabla \cdot b + |b|^2 + q\}u_\varepsilon , \, u_\varepsilon \eta_k^2 \right)$$

$$+ 2i \sum_{j=1}^{n} \left((b_j u)_\varepsilon - b_j u_\varepsilon , \, \partial_j(u_\varepsilon \eta_k^2) \right) + 2i(u_\varepsilon \eta_k, u_\varepsilon b \cdot \nabla \eta_k) \, .$$

The first two terms on the right-hand side are real and can be estimated from below by

$$\|u_\varepsilon \eta_k\|^2 - \frac{\max |\eta'|}{k^2} \|u_\varepsilon\|^2 \, .$$

The sum of the other terms must be real too and tends to

$$2i(\mathrm{im}(\eta_k \nabla u, \, u \nabla \eta_k) + \int |u|^2 \eta_k b \cdot \nabla \eta_k)$$

as $\varepsilon \to 0$, which thus must be 0. Hence we arrive at

$$\|u\,\eta_k\|^2 \le \frac{\max |\eta'|^2}{k^2} \|u\|^2 \, ,$$

and letting $k \to \infty, u = 0$ follows. $\qquad\qquad\qquad\square$

Another way to establish essential self-adjointness of T is by imposing global conditions on q_- such as $q_- \in K$ or $K + O(|x|^2)$ (i.e. $q_- = q_1 + q_2$ with $q_1 \in K$, and $(1 + |\cdot|)^{-2}q_2$ is bounded). Theorem 2.1 allows to consider the even larger class K_ρ with $\rho(x) = (1+|x|)^{-1}$, although T will not be bounded from below in that case.

Corollary 2.2 *Let* $q \in L_{2,loc}$ *with* $q_- \in K_{(1+|\cdot|)^{-1}}$, $b \in C^1$. *Then* T *is essentially self-adjoint in* L_2.

Proof. Let us first assume that $q_- \in K$. Then q_- is relatively form bounded with respect to $-\Delta$ ([2], Lemma 3.2) and consequently also with respect to $-(\nabla - ib)^2$ with the same bound ([4], Lemma 2.3), namely 0. Hence for all $\varphi \in C_0^\infty$:

$$(\varphi, T\varphi) = \| \, |(\nabla - ib)\varphi| \, \|^2 + (\varphi, q\varphi) \ge \mathrm{const} \, \|\varphi\|^2 \, ,$$

i.e. T is bounded from below. By Theorem 2.1 T is essentially self-adjoint.

The transition to general $q_- \in K_{(1+|\cdot|)^{-1}}$ by cutting q_- off outside balls and recourse to the first case is done as in ([2], Section 3.2), where Δ has to be replaced by $(\nabla - ib)^2$ in Lemma 3.5; the necessary changes are straightforward.   □

3 Bounds on Eigensolutions and the Spectrum

A classical subject of spectral theory of Schrödinger operators T is the discussion of connections between the behavior at infinity of eigensolutions for λ and the position of λ in the spectrum $\sigma(T)$. Apart from extreme cases, the discrete spectrum $\sigma_d(T)$ is associated with exponentially decaying eigenfunctions, whereas a λ in the essential spectrum $\sigma_e(T)$ has only (polynomially) bounded eigensolutions. We will make this precise with the aid of a method of Emmanuil Eh. Shnol', based on the following lemma, which is an easy extension of the well-known Weyl criterion for the essential spectrum:

Lemma 3.1 *Let T be a self-adjoint operator in a Hilbert space; $\lambda \in \mathbb{R}$. Then for any sequence $(u_k)_{k \in \mathbb{N}} \subset D(T)$ with $\forall k \in \mathbb{N}$: $\|u_k\| = 1$ and $u_k \xrightarrow{w} 0$, as $k \to \infty$:*

$$\mathrm{dist}(\lambda, \sigma_e(T)) \leq \liminf_{k \to \infty} \|(T - \lambda)u_k\|.$$

For the proof see ([1], p. 174).

We will now assume $T = -(\nabla - ib)^2 + q$ with $q \in L_{2,loc}$ and $b \in C^1$ throughout. Starting from an eigenfunction $u \in L_2$ for $\lambda \in \sigma_d(T)$ (a polynomially bounded eigensolution $u \in L_{2,loc} \backslash L_2$ for a $\lambda \in \mathbb{R}$) one can construct the sequence (u_k) by cutting off inside (outside) balls of increasing diameters. The bounds on $\mathrm{dist}(\lambda, \sigma_e(T))$ obtained from Lemma 3.1 can then be used to derive upper bounds for u (prove $\lambda \in \sigma_e(T)$).

Theorem 3.2 *Let* $q_- \in K_{(1+|\cdot|)^{-\gamma}}$ *with a* $\gamma \in [0,1]$. *Then for* $\lambda \in \sigma_d(T)$ *there is a* $\mu > 0$ *such that for any eigenfunction* u *for* λ:

$$u(x) = \begin{cases} O(e^{-\mu|x|^{1-\gamma}}) & , \text{ if } 0 \le \gamma < 1; \\ O(|x|^{(n-\mu)/2)}) & , \text{ if } \gamma = 1. \end{cases}$$

Theorem 3.3 *Let* $q_- \in K + o(|x|^2)$ *(i.e.* $q_- \in K + O(|x|^2)$ *and* $|x|^{-2}q_2(x) \to 0$, *as* $|x| \to \infty$ *). If for a* $\lambda \in \mathbb{R}$ *there is a polynomially bounded solution* $u \in L_{2,loc} \backslash L_2$ *of* $Tu = \lambda u$, *then* $\lambda \in \sigma_e(T)$.

The technical details of the proofs depend on the regularity results of Corollary 1.3, on the observation that

$$\Delta(|u|^2) = 2(q - \lambda)|u|^2 + 2|(\nabla - ib)u|^2$$

([4], Lemma 3.9) and on form boundedness. We refer to ([2], Section 4.2) and ([4] Section 3.2), respectively.

In the proof of Theorem 3.2 the mean value inequality Theorem 1.5 enters in a step where L_2-bounds on u are transferred into the desired pointwise bounds. This procedure is also used in proving a kind of converse of Theorem 3.3, namely the fact that $\sigma(T)$ is the closure of the set of those $\lambda \in \mathbb{R}$ for which there is a polynomially bounded non-trivial eigensolution. One starts from an expansion in generalized eigenfunctions u which lie in some weighted L_2-spaces (see ([4], Section 3.1) for details). This L_2-bound can then be turned into a pointwise bound by Theorem 1.5. We thus arrive at:

Theorem 3.4 *Let* $q_- \in K + o(|x|^2)$. *Then*

$$\sigma(T) = \overline{\{\lambda \in \mathbb{R} : \exists s > 0 \exists u \ne 0, (1 + |\cdot|)^{-s}u \in L_\infty(\mathbb{R}^n) : Tu = \lambda u\}}.$$

The fact that $q_-(x) = O(|x|^2)$ is excluded here and turns up as an exception in Theorem 3.2 is explained by the existence of an example due to Halvorsen, where 0 is a discrete eigenvalue with an only polynomially decaying eigenfunction and where there is a bounded eigensolution to every $\lambda \in \mathbb{R}$, including those in the neighborhood of 0 which are not in the spectrum. Halvorsen's example is in $\mathbb{R}^1$, and it is an open question if this phenomenon extends to higher dimensions (see the discussion in ([2], Chapter 5)).

Acknowledgement

My travel to the conference in Atlanta was supported by the Deutsche Forschungsgemeinschaft.

Bibliography

[1] A. M. Hinz, *Asymptotic behavior of solutions of* $-\triangle v + qv = \lambda v$ *and the distance of* λ *to the essential spectrum*, Math. Z. 194 (1987), p. 173-182.

[2] A. M. Hinz, *Regularity of solutions for singular Schrödinger equations*, Rev. Math. Phys. 4 (1992), p. 95-161.

[3] A. M. Hinz and H. Kalf, *Subsolution estimates and Harnack's inequality for Schrödinger operators*, J. Reine Angew. Math. 404 (1990), p. 118-134.

[4] A. M. Hinz and G. Stolz, *Polynomial boundedness of eigensolutions and the spectrum of Schrödinger operators*, Math. Ann. (to appear).

[5] T. Kato, *Schrödinger operators with singular potentials*, Israel J. Math. 13 (1972), p. 135-148.

Linearization of Ordinary Differential Equations

Nail H. Ibragimov
Institute of Mathematical Modelling
Russian Academy of Sciences
Moscow 125047, Russia

To Bill Ames on the occasion of his 65th birthday.

Abstract

Lie group approach is discussed to linearization of second and first order ordinary differential equations. For first order equations we use changes of the dependent variable only while for second order equations general changes of dependent and independent variables are considered.

1 Second Order Equations

One can extract, from several results of S. Lie [1], [2], the following statement [3]:

Theorem 1 *The following assertions are equivalent:*
(i) a second order ordinary differential equation

$$y'' = f(x, y, y')$$ (1)

Differential Equations with
Applications to Mathematical
Physics

can be linearized by a change of variables $\overline{x} = \phi(x,y)$, $\overline{y} = \psi(x,y)$; *(ii) equation* (1) *has the form*

$$y'' + F_3(x,y)y'^3 + F_2(x,y)y'^2 + F_1(x,y)y' + F(x,y) = 0 \quad (2)$$

with coefficients F_3, F_2, F_1, F *satisfying the integrability conditions of an auxiliary overdetermined system*

$$\frac{\partial z}{\partial x} = z^2 - Fw - F_1 z + \frac{\partial F}{\partial y} + FF_2,$$

$$\frac{\partial z}{\partial y} = -zw + FF_3 - \frac{1}{3}\frac{\partial F_2}{\partial x} + \frac{2}{3}\frac{\partial F_1}{\partial y},$$

$$\frac{\partial w}{\partial x} = zw - FF_3 - \frac{1}{3}\frac{\partial F_1}{\partial y} + \frac{2}{3}\frac{\partial F_2}{\partial x},$$

$$\frac{\partial w}{\partial y} = -w^2 + F_2 w + F_3 z + \frac{\partial F_3}{\partial x} - F_1 F_3; \quad (3)$$

(iii) equation (1) *admits an* 8*-dimensional Lie algebra;*
(iv) equation (1) *admits a* 2*-dimensional Lie algebra with a basis*

$$X_\alpha = \xi_\alpha(x,y)\frac{\partial}{\partial x} + \eta_\alpha(x,y)\frac{\partial}{\partial y}, \qquad \alpha = 1,2,$$

such that their pseudoscalar product

$$X_1 \vee X_2 = \xi_1 \eta_2 - \eta_1 \xi_2 \quad (4)$$

vanishes.

Example 1. The equation

$$y'' = e^{-y'}$$

is not linearized since it is not of the form (2).

Example 2. Let's consider equations of the form

$$y'' = f(y') \quad (5)$$

from Table 2, and inspect when they are linearized. In accordance with Theorem 1(ii) it is necessary that the function $f(y')$ is a polynom of the third degree, i.e., the equation (5) has the form

$$y'' + A_3 y'^3 + A_2 y'^2 + A_1 y' + A_0 = 0 \quad (6)$$

with constant coefficients A_i. One can easily verify that the auxiliary system (3) for Eq. (6) is integrable. Therefore, Eq. (6) is linearized for arbitrary coefficients A_i.

Example 3. Let's take, from Table 2, equations of the form

$$y'' = \frac{1}{x} f(y').$$ (7)

When are they linearized? Again, by Theorem 1(ii) we have to consider only equations of the form

$$y'' + \frac{1}{x}(A_3 y'^3 + A_2 y'^2 + A_1 y' + A_0) = 0$$

with constant coefficients A_i. In this case we have from the integrability conditions of the corresponding system (3) the following equations:

$$A_2(2 - A_1) + 9A_0 A_3 = 0, \qquad 3A_3(1 + A_1) - A_2^2 = 0.$$

We put $A_3 = -a$, $A_2 = -b$ and obtain $A_1 = -\left(1 + \frac{b^2}{3a}\right)$, $A_0 = -\left(\frac{b}{3a} + \frac{b^3}{27a^2}\right)$. Hence, Eq. (7) is linearized iff it is of the form (see also [4])

$$y'' = \frac{1}{x}\left[ay'^3 + by'^2 + \left(1 + \frac{b^2}{3a}\right)y' + \frac{b}{3a} + \frac{b^3}{27a^2}\right].$$ (8)

A linearizing change of variables can be found via statement (iv) of Theorem 1.

For example, we find a linearization of Eq. (8) in the case $a = 1$, $b = 0$, i.e., of the equation

$$y'' = \frac{1}{x}(y' + y^3).$$ (9)

This equation admits L_2 with the basis

$$X_1 = \frac{1}{x}\frac{\partial}{\partial x}, \qquad X_2 = \frac{y}{x}\frac{\partial}{\partial x},$$ (10)

which satisfies the condition $X_1 \vee X_2 = 0$ of Theorem 1(iv). The operators (10) are of the type II from Table 2. Therefore a linearization is obtained by turning to the canonical variables

$$\overline{x} = y, \qquad \overline{y} = \frac{1}{2}x^2$$

in which the operators (10) become

$$\overline{X}_1 = \frac{\partial}{\partial \overline{y}}, \qquad \overline{X}_2 = \overline{x}\frac{\partial}{\partial \overline{y}},$$

in accordance with Table 2. Then, excluding the special solution $y = \text{const.}$, we have the transformed equation (9):

$$\overline{y}'' + 1 = 0.$$

Example 4. We now take equations

$$y'' = F(x, y) \qquad (11)$$

and verify that the

Question: *When a nonlinear equation of the form* (11) *is linearized?* has the

Answer: *Never.*

Indeed, our equation (11) is a particular case of Eq. (2) with coefficients $F_1 = F_2 = F_3 = 0$. The system (3) is

$$z_x = z^2 + Fw - Fy, \quad w_x = zw,$$
$$z_y = -zw, \qquad\qquad w_y = -w^2,$$

and one of the integrability conditions, namely

$$z_{xy} = z_{yx}$$

yields

$$F_{yy} = 0.$$

It follows that Eq. (11), where $F(x, y)$ is nonlinear in y, is not linearizable.

Example 5. Here we discuss in detail a construction of a linearization. One can readily find that the equation

$$y'' = \left(y' - \frac{y}{x} \right)^3 f \left(\frac{y}{x} \right) \tag{12}$$

with an arbitrary function f admits the 2–dimensional Lie algebra spanned by

$$X_1 = x^2 \frac{\partial}{\partial x} + xy \frac{\partial}{\partial y}, \qquad X_2 = xy \frac{\partial}{\partial x} + y^2 \frac{\partial}{\partial y}. \tag{13}$$

This algebra belongs to the type II of Table 2. Therefore Eq. (12) can be linearized and a linearizing change of variables $\bar{x} = \phi(x,y)$, $\bar{y} = \psi(x,y)$ is obtained from the conditions

$$X_1(\phi)\frac{\partial}{\partial \bar{x}} + X_1(\psi)\frac{\partial}{\partial \bar{y}} = \frac{\partial}{\partial \bar{y}}, \qquad X_2(\phi)\frac{\partial}{\partial \bar{x}} + X_2(\psi)\frac{\partial}{\partial \bar{y}} = \bar{x}\frac{\partial}{\partial \bar{y}}. \tag{14}$$

We have from (14) the following four equations to determine ϕ, ψ:

$$X_1(\phi) = 0, \qquad X_1(\psi) = 1; \tag{15}$$

$$X_2(\phi) = 0, \qquad X_2(\psi) = \phi. \tag{16}$$

The general solution of Eqs. (15) is

$$\phi = g\left(\frac{y}{x}\right), \qquad \psi = -\frac{1}{x} + h\left(\frac{y}{x}\right).$$

By these functions the first Eq. (16) is satisfied identically while the second one gives $\phi = y/x$. We choose $h = 0$ to obtain the following change of variables:

$$\bar{x} = \frac{y}{x}, \qquad \bar{y} = -\frac{1}{x}.$$

After this transformation the equation (12) becomes

$$\bar{y}'' + f(\bar{x}) = 0.$$

2   First Order Equations

In the case of first–order equations Theorem 1 is replaced by the following.

Theorem 2 *Given a first–order ordinary differential equation*

$$y' = f(x, y) \qquad (17)$$

one can by means of an appropriate change of variables

$$\overline{x} = \phi(x, y), \qquad \overline{y} = \psi(x, y) \qquad (18)$$

transform (17) *into any given equation*

$$\overline{y}' = g(\overline{x}, \overline{y}). \qquad (19)$$

We consider here, instead of general changes (18) of both independent and dependent variables, transformations of the dependent variable only:

$$\overline{y} = \psi(y). \qquad (20)$$

If Eq. (17) is linear, then after transformation (20) we have, in general, a nonlinear equation (19). This equation will be a particular case of equations possessing a fundamental system of solutions, or a nonlinear superposition principle ([5]–[9]). Further, any first–order ODE possessing a nonlinear superposition can be written after a transformation (20) in the form of a Riccati equation

$$y' = P(x) + Q(x)y + R(x)y^2. \qquad (21)$$

So, the question is when is Eq. (21) linearized by a transformation of the form (20)? We formulate an answer as follows ([10]):

Theorem 3 *If the Riccati equation* (21) *possesses one of the following four properties, then it should possess all of them:*
(i) Eq. (21) *is linearized by a transformation* (20):
(ii) Eq. (21) *can be written in the form*

$$y' = T_1(x)\xi_1(y) + T_2(x)\xi_2(y) \qquad (22)$$

so that the operators

$$X_1 = \xi_1(y)\frac{d}{dy}, \qquad X_2 = \xi_2(y)\frac{d}{dy} \qquad (23)$$

span a 2–dimensional Lie algebra, i.e.,

$$[X_1, X_2] = \alpha X_1 + \beta X_2$$

(if $[X_1, X_2] = 0$ we have one–dimensional algebra and the variables in the Riccati equation are separated);
  (iii) Eq. (21) is either of the form

$$y' = Q(x)y + R(x)y^2 \qquad (24)$$

or

$$y' = P(x) + Q(x)y + k[Q(x) - kP(x)]y^2 \qquad (25)$$

with any coefficients $P(x), Q(x), R(x)$ and a certain constant k (in general, complex);
  (iv) Eq. (21) admits a constant (in general, complex) solution.

Remark. Eq. (25) has the constant solution $y = -1/k$. Therefore a linear equation being a particular case of Eq. (25) with $k = 0$, can be considered as a Riccati equation having the point at infinity as its constant solution.

Example 1. The equation

$$y' = x + y^2$$

is neither of the form (24) nor (25). Hence it cannot be linearized. We also notice that it is of the form (22) with coefficients $T_1 = x$, $\xi_1 = 1$; $T_2 = 1$, $\xi_2 = y^2$ so that operators (23) are

$$X_1 = \frac{d}{dy}, \qquad X_2 = y^2\frac{d}{dy}.$$

The two–dimensional vector space spanned by these operators is not a Lie algebra since the commutator

$$[X_1, X_2] = 2y\frac{d}{dy}$$

is not a linear combination of X_1 and X_2.

Example 2. The equation

$$y' = x + (x + \sqrt{2})^2 y + 2\sqrt{2}(2 + x^2)y^2 \tag{26}$$

is not of the form (22). But, it would be erroneous to make a conclusion that this equation cannot be linearized. Indeed it has the following constant solution:

$$y = -\frac{1}{2\sqrt{2}}$$

and thus Eq. (26) is linearizable. This is not in a contradiction with Theorem 3(ii). In fact one can represent Eq. (26) in the form (22) as follows:

$$y' = x(1 + 2\sqrt{2}y) + (2 + x^2)(y + 2\sqrt{2}y^2). \tag{27}$$

The corresponding operators (23) for Eq. (27) are equal to

$$X_1 = (1 + 2\sqrt{2}y)\frac{d}{dy}, \qquad X_2 = (y + 2\sqrt{2}y^2)\frac{d}{dy}$$

and form a 2–dimensional Lie algebra since

$$[X_1, X_2] = X_1 + 2\sqrt{2}X_2.$$

Example 3. Now we discuss details of a linearization. Consider the equation

$$y' = P(x) + Q(x)y + [Q(x) - P(x)]y^2 \tag{28}$$

which is of the form (25) with $k = 1$. It is written in the form (22) with $T_1 = P$, $T_2 = Q$, $\xi_1 = 1 - y^2$, $\xi_2 = y + y^2$. Hence the operators (23) are

$$X_1 = (1 - y^2)\frac{d}{dy}, \qquad X_2 = (y + y^2)\frac{d}{dy}. \tag{29}$$

They span L_2 since

$$[X_1, X_2] = X_1 + 2X_2.$$

To find the linearizing transformation we first choose the new basis of L_2 as follows:

$$\overline{X}_1 = X_1 + 2X_2 = (1+y)^2 \frac{d}{dy}, \qquad \overline{X}_2 = X_2. \qquad (29')$$

Then $[\overline{X}_1, \overline{X}_2] = \overline{X}_1$ and therefore we seek for a transformation (20) such that the operators $(29')$ become

$$\overline{X}_1 = \frac{d}{d\overline{y}}, \qquad \overline{X}_2 = \overline{y}\frac{d}{d\overline{y}}.$$

This transformation is found from the equation

$$\overline{X}_1(\overline{y}) \equiv (1+y)^2 \frac{d\overline{y}}{dy} = 1$$

and is given by

$$\overline{y} = -\frac{1}{1+y}. \qquad (30)$$

After this Eq. (28) becomes

$$\overline{y}' = Q(x) - P(x) + [Q(x) - 2P(x)]\overline{y} \qquad (28')$$

Table 1. Lie Group Classification of Second Order Equations

Group	Basis of Lie Algebra	Equation
G_1	$X_1 = \frac{\partial}{\partial x}$	$y'' = f(y, y')$
G_2	$X_1 = \frac{\partial}{\partial x}, X_2 = \frac{\partial}{\partial y}$	$y'' = f(y')$
	$X_1 = \frac{\partial}{\partial y}$, $X_2 = x\frac{\partial}{\partial x} + y\frac{\partial}{\partial y}$	$y'' = \frac{1}{x}f(y')$
G_3	$X_1 = \frac{\partial}{\partial x} + \frac{\partial}{\partial y}$, $X_2 = x\frac{\partial}{\partial x} + y\frac{\partial}{\partial y}$, $X_3 = x^2\frac{\partial}{\partial x} + y^2\frac{\partial}{\partial y}$	$y'' + 2\frac{y' + Cy'^{3/2} + y'^2}{x - y} = 0$
	$X_1 = \frac{\partial}{\partial x}$, $X_2 = 2x\frac{\partial}{\partial x} + y\frac{\partial}{\partial y}$, $X_3 = x^2\frac{\partial}{\partial x} + xy\frac{\partial}{\partial y}$	$y'' = Cy^{-3}$
	$X_1 = \frac{\partial}{\partial x}, X_2 = \frac{\partial}{\partial y}$, $X_3 = x\frac{\partial}{\partial x} + (x + y)\frac{\partial}{\partial y}$	$y'' = Ce^{-y'}$
	$X_1 = \frac{\partial}{\partial x}, X_2 = \frac{\partial}{\partial y}$, $X_3 = x\frac{\partial}{\partial x} + ky\frac{\partial}{\partial y}$	$y'' = Cy'^{(k-2)/(k-1)}$, $k \neq 0, \frac{1}{2}, 1, 2$
G_8	$X_1 = \frac{\partial}{\partial x}, X_2 = \frac{\partial}{\partial y}$, $X_3 = x\frac{\partial}{\partial y}, X_4 = x\frac{\partial}{\partial x}$, $X_5 = y\frac{\partial}{\partial x}, X_6 = y\frac{\partial}{\partial y}$, $X_7 = x^2\frac{\partial}{\partial x} + xy\frac{\partial}{\partial y}$, $X_8 = xy\frac{\partial}{\partial x} + y^2\frac{\partial}{\partial y}$	$y'' = 0$

Table 2. Canonical Form of 2–Dimensional Lie Algebras
and Invariant Second Order Equations

Type	Structure of L_2	Basis of L_2 in Canonical Variables	Equation
I	$[X_1, X_2] = 0,$ $X_1 \vee X_2 \neq 0$	$X_1 = \frac{\partial}{\partial x}, X_2 = \frac{\partial}{\partial y}$	$y'' = f(y')$
II	$[X_1, X_2] = 0,$ $X_1 \vee X_2 = 0$	$X_1 = \frac{\partial}{\partial y}, X_2 = x\frac{\partial}{\partial y}$	$y'' = f(x)$
III	$[X_1, X_2] = X_1,$ $X_1 \vee X_2 \neq 0$	$X_1 = \frac{\partial}{\partial y},$ $X_2 = x\frac{\partial}{\partial x} + y\frac{\partial}{\partial y}$	$y'' = \frac{1}{x}f(y')$
IV	$[X_1, X_2] = X_1,$ $X_1 \vee X_2 = 0$	$X_1 = \frac{\partial}{\partial y}, X_2 = y\frac{\partial}{\partial y}$	$y'' = f(x)y'$

Bibliography

[1] Lie, S., *Klassifikation und Integration von gewöhnlichen Differentialgleichungen zwischen x, y, die eine Gruppe von Transformationen gestatten*, Arch. for Math. (1883), Bd. VIII, (1884), Bd. IX; in Gesammelte Abh., Bd. 5.

[2] Lie, S., *Vorlesungen über Differentialgleichungen mit bekannten infinitesimalen Transformationen* (Bearbeitet und herausgegeben von Dr. G. Scheffers) (1891) Leipzig: B. G. Teubner.

[3] Ibragimov, N. H., *Essays in the group analysis of ordinary differential equations* (1991) Moscow: Znanie, see also: *Group analysis of ordinary differential equations and new observations in mathematical physics*, Uspekhi Mat. Nauk, to appear.

[4] Mahomed, F. M., Leach, P. G. L., *Lie algebras associated with scalar second-order ordinary differential equations*, J. Math. Phys. (1989), v. 30, no. 12.

[5] Vessiot, E., *Sur une classe d'équations différentielles*, Ann. Sci. École Norm. Sup. (1893), T. 10.

[6] Guldberg, A., *Sur les équations différentielles ordinaire qui possèdent un système fundamental d'integrales*, C. R. Paris (1893), T. 116.

[7] Lie, S., *Vorlesungen über continuierliche Gruppen mit geometrischen and anderen Anwendungen*, (Bearbeitet and herausgegeben von Dr. G. Scheffers) (1893) Leipzig: B. G. Teubner.

[8] Ames, W. F., *Nonlinear superposition for operator equations*. In *Nonlinear Equations in Abstract Spaces*, ed. V. Lakshmikantham (1978) New York: Academic Press.

[9] Anderson, R. L., *A nonlinear superposition principle admitted by coupled Riccati equations of the projective type*, Lett. Math. Phys. (1989), v. 4.

[10] Ibragimov, N. H., *Primer on the group analysis* (1989) Moscow: Znanie.

Expansion of Continuous Spectrum Operators in Terms of Eigenprojections

Robert M. Kauffman
Dept. of Mathematics
University of Alabama at Birmingham
Birmingham, AL 35294

1 Introduction

Eigenfunction expansions are at the heart of the picture of quantum mechanics which was developed by Dirac. The idea is to expand states which change with time as they evolve under the Schrödinger equation $d\Psi/dt = iH\Psi$ in terms of those which do not, in the sense that they give the same expectation values for all observables. However, in quantum mechanics, observables in the physical sense correspond to operators in a Hilbert space. The operator which maps the initial condition $\Psi(0)$ for the Schrödinger equation to the solution $\Psi(t)$ at time t is denoted by e^{iHt}. Since this is the fundamental operator of quantum mechanics, it makes sense to expand it in terms of simple operators; the most natural way of doing this is to expand in terms of operators of the form $e^{i\lambda t}P_\lambda$, where P_λ is a projection onto a one-dimensional space of eigenfunctions with eigenvalue λ. This turns the operator $\exp(iHt)$ of time evolution into a diagonal matrix; unfortunately, it in general has uncountably many entries. For many physical problems, such as those connected with scattering

Differential Equations with
Applications to Mathematical
Physics

theory, H is not known, or is only known up to a small perturbation; the object is to find it, from measurements involving the time evolution of the physical system. If $\exp(iHt)$ is considered as a matrix with uncountably many entries, it would take uncountably many measurements in general to find it. Ideally, the matrix would have only finitely many entries; this is of course not possible unless we perform an approximation. In this note we discuss how to approximate $\exp(iHt)$, where H is a self-adjoint operator with arbitrary spectrum, in terms of eigenprojections of multiplicity one. These terms of course must be defined rigorously as part of the program. We concentrate on the approximation of spectral projections by finitely many eigenprojections, since once this is done the spectral theorem can be used to do the rest.

Our approach is self-contained, and involves developing the theory of continuous spectrum eigenfunctions afresh and paying very careful attention to convergence; in fact, new results on convergence are contained in the paper. Outside of related papers by the author [4], with Edmunds [1] and with Hinton [3], it is probably closest in spirit to the recent paper of [5], though it also harks back to work of Gelfand and others in the 1950's. The purpose of our approach is to give a very concrete answer to the question of what the eigenfunctions are and how the expansion converges. This paper gives new convergence results, which hold even in situations where no reasonable a priori estimates on the domain of the self-adjoint operator are available; one such situation would be the Laplace-Beltrami operator on a semi-Riemannian manifold. However, even when the a priori estimates needed to apply the results of [1], [3], and [4] hold for the operator in question, the results of Theorem 11 and Theorem 13 are not implied by these other results. The difference is that the convergence we study is uniform on the proper hull of appropriate sets; the concepts of hull and proper hull are given in Definition 6 and are introduced in this paper.

2 Definitions and Results

Eigenfunctions will be defined as elements of the dual space of a topological vector space W, which we call the space of *attainable states*. This space is a background space, which is in the domain of any reasonable self-adjoint operator. Since in quantum mechanics, there are good reasons for wishing the self-adjoint operators to be a ring, it is natural to expect the attainable states to be a subset of the C^∞ functions in many applications. This indicates that W is more likely to be a topological vector space than a Banach space.

    Spectral projection operators arise as operators from W into W' with range contained in the eigenfunctions of the self-adjoint operator H being studied. These are in turn defined to be solutions in W' to the equation $H'F = \lambda F$. It is interesting to observe that one of the most difficult convergence questions arises from the decomposition of the entire Hilbert space into a direct sum of cyclic subspaces. A *cyclic subspace* $\Re_f$ is the linear span of $\{e^{iHt}f : t \in \mathbf{R}\}$, where f is a fixed vector in the Hilbert space. Thinking of f as an impulse, the decomposition into orthogonal cyclic subspaces breaks the Hilbert space into invariant subspaces corresponding to orthogonal impulses; on each subspace the possibly non-normalizable eigenfunctions corresponding to a given eigenvalue have multiplicity one. The projection onto a cyclic subspace then seems to have physical meaning. However, the space W of attainable states is not in general closed under projections onto subspaces $\Re_f$; or under the group e^{iHt}. Especially this latter property is a major physical defect. It is desirable to have a larger subspace than W which is closed under these operations, but which is small enough that everything still converges. The hull of W, introduced in Definition 6, has these properties.

Definition 1 *A locally convex topological vector space is said to be a* **nuclear space** *if, for any convex balanced neighborhood V of 0, there exists another convex balanced neighborhood $U \subseteq V$ of 0 such that the canonical mapping $T : X_U \to X_V$ is nuclear. A* **nuclear operator** *from a locally convex topological vector space X into a*

Banach space Y *is an operator of the form*

$$Tx = s - \lim_{n \to \infty} \sum_{j=1}^{n} c_j f_j(x) y_j$$

where $\{f_j\}$ *is an equicontinuous sequence of continuous linear functionals on* X, $\{y_j\}$ *is a bounded sequence of elements of* Y, *and* $\{c_j\}$ *is a sequence of non-negative real numbers such that* $\sum_{j=1}^{\infty} c_j < \infty$. *The spaces* X_U *and* $\hat{X}_U$ *are defined as follows: let* U *be a convex balanced neighborhood of* 0 *in* X. *Let* κ_U *be the Minkowski functional on* U. *Let* $N_U = \{x \in X : \lambda x \in U \ \forall \ \lambda > 0\}$. *Then* N_U *is a closed subspace of* X, *and the quotient space* $\frac{X}{N_U}$ *is a normed linear space* X_U *under the norm induced by* κ_U. $\hat{X}_U$ *is the completion of* X_U.

Definition 2 *Let* Ω *be a separable Hilbert space. Let* H *be a (possibly unbounded) self-adjoint operator in* Ω. *A space* W *of* **attainable states** *for* H *is defined to be a locally convex topological vector space with the following properties:*

1. H *takes* W *continuously into* W;

2. W *is a nuclear space;*

3. W *is a dense subspace of* Ω, *such that the injection from* W *into* Ω *is continuous;*

4. W *is the inductive limit of a finite or infinite sequence* $\{V_n\}$ *of separable Frechet spaces such that* $\{V_n\}$ *is algebraically and topologically contained in* V_{n+1}.

Definition 3 *The space of* **idealized states** *is defined to be the dual space* W' *of the space of attainable states.* W' *is given the topology* $\beta(W, W')$, *where a subbase for the neighborhoods of* 0 *in* W' *is defined to be sets of the form* $A^\circ = \{F \in W' : |F(x)| \leq 1 \ \forall \ x \in A\}$, *where* A *ranges over the balanced convex bounded subsets of* W.

Note: We naturally embed Ω into W'; this causes complex conjugates to appear in various formulae.

Remark: The standard topological vector spaces of analysis, such as the rapidly decreasing functions, $C_0^\infty(R^n)$, and many others, satisfy the hypotheses of Definition 2; see [6], page 74.

Theorem 4 *A locally convex topological vector space X is nuclear if and only if for any convex balanced neighborhood V of 0, the natural mapping I_V from X into $\hat{X}_V$ is nuclear.*

Proof: This is Theorem 1, p. 291, [7].

Theorem 5 *Every space W satisfying the hypotheses of Definition 2 is a Montel space, which is by definition a separated barrelled space such that closed and bounded subsets are compact.*

Remark: The proof is not difficult, and will be omitted.

Definition 6 *Let $\phi \in W$; let $\{e_i\}$ be an orthonormal set in Ω; assume that the cyclic subspaces $\Re_{e_i}$ generated by e_i have the property that $\Re_{e_i} \perp \Re_{e_j}$ for $i \neq j$. Let P_i be the projection onto $\Re_{e_i}$. Let H be as in Definition 2; let $\Delta \to P(\Delta)$ be the spectral measure for H. Let $\sigma_{e_i}(\Delta) = [P(\Delta)e_i, e_i]$. By the spectral theorem there exists a unique isometry T_i taking the range of P_i into $L_2(\sigma_{e_i})$ such that $T_i e_i(\lambda) \equiv 1$ and such that for any $g \in domain(H)$, $T_i(HP_ig)(\lambda) = \lambda T_i g(\lambda)$. An element $e \in L_2$ is said to be in the* **hull** *$h(\phi)$ of $\phi \in W$ if $\forall\, i$, $T_i P_i e(\lambda) = \beta_{e,i}(\lambda)T_i\phi(\lambda)$ for some Borel measurable function β_i of modulus one; the* **proper** *hull is the set of elements of the hull where the functions β_i are equicontinuous when restricted to compact sets. The hull $h(A)$ of a set A is $\{h(\phi) : \phi \in A\}$; the proper hull of A is defined analogously.*

Lemma 7 *Let $e \in \Omega$. There exists a neighborhood U_0 of the origin in W, and a positive constant β, with the following property: for any disjoint family $\{\xi(r)\}_{r=1}^s$ of subsets of $\mathbf{R}$, and any set $\{\theta_{r,i}\}$ of elements of U_0,*

$$\sum_{r=1}^s \sum_{i=1}^k |[P(\xi(r))\theta_{r,i}, P_i e]| < \beta \left\| \sum_{r=1}^s P(\xi(r))e \right\|.$$

Proof: Since the embedding from W into Ω is continuous, it follows that if V is the intersection of the unit ball of Ω with W, then V is a neighborhood in W. If we let N_V denote the subspace of W consisting of elements which are contained in all multiples of V, then N_V is the trivial subspace. The Minkowski functional κ_V is the norm of Ω, and the space X_V defined in Definition 1 is the normed linear space formed by giving W this norm. The mapping I_V is the identity mapping from W into X_V. We then see that

$$\sum_{i=1}^{k}\sum_{r=1}^{s}|[P(\xi(r))\theta_{r,i},P_ie]| \;=\; \sum_{r,i}|[I_V\theta_{r,i},P(\xi(r))P_ie]|$$

$$=\; \sum_{r,j,i}\left|\left[\sum_{j=1}^{\infty}c_j\alpha_j\theta_{r,i}\beta_j,P(\xi(r))P_ie\right]\right|$$

$$=\; \sum_{r,j,i}\sum_{j=1}^{\infty}[c_j\alpha_j(\theta_{r,i})\beta_j,b_{r,j,i}P(\xi(r))P_ie]$$

$$\leq\; \gamma\sum_{j=1}^{\infty}|c_j|\left\|\sum_{i=1}^{k}\alpha_j(\theta_{r,i})b_{r,j,i}P(\xi(r))P_ie\right\|$$

for some summable sequence c_j of complex numbers, and for some equicontinuous sequence α_j of elements of W', some set $\{b_{r,j,i}\}$ of complex numbers of modulus one, and some bounded sequence $\{\beta_j\}$ with norm less than γ of elements of the normed space X_V, which of course is just $\Omega \cap W$. In fact, the last inequality is proved as follows: since $\{\alpha_j\}$ is equicontinuous, there exists a neighborhood U of the origin in W such that $|\alpha_j(x)| \leq 1$ for all $x \in U$. But, if $P_{i,r} = P(\xi(r))P_i$, then $\{P_{i,r}\}$ is a set of mutually orthogonal projections, since P_i commutes with $P(\xi(r))$. Hence, if $\{\theta_j\}$ is chosen from $U \cap V$, the conclusion is established, where $\beta = \gamma\sum_{j=1}^{\infty}|c_j|$.

Definition 8 *Assume the following for the rest of the paper. Let $\{e_i\}$ be an orthonormal set in Ω such that the cyclic subspaces $\Re_{e_i}$ generated by e_i have the property that $\Re_{e_i}\perp\Re_{e_j}$ for $i \neq j$, and such that for all $j > 1$ σ_{e_j} is absolutely continuous with respect to σ_{e_1}; using the spectral theorem such an orthonormal set may be selected.*

Let

$$d\sigma_{e_i} = \delta_i(\lambda) d\sigma_{e_1}.$$

Note that W has a countable dense subset. Let S_1 be a countable dense subset of W, which is also a subspace over the rational numbers. Let $S = S_1 + HS_1$. For each $\phi \in S$, $T_i P_i \phi(\lambda)$ is well defined for all i, except on a set of λ which has measure 0 with respect to σ_{e_i}. Define F_{λ,e_i} to be zero on the exceptional set, which may be chosen independently of ϕ. On the complement of this set, define F_{λ,e_i} for each i by $F_{\lambda,e_i}(\phi) = T_i P_i \phi(\lambda)$; this defines a linear functional on S, or more precisely a function from S into the real numbers which is linear over the field of rational numbers. We extend this functional to all of W.

Lemma 9 *For almost every λ with respect to σ_1, there exists a unique element F_{λ,e_i} of W' which agrees with the previously defined functional F_{λ,e_i} on S, and which has the following properties:*

1. *$H' F_{\lambda,e_i} = \lambda F_{\lambda,e_i}$;*

2. *for each $\phi \in W$, there exists a set Δ depending on ϕ, such that $P(\Delta) = I$ (the identity operator), and such that for all $\lambda \in \Delta$, $F_{\lambda,e_i}(\phi) = T_i P_i \phi(\lambda) \; \forall \; i$;*

3. *the function $\alpha_i : \alpha_i(\lambda) = F_{\lambda,e_i}$ is a measurable function from $\mathbf{R}$ into W' with respect to σ_1, in the sense that $\forall \; \epsilon > 0$, $\exists$ a closed set Δ_ϵ such that $\sigma_{e_1}(\mathbf{R} \backslash \Delta_\epsilon) < \epsilon$, and such that the restriction of α_i to Δ_ϵ is a continuous function from $\mathbf{R}$ to W'.*

Proof: We extend F_{λ,e_i} from S to W by continuity. We show that there exists a neighborhood U of zero in W such that for almost every λ with respect to σ_{e_i}, F_{λ,e_i} is bounded on $U \cap S$. In fact, take $U = U_0$, where U_0 is the neighborhood defined in Lemma 7. Let

$$\gamma_U(F_{\lambda,e_i}) = \sup_{\theta \in U \cap S} |F_{\lambda,e_i}(\theta)|.$$

Note that $\gamma_U(F_{\lambda,e_i}) = \sup\{F_{\lambda,e_i}(\theta) : \theta \in S \cap U\}$. It follows from Lemma 7 that defining $\hat{\gamma}_U$ by

$$\hat{\gamma}_U(i, \lambda) = \sup_{\phi \in U} |T_i P_i \phi(\lambda)|,$$

$\hat{\gamma}_U(i, \cdot) \in L_1(\sigma_{e_i})$. Thus $\sigma_{e_i}\{\lambda : \gamma_U(i, \lambda) = \infty\} = 0$. Hence for almost every λ with respect to σ_{e_i}, $\gamma_U(F_{\lambda, e_i})$ is finite. It is now elementary to extend F_{λ, e_i} uniquely to be an element of W'; the fact that $H'F_{\lambda, e_i} = \lambda F_{\lambda, e_i}$ follows from the fact that for all $\phi \in S_1$, $T_i(HP_i\phi)(\lambda) = \lambda T_i(P_i\phi)(\lambda)$. It is easy to see that for any element $\phi \in W$, for almost every λ with respect to σ_{e_i}, $T_i(P_i\phi)(\lambda) = F_{\lambda, e_i}(\phi)$, although the exceptional set can now depend on ϕ. It follows that $F_{\lambda, e_i}(\phi)$ is a Borel measurable function for each $\phi \in W$. It is also clear that, except for a set of measure 0, $\{F_{\lambda, e_i}\}$ is contained in a bounded subset of W', in the given topology $\beta(W', W)$. A Montel space is reflexive; see page 74 of [6]. Hence, in the terminology of [2], page 558, the function $\alpha(\lambda) = F_{\lambda, e_i}$ is scalarwise measurable from $\mathbf{R}$ into W'. Since the functions F_{λ, e_i} are in W', they are also in the dual space of each of the Frechet spaces in the inductive limit which forms W. By Proposition 8.15.3, page 575, [2] it follows that the function α is continuous on a closed set whose complement has arbitrarily small measure with respect to σ_{e_i}, as a function with range contained in the weak dual of each Frechet space. Picking the sets of measure 0 corresponding to each Frechet space, we see that α is measurable considered as a function with range in W', where W' is given the weak topology. But on closed, bounded subsets of the Montel space W', the injection from the given topology into the weak topology is a continuous one-to-one function defined on a compact Hausdorff space, which is therefore a homeomorphism. It follows that α is a measurable function with values in W', under the given topology. The lemma is proved.

Definition 10 *A series $\sum_{i=1}^{\infty} F_i$ of elements of W' will be said to converge* **absolutely** *if, for every continuous seminorm ρ, the series $\sum_{i=1}^{\infty} \rho(F_i)$ converges.*

Theorem 11 *Let δ_i be as in Definition 8. There exists a convex, balanced neighborhood V of 0 in W such that if $\rho_V(F) = \sup_{\theta \in V} |F(\theta)|$, then for almost every λ with respect to spectral measure there exists an element F_{λ, e_i} of W' for each i such that $H'F_{\lambda, e_i} = \lambda F_{\lambda, e_i}$ and such that the following properties hold:*

1. *define the measure* Γ *on* $\mathbf{R} \times N$, *where* N *denotes the natural numbers, by* $\Gamma = \sigma_{e_1} \times \mu_N$, *where* μ_N *denotes counting measure; then for any* $\phi \in W$, *the function* $f_\phi \in L_1(\Gamma)$, *where*

$$f_\phi(\lambda, i) = \delta_i(\lambda) F_{\lambda, e_i}(\phi) \rho_V(F_{\lambda, e_i});$$

2. *for almost every* λ *with respect to* σ_{e_1}, $\sum_{i=1}^{\infty} \delta_i(\lambda) F_{\lambda, e_i}(\phi) \bar{F}_{\lambda, e_i}$ *converges absolutely in* W' *for every* ϕ *in* $h(W)$;

3. *for every Borel set* Δ *and every* $\phi \in h(W)$,

$$P(\Delta)\phi = \int_\Delta \left(\sum_i F_{\lambda, e_i}(\phi) \bar{F}_{\lambda, e_i} \right) \delta_i(\lambda) \, d\sigma_1(\lambda).$$

4. *for almost every* λ *with respect to* σ_{e_i}, *there is a sequence* Δ_n *of Borel sets such that* Δ_n *is supported in* $(\lambda - \frac{1}{n}, \lambda + \frac{1}{n})$ *and the sequence* $\Psi_n = P(\Delta_n) e_i / \sigma_{e_i}(\Delta_n)$ *converges to* F_{λ, e_i} *in* W', *so that by the continuity of* H' *as a linear transformation of* W' *into itself,* $H' \Psi_n$ *converges in* W' *to* $\lambda F_{\lambda, e_i}$.

Remark: The above formulae show how to spectrally decompose projection operators. From these, one can spectrally decompose all functions of H.

Proof of the Theorem: Note that $\rho_V(F_{\lambda, e_i})$ is the supremum of countably many Borel measurable functions of λ, and is thus measurable. The first assertion follows from Lemma 7, upon selecting $\theta_{r,i}$ carefully; the method of proof is that of assertion iii), Lemma 1.6, [4]. The second assertion follows from Fubini's theorem. Note that by the spectral theorem and the definition of F_{λ, e_i}, for any $\phi, \theta \in W$,

$$[P(\Delta)\phi, \theta] = \sum_i \int_\Delta F_{\lambda, e_i}(\phi) \bar{F}_{\lambda, e_i}(\theta) \, d\sigma_{e_i}(\lambda).$$

The third assertion follows immediately. The fourth assertion follows from the formula

$$P(\Delta)e_i = \int_\Delta \bar{F}_{\lambda, e_i} \, d\sigma_{e_i} \tag{1}$$

together with assertion 3 of Lemma 9. Equation 1 follows from the third assertion by passing to the closure and noting that $T_i P_i e_i = 1$ by the spectral theorem.

Lemma 12 *Let V be a bounded convex balanced subset of W. Then $\forall\ \epsilon > 0\ \exists\ N > 0$ and compact subsets $\{\Delta_i\}_{i=1}^{N-1}$ of $\mathbf{R}$ such that the function $\alpha(\lambda) = F_{\lambda,e_i}$ is a continuous function from Δ_i into W' and $\forall \phi \in V$:*

1. $\sum_{i=N}^{\infty} \int \rho_V(F_{\lambda,e_i}) |F_{\lambda,e_i}(\phi)|\ d\sigma_{e_i} < \epsilon;$

2. $\sum_{i=1}^{N-1} \int_{\mathbf{R}\setminus\Delta_i} \rho_V(F_{\lambda,e_i}) |F_{\lambda,e_i}(\phi)|\ d\sigma_{e_i} < \epsilon.$

Proof: V is compact in W and therefore in $L_2(\Omega)$. It follows that $\forall\ \epsilon > 0\ \exists\ N > 0\ \ni\ \sum_{i=N}^{\infty} \|P_i\phi\|^2 < \epsilon\ \forall\ \phi \in V$. The first conclusion follows from picking $\theta_{r,i}$ carefully and using Lemma 7, together with the preceding theorem. (Recall that since V is bounded, V is contained in some multiple of the neighborhood U of Theorem 11.) The second conclusion follows in the same fashion.

Theorem 13 *For any bounded convex balanced subset V of W, if $ph(V)$ denotes the proper hull of V, and $\rho_V(F) = \sup_{\theta \in V} |F(\theta)|$, and Δ is any Borel set, then for every $\epsilon > 0$ there exists a subset J of the positive integers and for each $j \in J$ a finite set $\{\lambda_{i,j} : i \leq n(j)\}$ of real numbers and $\{\alpha_{i,j}\}$ of positive real numbers such that for every $\phi \in ph(V)$,*

$$\rho_V\left\{P(\Delta)\phi - \sum_{j \in J}\sum_{i \leq n(j)} \alpha_{i,j} F_{\lambda_{i,,j},e_j}(\phi)\bar{F}_{\lambda_{i,j},e_j}\right\} < \epsilon.$$

Remark: We need to use the proper hull instead of the hull to control the sets of measure zero, and pick the $\lambda_{i,j}$ independently of ϕ.

Proof: We may use the preceding lemma to cut down to a finite set of e_j and compact sets Δ_j on which α is continuous. The integral then becomes a Riemann integral. (This is the method of proof of the implication ii) $\Rightarrow$ iii) of Theorem 3.3 of [4]; more details are given there.)

Remark: The preceding theorem shows the importance of using the largest possible space W. For example, if $W = C_0^{\infty}(R^n)$, bounded subsets of W must be supported in some fixed compact subset of R^n; however, bounded subsets of the rapidly decreasing functions are much larger. Larger bounded sets give better convergence. The obstacle to using large spaces W is that H must take W continuously into itself, in order to make the eigenfunctions F_{λ,e_i} satisfy the equation

$$H'F_{\lambda,e_i} = \lambda F_{\lambda,e_i}. \tag{2}$$

It is this last equation which gives legitimacy to the eigenfunctions, because it leads to conclusion 4 of Theorem 11. When H is a partial differential operator arising from a hypoelliptic differential expression, equation 2 leads to regularity results and Sobolev inequalities for the eigenfunctions F_{λ,e_i} .

Bibliography

[1] D. E. Edmunds and R. M. Kauffman, *Eigenfunction Expansions and Semigroups in $L_1(\mathbf{R})$*, (in preparation).

[2] R. E. Edwards, *Functional Analysis, Theory and Applications*, New York: Holt, Rinehart and Winston, 1965.

[3] D. B. Hinton and R. M. Kauffman, *Discreteness of Some Continuous Spectrum Eigenfunction Expansions* (to appear).

[4] R. M. Kauffman, *Finite Eigenfunction Approximations for Continuous Spectrum Operators*, Int. J. Math. and Math. Sci. (to appear).

[5] T. Poerschke, G. Stolz and J. Weidmann, *Expansions in Generalized Eigenfunctions of Selfadjoint Operators*, Math. Z. **202** (1989) 397-408.

[6] A. P. Robertson and W. J. Robertson, *Topological Vector Spaces*, second edition, Cambridge Tracts in Mathematics **53**, Cambridge: Cambridge University Press, 1973.

[7] K. Yosida, *Functional Analysis*, second edition, Grund. Math. Wiss. **123**, Berlin-New York: Springer-Verlag, 1968.

On Unique Continuation Theorem for Uniformly Elliptic Equations with Strongly Singular Potentials

Kazuhiro Kurata
Gakushuin University

Abstract

We prove unique continuation properties for solutions of uniformly elliptic equations: $-div(A(x)\nabla u) + \mathbf{b}(x) \cdot \nabla u + (V(x) + W(x))u = 0$ with Lipschitz continuous $A(x)$ and singular $\mathbf{b}(x)$, $W(x)$ and $V(x)$.

The principal assumptions on $\mathbf{b}(x)$, $W(x)$ and $V(x)$, in our theorems, are $V, (2V + x \cdot \nabla V)^-, W^+, (|x|W^+)^2 \in Q_t(\Omega)$, $|W^-(x)| \leq C/|x|^2, |\mathbf{b}(x)| \leq C/|x|$ for some constant $C > 0$, where $Q_t(\Omega) = K_n(\Omega) + F_t(\Omega)$ for some $1 < t \leq n/2$, $V^- = \max(0, -V), V^+ = \max(0, V)$. Here $K_n(\Omega)$ is the Kato class and $F_t(\Omega)$ is the Fefferman-Phong class.

1 Introduction

We consider the second order uniformly elliptic equation with real coefficients:

$$Lu = -div(A(x)\nabla u) + \mathbf{b}(x) \cdot \nabla u + V(x)u + W(x)u = 0 \quad (1)$$

Differential Equations with
Applications to Mathematical
Physics

in a domain $\Omega \subset R^n$ ($n \geq 3$). Here $A(x) = (a_{ij}(x))_{1 \leq i,j \leq n}$ is a symmetric matrix which satisfies, for some $\lambda \in (0,1]$ and $\Gamma > 0$,

$$\lambda |\xi|^2 \leq \sum_{i,j=1}^{n} a_{ij}(x)\xi_i \xi_j \leq \lambda^{-1} |\xi|^2, \quad x \in \Omega, \quad \xi \in \mathbf{R}^n, \qquad (2)$$

$$|a_{ij}(x) - a_{ij}(y)| \leq \Gamma |x - y|, \quad i,j = 1,2,\cdots,n \quad x,y \in \Omega, \qquad (3)$$

and $\mathbf{b}(x) = (b_i(x))_{1 \leq i \leq n}$. The following two types of unique continuation property for solutions of (1) are well known for bounded coefficients $\mathbf{b}, V, W$.

(W) Let $u \in W_{loc}^{1,2}(\Omega)$ be a weak solution of $Lu = 0$ in Ω and $u \equiv 0$ on some open subset Ω' of Ω, then $u \equiv 0$ in Ω.

(S) Let $u \in W_{loc}^{1,2}(\Omega)$ be a weak solution of $Lu = 0$ in Ω and u vanishes of infinite order at a point $x_o \in \Omega$ in the sense $r^{-m} \int_{B_r(x_o)} u^2 \, dy \to 0$ as $r \to 0$ for every $m > 0$ at a point $x_o \in \Omega$, then $u \equiv 0$ in Ω.

Recently these results are extended to various classes of unbounded coefficients. When $A(x) \equiv (\delta_{ij})$, see e.g. [10], [7], [12], [3], [2]. In particular, Jerison and Kenig [7] showed the property (S) for $W \in L_{loc}^{n/2}(\Omega)$ ($\mathbf{b}, V \equiv 0$) and Stein [12] extended this result to the weak-$L^{n/2}$ class. Sawyer [10] and Fabes, Garofalo and Lin [3] studied it for W of the Kato class $K_n(\Omega)$ and Chanillo and Sawyer [2] for W in the Fefferman and Phong class F_t with $t > (n-1)/2$.

As is well-known (cf. [9]), in general, the Hölder continuity of the coefficients $a_{ij}(x)$ does not suffice for solutions of (1) to have the property (W). Therefore, the regularity condition (3) is optimal. Under general conditions (2), (3), the unique continuation theorem for (1) was shown under different assumptions on $\mathbf{b}$ and V, W by [1], [6], [4] and [5].

Hörmander proved the property (W) for (1), when $n > 4$, $V = 0, W \in L_{loc}^p(\mathbf{R}^n)$, $p \geq (4n-2)/7$ and $\mathbf{b} \in L_{loc}^q(\mathbf{R}^n)$, $q > (3n-2)/2$; (S) at the origin, when $n \geq 3$, $V = 0, |\mathbf{b}(x)| \leq C/|x|^{1-\delta}$, $|W(x)| \leq C/|x|^{2-\delta}$ for some $\delta > 0$. When $A(x) \in C^\infty(\Omega)$, Sogge [11] proved (S) for (1), if $|\mathbf{b}(x)| \leq C/|x|^{1-\delta}$ for some $\delta > 0$ and $V = 0, W \in$ w$-L_{loc}^{n/2}$ (see also [13]).

The standard approach for the unique continuation problem is to establish an appropriate Carleman estimate and almost all results was shown by this method ([1], [2], [6], [7], [10], [12], [11], [13]). Garofalo and Lin found a new approach to this problem and partially improved the result of [6]: they proved (S) at the origin, when $n \geq 3$, $V = 0, |b(x)| \leq Cf(|x|)/|x|$, $|W(x)| \leq Cf(|x|)/|x|^2$ with $\int_0^{r_0} f(t)/t \, dt < +\infty$.

In this paper we shall extend these results to several directions under an additional assumption on the quantity $(2V + x \cdot \nabla V)^-$. In particular, we shall generalize the results of [4], [5] and show (W) for (1) with $b \equiv 0$ under the assumptions (i) $V, |x||\nabla V|, W^+, (|x|W^+)^2 \in K_n(\Omega) + F_t(\Omega)$ for some $1 < t \leq n/2$, (ii) $|W^-(x)| \leq \delta/|x|^2$ for sufficientlly small $\delta > 0$, (iii) certain smallness condition on V^-; (S) under additional technical conditions. We also deal with the case $b \not\equiv 0$ and basically our assumption on b is the same as in [5].

2 Main Results

To state our results we first recall the definitions of $K_n(\Omega)$ and $F_t(\Omega)$. $V \in L^1_{loc}(\mathbf{R}^n)$ is said to be of the Kato class K_n if

$$\lim_{r \to 0} \eta^K(r; V) = 0, \quad \eta^K(r; V) = \sup_{x \in \mathbf{R}^n} \int_{B_r(x)} \frac{|V(y)|}{|x - y|^{n-2}} \, dy, \quad (4)$$

where $B_r(x) = \{y \in \mathbf{R}^n | |x - y| < r\}$ for $r > 0$. For $1 \leq t \leq n/2$, $V \in L^t_{loc}(\mathbf{R}^n)$ is said to be of the Fefferman and Phong class F_t if

$$\|V\|_{F_t} = \sup_{x \in \mathbf{R}^n, r > 0} r^2 \left(\frac{1}{|B_r(x)|} \int_{B_r(x)} |V|^t \, dy \right)^{1/t} < +\infty. \quad (5)$$

We say $V \in K_n(\Omega)$ (resp. $V \in F_t(\Omega)$) if $\chi_\Omega V \in K_n$ (resp. $\chi_\Omega V \in F_t$), where χ_Ω is the characteristic function of Ω. We note that $F_{n/2} = L^{n/2}(\mathbf{R}^n) \subset F_t \subset F_s$ for $1 \leq s \leq t \leq n/2$ and weak-$L^{n/2}(\mathbf{R}^n) \subset F_t$ for every $t \in: 1, n/2)$; $V \in K_n(\Omega)$ implies $V \in F_1(\Omega)$; and that $L^{n/2}(\Omega)$ and $K_n(\Omega)$ are incomparable for $n \geq 3$.

We introduce some functional spaces. For $1 < t \leq n/2$, we define the function space $Q_t(\Omega)$ by $Q_t(\Omega) = \{V = V_1 + V_2; V_1 \in$

$K_n(\Omega),\ V_2 \in F_t(\Omega)\}$ and set

$$\eta(r;x;V) = \inf_{V=V_1+V_2 \in Q_t(\Omega)} \{\eta^K(r; \chi_{B_r(x)\cap\Omega}V_1) + \parallel \chi_{B_r(x)\cap\Omega}V_2 \parallel_{F_t}\}$$

(6)

For $1 < t \le n/2$ and $\epsilon > 0$, $V \in Q_t(\Omega)$ is said to be in $M(\Omega; t, \epsilon)$ if;

(a) V^- satisfies $\lim_{r\to 0}(\sup_{x\in\Omega} \eta(r;x;V^-)) \le \epsilon$.

(b)   For every $x_o \in \Omega$, there exists $r_o > 0$ such that $|x - x_o||\nabla V(x)| \in Q_t(B_{r_o}(x_o) \cap \Omega)$.

For **b**, V, (for W, see REMARK 2), we assume
ASSUMPTION (A.1):

(i) $V \in M(\Omega; t, \epsilon)$ for a sufficiently small $\epsilon = \epsilon(n, t, \lambda, \Gamma)$.

(ii)   For every $x_o \in \Omega$

$$\lim_{r\to 0} \eta(r; x_o; (2V + (x - x_o) \cdot \nabla V)^-) = 0,$$

$$\lim_{k\to\infty} \eta(r_o; x_o; V^-\chi_{\{V->k\}}) = 0.$$

(7)

(iii)   When $\mathbf{b}(x) \not\equiv 0$, for every $x_o \in \Omega$, there exists $f : (0, r_o) \to \mathbf{R}^+$ and $C > 0$ such that f is nondecreasing on $(0, r_o)$, $\lim_{r\to 0} f(r) = 0$, and for every $x \in B_{r_o}(x_o) \cap \Omega$

$$|\mathbf{b}(x)| \le C\frac{f(|x - x_o|)}{|x - x_o|}, \qquad |V^-(x)| \le \frac{C}{|x - x_o|^2}$$

(8)

To obtain the property (S) for L, we require an additional
ASSUMPTION (A.2): For every $x_o \in \Omega$

$$\int_0^{r_o} \frac{f(r)}{r}\,dr < +\infty, \quad \int_0^{r_o} \frac{\eta(r; x_o; (2V + (x - x_o) \cdot \nabla V)^-)}{r}\,dr < +\infty.$$

(9)

Theorem 1 *Suppose that* (A.1) *and* (A.2) *are satisfied. Then L has the property* (S) *in Ω for $W^{1,2}_{loc}(\Omega)$-solutions.*

Theorem 2 *Suppose that* (A.1) *is satisfied. Then* L *has the following property* (P) *in* Ω *for* $W_{loc}^{1,2}(\Omega)$-*solutions: If* $u \in W_{loc}^{1,2}(\Omega)$ *is a weak solution of* (1) *in* Ω *and satisfies, for some* $x_o \in \Omega$ *and* $A, \alpha > 0$,

$$\int_{B_r(x_o)} u^2 \, dx = O\left(\exp\left(-\frac{A}{r^\alpha}\right)\right) \tag{10}$$

as $r \to 0$, *then* $u \equiv 0$ *in* Ω.

To obtain the property (W) for L, we can weaken our conditions.
ASSUMPTION (C):

(i) $V \in M(\Omega; t, \epsilon)$ for some $1 < t \leq n/2$ and a sufficiently small $\epsilon = \epsilon(n, t, \lambda, \Gamma)$.

(ii) When $\mathbf{b}(x) \not\equiv 0$, for every $x_o \in \Omega$, there exist $r_o > 0$, $C > 0$ and a sufficiently small $\epsilon(n, \lambda, \Gamma) > 0$ such that

$$|\mathbf{b}(x)| \leq \frac{\epsilon(n, \lambda, \Gamma)}{|x - x_o|}, \quad |V^-(x)| \leq \frac{C}{|x - x_o|^2}, \quad x \in B_{r_o}(x_o) \cap \Omega. \tag{11}$$

Theorem 3 *Suppose that* (2), (3) *and* Assumption (C) *is satisfied. Then* L *has the property* (W) *in* Ω *for* $W_{loc}^{1,2}(\Omega)$-*solutions.*

We should mention several remarks on Theorems 1, 2 and 3.

REMARK 1: We obtain Theorems 1, 2 and 3 by strong quantitative estimates; for example, under (A.1) and (A.2), for weak solutions u of (1), we have

$$\int_{B_{2r}(x_o)} u^2 \, dx \leq C_1 \int_{B_r(x_o)} u^2 \, dx \tag{12}$$

for $0 < r \leq r^*/2$, where C_1 depends on u, n, t, λ, Γ, and the local properties of $\mathbf{b}$, V, and W at x_o and $r^*(\leq r_o)$ on n, t, λ, Γ, and the local properties of $\mathbf{b}$, V, and W at x_o. For the details, see [8, Theorem 1.1, 1.2, 1.6].

REMARK 2: When $W \not\equiv 0$, the property (W) for L also holds under $W^+, (|x-x_o|W^+)^2 \in Q_t(B_{r_o}(x_o) \cap \Omega)$ and $|W^-(x)| \leq \frac{\delta(n,\lambda,\Gamma)}{|x-x_o|^2}$ for a sufficiently small $\delta(n, \lambda, \Gamma) > 0$; (P) under $\lim_{r \to 0} \eta(r; x_o; W^+ + (|x -$

$x_o|W^+)^2) = 0$ and $|W^-(x)| \leq C\frac{f(|x-x_o|)}{|x-x_o|^2}$; (S) under $\int_0^{r_o}(\eta(r;x_o;W^+)$
$+\sqrt{\eta(r;x_o;(|x-x_o|W^+)^2)})/r\,dr < +\infty$.

REMARK 3: For solutions $u \in W_{loc}^{2,2}(\Omega)$, the condition $\lim_{k\to\infty}\eta(r_o;x_o;V^-\chi_{\{V^->k\}}) = 0$ in (A.1) can be removed, because we do not need (STEP 2) in the proof of Theorems (see section 3).

REMARK 4: When $2V + (x - x_o) \cdot \nabla V \geq 0$ a.e. $x \in B_{r_o}(x_o)$, then the property (S) is satisfied without the smallness assumption $\lim_{r\to 0}\sup_{x\in\Omega}\eta(r;x;V^-) \leq \epsilon$ on V^- in some special case (see [8, Theorem 1.5]). However, in general, this smallness condition cannot be removed (see [14]).

REMARK 5: For $A(x) \equiv (\delta_{ij})$, the condition (b) in the definition of $M(\Omega;t;\epsilon)$ can be relaxed by $(x - x_o) \cdot \nabla V(x) \in Q_t(B_{r_o}(x_o)) \cap \Omega$.

Theorems 1, 2, 3 extend the results in [4], [5] which assumes $V \equiv 0$ and stronger pointwise condition (see section 1) on W in our terminology; Theorem 1 is a partial extension of the result in [2] to general $A(x)$; the property (P) is studied in [5] and [3], and Theorem 2 extends their results to the operator L with more singular V.

Let us clarify how do our theorems extend the previous works by using the following example.

EXAMPLE 1: Let $V(x) = K_\delta\frac{|\log(1/|x|)|^{-\delta}}{|x|^2} - L_\gamma\frac{|\log(1/|x|)|^{-\gamma}}{|x|^2}$, where $K_\delta, L_\gamma, \delta, \gamma \geq 0$ are constants. When $\gamma > 0$, V satisfies (A.1) and (A.2) and Theorem 1 yields the property (S) for general A and $\mathbf{b}$ satisfying (2)-(3), (8) with $\int_0^{r_o} f(t)/t\,dt < +\infty$. The results in [4], [5] only assure (W) (see [5, Theorem 1.3]) for general elliptic equations with this potential V in the case $0 \leq \delta, \gamma \leq 1$; the ones in [2] are applicable for $\delta, \gamma \geq 0$ and sufficiently small L_0, K_0, but those are restricted in the special case $A(x) = (\delta_{ij})_{1\leq i,j\leq n}$ and $\mathbf{b} \equiv 0$ (cf. [11]).

EXAMPLE 2: Consider the operator $L = -\sum_{i=1}^N \Delta_{x_i} + V$ in $\Omega \subset \mathbf{R}^n$ and $V = -\sum_{i=1}^N \frac{1}{|x_i - R_i|}$, where $x_i, R_i \in \mathbf{R}^\nu$, $\nu \geq 3$, $i = 1, \cdots, N$ ($N \geq 1$) and $n = \nu N$. Since $V, (x - x_o) \cdot \nabla V \in K_n(\Omega)$ we can apply Theorem 3, and Theorem 2 for solutions $u \in W_{loc}^{2,2}(\Omega)$ of $Lu = 0$. However, for $N > 1$ previous results do not yield unique continuation property for L.

3 Sketch of the Proof of Theorems

The basic idea of the proof is to combine Garofalo and Lin's variational method and the inequality:

$$\int_{B_r(y)} |V| u^2 \, dx \leq C \|V \chi_{B_r(y)}\|_{F_t} \left(\frac{1}{r} \int_{\partial B_r(y)} u^2 \, dS + \int_{B_r(y)} |\nabla u|^2 \, dx \right), \tag{13}$$

for $V \in F_t$ $(1 < t \leq n/2)$ and for every $r > 0, y \in \mathbf{R}^n$, and $u \in C^\infty(\mathbf{R}^n)$. See [8] for the extension of this inequality and its proof. (STEP 1) First we use the geometric reduction procedure of [1]. We fix a point x_o in Ω, then the equation (1) is reduced to

$$- div_M(\mu(x)\nabla_M u) + \mathbf{b}_M \cdot \nabla_M u + V_M(x)u + W_M(x)u = 0, \quad (14)$$

where M is the Riemannian manifold $(B_{r_o}(x_o), G)$ for sufficiently small $r_o > 0$ and $\mathbf{b}_M = G(\mathbf{b}/\sqrt{g}), V_M = V/\sqrt{g}, W_M = W/\sqrt{g}$, and $\mu(x)$ is a Lipschitz function satisfying $\mu(x_o) = 1, |\frac{\partial \mu}{\partial r}(r,t)| \leq \Lambda_1, \mu_1 \leq \mu(x) \leq \mu_2$ for some positive constants Λ_1, μ_1, and μ_2 which depend only on n, λ, and Γ. Here $G(x) = (g_{ij}(x))_{1 \leq i,j \leq n}$ is determined by $A(x)$ as follows:

$$\bar{g}_{ij}(x) = a^{ij}(x)(det A(x))^{1/(n-2)},$$

$$r(x)^2 = \sum_{i,j} \bar{g}_{ij}(x_o)(x - x_o)_i(x - x_o)_j,$$

$$P(x) = \sum_{k,l} \bar{g}^{kl}(x) \frac{\partial r}{\partial x_k}(x) \frac{\partial r}{\partial x_l}(x),$$

$$g_{ij}(x) = P(x)\bar{g}_{ij}(x),$$

where $g = |det(G)|$, $A^{-1}(x) = (a^{ij}(x))_{1 \leq i,j \leq n}$, $(\bar{g}^{ij}) = (\bar{g})_{ij}^{-1}$, and div_M, ∇_M are the intrinsic divergence and gradient in the metric G. Note that $r(x)$ is the geodesic distance from x_o to x in the metric g_{ij} (cf. [1, p. 427]).

Therefore, to prove Theorems we may assume $x_o = O$ and study the local properties of solutions for the equation (14) on $M = (B_{r_o}, G)$, $B_{r_o} = B_{r_o}(O)$.

(STEP 2) We approximate the solution u of (14) by the ones of the following boundary value problem:

$$-div_M(\mu\nabla_M v) + (\mathbf{b}_k)_M \cdot \nabla_M v + (V_k)_M v + (W_k)_M v = 0 \quad \text{in } B_{R_1},$$

$$v = u \quad \text{on } \partial B_{R_1}, \tag{15}$$

where f_k, for each $k > 0$, is defined by $f_k(x) = f(x)$ (if $|f(x)| \le k$), $= k$ (if $f(x) > k$), $= -k$ (if $f(x) < -k$) for any function f, and $\mathbf{b}_k = ((b_j)_k)_{1 \le j \le n}$.

There exists a sufficiently small $R_1 > 0$ depending only on n, t, λ, Γ, and the local properties of $\mathbf{b}$ and V at O such that the problem (15) has a unique weak solution $v = u_k \in W^{2,2}_{loc}(B_{R_1})$ and u_k satisfies $\|u_k - u\|_{H^{1,2}(B_{R_1})} \to 0$ as $k \to +\infty$ (see [8, Lemma 4.2]).

(STEP 3) Define

$$I(r) = \int_{B_r} (\mu|\nabla_M u|^2 + (\mathbf{b}_M \cdot \nabla_M u)u + (V_M + W_M)u^2)\, dv_M, \tag{16}$$

$$H(r) = \int_{\partial B_r} \mu u^2\, dS_M, \quad N(r) = \frac{rI(r)}{H(r)}, \quad D(r) = \int_{B_r} \mu|\nabla_M u|^2\, dv_M, \tag{17}$$

$$\begin{aligned}(r;W) \;=\; & \int_{\partial B_r} W_M u^2\, dS_M - \frac{n-2}{r}\int_{B_r} W_M u^2\, dv_M \\ & -\frac{2}{r}\int_{B_r}(x \cdot \nabla_M u)W_M u\, dv_M,\end{aligned}$$

and $H_k(r), I_k(r)$ and $N_k(r)$ by using (16), (17) for functions $\mathbf{b}_k, V_k$, W_k, and u_k instead of $\mathbf{b}, V, W$, and u. We use the following identity (see [8, Lemma 4.1]) for u_k obtained in (STEP 2).

Let $u \in W^{2,2}(B_{r_o})$ satisfy (14) a.e. on B_{r_o}. Assume that $V, W, |\mathbf{b}|^2 \in Q_t(\Omega)$ for some $1 < t \le n/2$, and there exists $r_o > 0$ such that $|x||\nabla V(x)| \in Q_t(B_{r_o} \cap \Omega)$. Then, for a sufficiently small $R_1 = R_1(n, \lambda, \Gamma) > 0$ and for a.e. $r \in (0, R_1)$, we have

$$\begin{aligned}I'(r) \;=\; & \frac{n-2}{r}I(r) + 2\int_{\partial B_r} \mu u_\rho^2\, dS_M + J(r;W) + R_0(r) \\ & + \int_{\partial B_r}(\mathbf{b}_M \cdot \nabla_M u)u\, dS_M\end{aligned}$$

$$+\frac{1}{r}\int_{B_r}(2V_M + x \cdot \nabla_M V_M)u^2\, dv_M$$

$$-\frac{n-2}{r}\int_{B_r}(\mathbf{b}_M \cdot \nabla_M u)u\, dv_M$$

$$-\frac{2}{r}\int_{B_r}(x \cdot \nabla_M u)(\mathbf{b}_M \cdot \nabla_M u)\, dv_M, \tag{18}$$

where $u_\rho = (x/|x|) \cdot \nabla_M u$, $|R_0(r)| \le C(\int_{B_r} |V_M|u^2\, dv_M + D(r))$.

(STEP 4) Let (A.1) and (A.2) be satisfied. By using (13), [3, Lemma 1.1] and (18), we compute $N_k'(r)/N_k(r)$ and obtain

$$N_k(r) \le \frac{C_* \max(1, N_k(R_1))}{r^{C_* \eta(R_1; V^- \chi_{\{V^- > k\}})}} \exp\left(\left|\int_0^{r_o} \frac{Cf(r) + C_*\Theta(r)}{r}\, dr\right| + Cr_o\right), \tag{19}$$

$r \in (0, R_1)$, $k = 1, 2, \cdots$, for some $C, C_* > 0$, where $\eta(r; U) = \eta(r; 0; U)$ and

$$\Theta(r) = \eta(r; (2V + x \cdot \nabla V)^-) + \eta(r; W^+) + \sqrt{\eta(r; (|x|W^+)^2)}.$$

(STEP 5) By using $\frac{d}{dr}\left(\log(\frac{H_k(r)}{r^{n-1}})\right) = 2\frac{N_k(r)}{r} + O(1)$, we have

$$\frac{H_k(2r)}{H_k(r)} \le 2^{n-1}\exp\left(\frac{C\max(1, N_k(R_1))}{r^{C_* \eta(R_1; V^- \chi_{\{V^- > k\}})}}\right), \quad 0 < r \le R_1/2. \tag{20}$$

Taking $k \to +\infty$ of (20), we obtain

$$\int_{\partial B_{2r}} u^2\, dS \le C_2 \exp(C_3 \max(1, N(R_1))) \int_{\partial B_r} u^2\, dS, \quad 0 < r \le R_1/2, \tag{21}$$

where C_2 depends only on n, t, λ, Γ and C_3 on n, t, λ, Γ, and the local properties of $\mathbf{b}$, V, and W at O. This implies Theorem 1. Theorem 2 and Theorem 3 can be proved in the same way.

Bibliography

[1] N. Aronszajn, A. Krzywicki and J. Szarski, *A unique continuation theorems for the exterior differential forms on Riemannian manifolds*, Ark. Mat., 4, 1962, p. 417–453.

[2] S. Chanillo and A. E. Sawyer. *Unique continuation for* $\Delta + v$ *and the C.Fefferman - Phong class*, Tran. Amer. Math. Soc., 318, 1990, p. 275–300.

[3] E. Fabes, N. Garofalo and F. H. Lin, *A partial answer to a conjecture of B. Simon concerning unique continuation*, J. Fun. Ann., 88, 1990, p. 194–210.

[4] N. Garofalo and F. H. Lin, *Monotonicity properties of variational integrals.* A_p *weights and unique continuation*, Indiana Univ. Math. J., 35, 1986, p. 245–268.

[5] N. Garofalo and F. H. Lin, *Unique continuation for elliptic operators: A geometric-variational approach*, Comm. Pure Appl. Math., 40, 1987, p. 347–366.

[6] L. Hörmander, *Uniqueness theorems for second-order elliptic differential equations*, Comm. in PDE, 8, 1983, p. 21–64.

[7] D. Jerison and C. E. Kenig, *Unique continuation and absence of positive eigenvalues of Schrödinger operators*, Annals of Math., 121, 1985, p. 463–488.

[8] K. Kurata, *On unique continuation theorem for uniformly elliptic equations with strongly singular potentials*, Gakushuin Preprint Series No.2 , 1991.

[9] A. Plis, *On non-uniqueness in Cauchy problems for an elliptic second order differential operators*, Bull. Acad. Pol. Sci., 11, 1963, p. 65–100.

[10] E. T. Sawyer, *Unique continuation for Schrödinger operators in dimension three or less*, Ann. Inst. Fourier (Grenoble), 34, No. 3, 1984, p. 189–200.

[11] C. D. Sogge, *Strong uniqueness theorems for second order elliptic differential equations*, Amer. J. Math., 112, 1990, p. 943–984.

[12] E. M. Stein, *An appendix to "Unique continuation and absence of positive eigenvalues of Schrödinger operators"*, by D. Jerison and C. E. Kenig, Ann. of Math., 121, 1985, p. 489–494.

[13] T. H. Wolff, *Unique continuation for* $|\Delta u| \leq V|\nabla u|$ *and related problems,* Revista Math. Iberoamericana, 6, 1990, p. 155–200.

[14] T. H. Wolff, *Note on counterexamples in strong unique continuation problems,* Proc. A.M.S., 114, 1992, p. 351–356.

Topics in the Spectral Methods in Numerical Computation — Product Formulas

S. T. Kuroda
Department of Mathematics
Gakushuim University
Mejiro, Tokyo 171 Japan

1 Introduction

This article is a brief account on some aspects of the combined use of product formulas of Lie–Trotter type and the Fast Fourier Transform (FFT) for solving the Schrödinger evolution equation

$$i \frac{\partial}{\partial t} u(x,t) = (-\Delta + V(x))u(x,t) \tag{1}$$

and for computing eigenfunctions and eigenvalues of the operator $-\Delta + V$. Here, $V(x)$ is a real valued function. The use of the product formula for this purpose goes back to [4] and the combined use with the FFT is due to [2].

The idea of [2] is as follows. Let $A = -\Delta$ and $B = V$. Then $\exp(-itA)$ and $\exp(-itB)$ are multiplication operators, one in the Fourier (ξ-)space and the other in the configuration (x-)space, respectively. Therefore, products like $\{\exp(-i(t/n)A)\exp(-(t/n)B)\}^n$ can be computed easily by going back and forth between these spaces. The transformation between x- and ξ-spaces can be implemented very efficiently by means of the FFT, which requires only $O(N \log N)$ multiplications for a problem with N mesh points. Using this idea for solving (1), [2] develops a method, which may be called "numerical spectroscopy," for simultaneously computing all eigenvalues in a wide energy range. Motivated by [2], [5] proposed a method, which may be called "numerical resonant excitation" for computing a particular eigenfunction and the associated eigenvalue very accurately. [2] and [5] also contain ample numerical examples.

In this article we shall focus our attention on product formulas and shall exploit various product formulas which may be used for the numerical procedure mentioned above. Not only a formula itself but the order of error is of interest. In Section 2.2 we shall list a few formulas with the order of error (for bounded generators). Some of these formulas seem not to have been noticed in the literature. Possibilities of applying these formulas will also be discussed in Section 2.3. Some remarks given in the talk on the methods developed in [2] and [5], especially on a way of handling remote eigenvalues in [2] will be reported elsewhere.

2   Product Formulas

2.1   Preliminaries

In this section we consider an abstract evolution equation in a Banach space X. The equation and its solution with the initial data u_0 are written as

$$\frac{d}{dt} u(t) = Cu(t), \quad t > 0; \quad u(t) = \exp(tC)u_0. \tag{2}$$

We assume that the generator C and other operators appearing later are all bounded linear operators in X. The reason for assuming the boundedness is twofold. Firstly, it makes the error estimate simpler, and secondly, in applications, product formulas will be applied after discretization, i.e., in a finite–dimensional space.

The product formulas we shall discuss are written generally as

$$\exp(tC) = \lim_{n \to \infty} F(t/n)^n. \tag{3}$$

Here, $F(t)$ is an approximation of $\exp(tC)$ for small t. We may call it a *unit increment* of the product approximation. The order of error (for large n) in (3) is related to the order of error (for small t) in the unit increment. Namely, it can be seen by a standard argument (cf. [6], p. 295) that

$$\| \exp(tC) - F(t) \| = O(t^p), \qquad t \to \infty, \quad p > 1, \qquad (4)$$

implies

$$\| \exp(tC) - F(t/n)^n \| = t^p O\left(n^{-(p-1)} \right). \qquad (5)$$

2.2 List of Formulas

We are interested in two cases $C = A + B$ and $C = [A, B]$ and shall list several product formulas. Since F is related to A and B we write $F(t; A, B)$ instead of $F(t)$. In this list we use e^A instead of $\exp(A)$. No. 1–No. 5 and No. 8 are main formulas. No. 6 and No. 7 will be used in the proof of No. 3, p in the last column is p of $O(t^p)$ in (4).

	e^{tC}	$F(t; A, B)$	p
1	$e^{t(A+B)}$	$e^{tA}e^{tB}$	2
2	$e^{t(A+B)}$	$e^{tA/2}e^{tB}e^{tA/2}$	3
3	$e^{t(A+B)}$	$e^{tA/2}e^{tB}e^{tA/2}$	
		$+\frac{1}{6}(e^{tA/2}e^{tB}e^{-tA/2}e^{-2tB}e^{-tA/2}e^{tB}e^{tA/2} - I)$	4
4	$e^{t[A,B]}$	$e^{\sqrt{t}A}e^{\sqrt{t}B}e^{-\sqrt{t}A}e^{-\sqrt{t}B}$	3/2
5	$e^{t[A,B]}$	$\frac{1}{2}(e^{\sqrt{t}A}e^{\sqrt{t}B}e^{-\sqrt{t}A}e^{-\sqrt{t}B}$	
		$+e^{-\sqrt{t}A}e^{-\sqrt{t}B}e^{\sqrt{t}A}e^{\sqrt{t}B})$	2
6	$e^{t(A+B)}$	$e^{tA/2}e^{tB}e^{tA/2} + \frac{1}{6}t^3[A/2 + B, [A/2, B]]$	4
7	I	$e^{tA/2}e^{tB}e^{-tA/2}e^{-2tB}e^{-tA/2}e^{tB}e^{tA/2}$	
		$-t^3[A/2 + B, [A/2, B]]$	4
8	$e^{t(A+B)}$	$e^{xtA}, e^{utB}e^{ytA}e^{vtB}e^{ytA}e^{utB}e^{xtA}$	4

In No. 8 u, v, x, y are given as

$$u = \frac{1}{3}\left(2 + \sqrt[3]{2} + \frac{1}{\sqrt[3]{2}}\right) = 1.351\cdots, \quad v = 1 - 2u = -1.702\cdots,$$

$$x = \frac{1}{2}\left(1 + \sqrt{\frac{1}{6u}}\right) = 0.675\cdots, \quad y = \frac{1}{2} - x = -0.175\cdots.$$

No. 1 is the classical Lie formula and No. 2 is its symmetrized form used in [4] and [2]. No. 4 is also well–known ([8], p. 99; see also [1] [3] for more recent developments).

No. 3 is a fourth order formula. Its feature is that the unit increment F contains only operators which remain bounded even when A and B are unbounded. (There are fourth order formulas with F containing A, B outside exponential factors. One example is (2.10) of [4] which has terms with double commutators sandwiched by exponential factors. Or, even the Taylor expansion up to the third order term may be regarded as such a formula.)

A few words on the proof of No. 3 and No. 5. Our only tool is brutal computations of Taylor coefficients. First, we try to improve the order of error of No. 2 by 1 and obtain No. 6. On the other hand, as suggested by (2.2.10) of [8], which is a third order formula, we obtain No. 7. No. 6 and No. 7 contain the same double commutator. They cancel each other to give No. 3. The remainder of the approximation No. 4 is computed as $-2^{-1}t^{3/2}[A + B, [A, B]] + O(t^2)$. We can replace A, B by $-A, -B$ without changing $[A, B]$. Adding these two formulas we obtain No. 5.

When A and B are both skew–adjoint, F in No. 1 and No. 2 are unitary. No. 3 does not have this advantage. Recently we have found an order 4 formula in which F is a product of seven exponential factors, so that it is unitary in skew–adjoint case. That is No. 8. The proof of No. 8 requires systematic computation of Taylor coefficients.

Remark 1. The Baker–Campbell–Hausdroff formula expresses $\exp(tA_1)\cdots\exp(tA_p)$ as $\exp(C(t))$ where the coefficients of the expansion of $C(t)$ in t involves multi–commutators of A_k. The Zassenhaus formula and its generalization (cf. [7]) is a kind of product formula, but again multi–commutators appear in exponential functions. For our application it is important that only scalar multiples of A

or B appear in each exponential factor. It is true that, say, No. 2 is easily derived from the B–C–H formula, and possibly others, too. We found, however, that a simple minded manipulation of Taylor coefficients will be less (or at most equally) complicated for a quick derivation of higher order formulas.

A more systematic analysis of these formulas with estimates will be published elsewhere.

2.3 An Application

Formula No. 5 may be used to solve numerically a Schrödinger operator with variable higher order coefficients by the method mentioned in Section 1. In this subsection we pretend that formulas like No. 3 and No. 5 remain valid also for unbounded operators. In fact, under suitable assumptions on the smoothness of the coefficients, these formulas are valid if $O(t^p)$ is interpreted with respect to a suitable norm.

Assuming for simplicity that the second order terms have constant coefficients, we consider the operator

$$
\begin{aligned}
H &= \sum_{k=1}^{n} (-i\partial_k + b_k(x))^2 + q(x) \\
&= -\Delta - \sum_{k=1}^{n} (2ib_k(x)\partial_k + i\partial_k b_k - b_k^2) + q(x)
\end{aligned} \tag{6}
$$

acting in $L^2(\mathbf{R}^n)$. Here, $\partial_k = \frac{\partial}{\partial x_k}$ and b_k, q are real functions. We put

$$
B_k(x) = \int_{a_k}^{x_k} b_k(x_1, \ldots, \overset{k}{t}, \ldots, x_n) dt, \tag{7}
$$

$$
b(x) = (b_1(x), \ldots, b_n(x)), \qquad B(x) = (B_1(x), \ldots, B_n(x)). \tag{8}
$$

Then H is expressed as

$$
H = -\Delta - i \sum_{k=1}^{n} [\partial_k^2, B_k] + Q(x), \qquad Q(x) = b(x)^2 + q(x) \tag{9}
$$

We now apply No. 5. Using the notation

$$
F(t; A, B) = e^{tA} e^{tB} e^{-tA} e^{-tB} \tag{10}
$$

and noting that repetitions of No. 1 also give rise to an error of order $O(t^2)$, we obtain

$$
e^{-itH} = e^{it\Delta}e^{-itQ} \prod_{k=1}^{n} e^{-t[\partial_k^2, B_k]} + O(t^2)
$$

$$
= \frac{1}{2^n}e^{it\Delta}e^{-itQ} \prod_{k=1}^{n} \left\{ F\sqrt{-t}; \partial_k^2, B_k \right)
$$

$$
+F(-\sqrt{-t}; \partial_k^2, B_k) \right\} + O(t^2). \tag{11}
$$

The result of a numerical test of this formula for $n = 2$ is promising. The details will be left to future research.

Even when second order coefficients are variable, a similar formula can be derived, but it becomes rather complicated. We have not yet tested the feasibility of such a formula in the numerical computation.

Bibliography

[1] P. R. Chernoff, *Universally commutable operators are scalars,* Michigan Math. J. **20** (1973), 101–107.

[2] M. D. Feit, J. A. Fleck, Jr., and A Steiger, *Solutions of the Schrödinger equation by a spectral method,* J. Comput. Phys. **47** (1982), 412–433.

[3] J. Goldstein, *A Lie product formula for one parameter groups of isometries on Banach space,* Math. Ann. **186** (1970), 299-306.

[4] R. Grimm and R. G. Storer, *A new method for the numerical solution of the Schrödinger equation,* J. Comput. Phys. 4 (1969), 230–249.

[5] S. T. Kuroda and Toshio Suzuki, *A time–dependent method for computing eigenfunctions and eigenvalues of Schrödinger operators,* Japan J. Appl. Math. 7 (1990), 231–253.

[6] M. Reed and B. Simon, *Methods of Modern Mathematical Physics I; Functional Analysis,* Academic Press, 1970.

[7] M. Suzuki, *On the convergence of exponential operators — the Zassenhaus formula, BCH formula and systematic approximants,* Comm. Math. Phys. **57** (1977), 193–200.

[8] V. S. Varadarajan, *Lie Groups, Lie Algebras, and their Representations,* Prentice–Hall, 1974.

Atoms in the Magnetic Field of a Neutron Star

Elliott H. Lieb[1] and Jan Philip Solovej
Department of Mathematics
Princeton University
Princeton, NJ 08544

Abstract

The ground state energy of an atom of nuclear charge Ze and in a magnetic field B is evaluated exactly in the asymptotic regime $Z \to \infty$. We present the results of a rigorous analysis that reveals the existence of 5 regions as $Z \to \infty$: $B \ll Z^{4/3}$, $B \approx Z^{4/3}$, $Z^{4/3} \ll B \ll Z^3$, $B \approx Z^3$, $B \gg Z^3$. Different regions have different physics and different asymptotic theories. Regions 1,2,3,5 are described exactly by a simple density functional theory, but only in regions 1,2,3 is it of the semiclassical Thomas-Fermi form. Region 4 cannot be described exactly by any simple density functional theory; surprisingly, it can be described by a simple *density matrix* functional theory.

1 Introduction

In these talks we shall discuss the effect on matter, specifically the ground state of atoms, of a very strong magnetic field. Results obtained in collaboration with J. Yngvason will be summarized and details will appear elsewhere [9]. The physical motivation for studying extremely strong magnetic fields of the order of 10^{12}-10^{13} Gauss

---

[1]Work supported by U.S. National Science Foundation grant no. PHY90-19433.

Differential Equations with Applications
to Mathematical Physics
ISBN 0–12–056740–7

is that they are supposed to exist on the surface of neutron stars. This study was essentially begun in the early 70's with the work of Kadomtsev [5], Ruderman [12] and Mueller, Rau and Spruch [11]; see [1] and [2] for further references. The argument given to explain these strong fields is that in the collapse, resulting in the neutron star, the magnetic field lines are trapped and thus become very dense. The structure of matter in strong magnetic fields is, therefore, a question of considerable interest in astrophysics. Mathematically, the problem turns out to involve an interesting exercise in semiclassical analysis.

We use units in which $e = \hbar = 2m_e = 1$. The natural unit of length is $\hbar^2/2m_e e^2$, i.e., half the Bohr radius. The natural unit of magnetic field strength that we shall use is $(2m_e)^2 e^3 c/\hbar^3 = 9.4 \times 10^9$ Gauss. This is the field for which the magnetic length $\sqrt{c\hbar/eB}$ equals half the Bohr radius. Thus, in our units, $B \approx 10^2 - 10^3$ for some neutron stars.

The atomic nucleus of principal interest on the surface of a neutron star is presumably iron with $Z = 26$. This number is large and hence it is sensible to ask (rigorously) about the limit of the ground state energy of an atom as $Z \to \infty$. We shall calculate this limit exactly; its application to $Z = 26$ instead of $Z = \infty$ will entail some errors – for which we can give bounds.

2 Main Results

To give the quantum mechanical energy of a charged spin-$\frac{1}{2}$ particle in a magnetic field B, we have to make a choice of vector potential $A(x)$, satisfying $B = \nabla \times A$. The energy is then given by the Pauli Hamiltonian

$$H_A = ((p - A(x)) \cdot \sigma)^2 . \tag{1}$$

Here $p = -i\nabla$ and $\sigma = (\sigma_1, \sigma_2, \sigma_3)$, are the Pauli matrices. We can also write $H_A = (p - A)^2 - B \cdot \sigma$. We shall here concentrate on the case where B is constant, say $B = (0, 0, B)$, with $B \geq 0$. We choose $A = \frac{1}{2}B \times x$.

The Hamiltonian describing an atom with N electrons and nuclear charge Z (with fixed nucleus) in a constant magnetic field B is

$$H_N = \sum_{i=1}^{N} \left(H_A^{(i)} - Z|x_i|^{-1} \right) + \sum_{1 \leq i < j \leq N} |x_i - x_j|^{-1} . \qquad (2)$$

The operator H_N acts on the Hilbert space $\mathcal{H}_N = \bigwedge^{N} L^2(\mathbf{R}^3; \mathbf{C}^2)$ of antisymmetric (i.e., fermionic) spinor-valued functions. We are interested in $E(N, B, Z) = \inf \mathrm{spec}_{\mathcal{H}_N} H_N$, **the ground state energy** of H_N.

We want to let B and Z go to infinity. It is surprising, but true, that there are *five* different regimes in B and Z, depending on the relative magnitudes of B and Z. In the following $\rho(x)$ is the electron density in the ground state ψ:

$$\rho(x) = N \int \|\psi(x, x_2, \ldots, x_N)\|^2 d^3 x_2 \ldots d^3 x_N. \qquad (3)$$

The five regions are the following.

1. $B \ll Z^{4/3}$, Z *large*: The effect of the magnetic field is negligible. Standard Thomas-Fermi (TF) theory is exact as $Z \to \infty$, and therefore the electron density is spherical to leading order.

2. $B \sim Z^{4/3}$, Z *large*: The magnetic field becomes important but the density is still almost spherical and stable atoms are almost neutral (see [14]). A modified TF theory (depending on the constant $B/Z^{4/3}$), in which the energy, as in standard TF theory, is approximated by a functional of the density ρ alone, is exact as $Z \to \infty$. We call this functional the Magnetic Thomas-Fermi (MTF) functional (see Sect. 4 below).

3. $Z^{4/3} \ll B \ll Z^3$, Z *large*: The magnetic field is increasingly important. To leading order all electrons will be confined to the lowest Landau band. The modified TF theory is still exact as $Z \to \infty$. In fact, the modified TF theory simplifies somewhat in this region compared to the MTF functional from the previous region. We call the new functional the Strong Thomas-Fermi (STF) functional. The only difference between STF and standard TF theory is that the usual $\rho^{5/3}$ is replaced

by ρ^3/B^2, while in the MTF theory from the previous region the function that replaces $\rho^{5/3}$ is more complicated (see (8) below). The density is almost spherical and stable atoms are almost neutral. Furthermore, the atom is getting smaller. The atomic radius behaves like $Z^{1/5}B^{-2/5} = Z^{-1/3}(B/Z^{4/3})^{-2/5}$. The energy behaves like $Z^{9/5}B^{2/5} = Z^{7/3}(B/Z^{4/3})^{2/5}$.

4. $B \sim Z^3$, Z *large*: The modified TF theories are no longer applicable. Indeed, we shall in general not approximate the energy by functionals of the density ρ alone. The energy is approximated by a more complicated functional to be described below in Sect. 4 depending on a one particle density matrix. We call this functional the Density Matrix (DM) functional. When B/Z^3 is large enough this functional again reduces to a density functional. For the first time the atom is no longer spherical to leading order. The length scale of the atom behaves like Z^{-1} and the energy like Z^3.

5. $B \gg Z^3$, Z *large*: In this hyper-strong case the atom is essentially one-dimensional. We can find a new functional, the Hyper-Strong (HS) functional depending only on the one-dimensional density $\bar{\rho}$ obtained from ρ by integrating ρ over the directions perpendicular to the field **B**, i.e.,

$$\bar{\rho}(x_3) = \int\!\!\int \rho(x_1, x_2, x_3)dx_1 dx_2 \ .$$

The energy behaves like $Z^3[\ln(B/Z^3)]^2$ and the length scale along the magnetic field is $Z^{-1}[\ln(B/Z^3)]^{-1}$, while the radius perpendicular to the field is $Z^{-1}(B/Z^3)^{-1/2}$.

The mathematically more precise statements of these results involve two energy functions $E_{\mathrm{MTF}}(N, B, Z)$ and $E_{\mathrm{DM}}(N, B, Z)$. The energy E_{MTF} is obtained as the minimum of the magnetic Thomas-Fermi functional mentioned under 2 above, and E_{DM} is the minimum of the density matrix functional mentioned under 4. The exact definitions of these functionals are given in Sect. 4 below.

The energies E_{MTF} and E_{DM} correspond to unique minimizers for the respective functionals. We denote the densities for these minimizers by ρ_{MTF} and ρ_{DM} respectively.

In the case when $B = 0$ the energy $E_{\mathrm{MTF}}(N, 0, Z)$ is the energy of standard TF theory. It is known [8] (see also [6]) that TF theory is asymptotically *exact* as $Z \to \infty$ with N/Z fixed, i.e.,

$$E_{\mathrm{MTF}}(N, 0, Z)/E(N, 0, Z) \to 1 \quad \text{as } Z \to \infty.$$

Is the same true when $B \neq 0$? The answer, surprisingly, depends on the relative magnitudes of B and Z, according to the 5 regions outlined above.

THEOREM 1 *Let N/Z be fixed and suppose $B/Z^3 \to 0$ as $Z \to \infty$. Then*

$$E_{\mathrm{MTF}}(N, B, Z)/E(N, B, Z) \to 1 \quad \text{as } Z \to \infty.$$

This theorem covers the regions 1–3 above. For the regions 4 and 5 we have

THEOREM 2 *Let N/Z be fixed and suppose $B/Z^{4/3} \to \infty$ as $Z \to \infty$. Then*

$$E_{\mathrm{DM}}(N, B, Z)/E(N, B, Z) \to 1 \quad \text{as } Z \to \infty.$$

Notice that there is an overlap of the regions of validity of the two theorems. In fact, both theorems cover region 3 above.

The energy functions satisfy the scalings

$$E_{\mathrm{MTF}}(N, B, Z) = Z^{7/3} E_{\mathrm{MTF}}(N/Z, B/Z^{4/3}, 1)$$

and

$$E_{\mathrm{DM}}(N, B, Z) = Z^3 E_{\mathrm{DM}}(N/Z, B/Z^3, 1)$$

In region 2 there is a non-trivial parameter $B/Z^{4/3}$. Likewise in region 4 there is B/Z^3. In the other three regions these parameters enter in a trivial way since they are tending either to 0 or ∞.

Region 1 corresponds to $B/Z^{4/3} \to 0$ and $B/Z^3 \to 0$ in which case

$$E_{\mathrm{MTF}}(N/Z, B/Z^{4/3}, 1) \to E_{\mathrm{MTF}}(N/Z, 0, 1),$$

which is the energy of standard TF theory.

Region 3 corresponds to $B/Z^{4/3} \to \infty$, in which case we have the asymptotic expansion

$$E_{\text{MTF}}(N/Z, B/Z^{4/3}, 1) \approx (B/Z^{4/3})^{2/5} E_{\text{STF}}(N/Z) \quad \text{as } B/Z^{4/3} \to \infty \,,$$

where E_{STF} is an energy function obtained from the simplified TF theory described under 3 above.

The overlap of the regions of validity of Theorems 1 and 2 implies that

$$E_{\text{DM}}(N/Z, B/Z^3, 1) \approx (B/Z^3)^{2/5} E_{\text{STF}}(N/Z) \quad \text{as } B/Z^3 \to 0 \ .$$

Finally, region 5 corresponds to $B/Z^3 \to \infty$, where the following asymptotic formula holds

$$E_{\text{DM}}(N/Z, B/Z^3, 1) \approx [\ln(B/Z^3)]^2 E_{\text{HS}}(N/Z) \quad \text{as } B/Z^3 \to \infty \ ,$$

where E_{HS} is an energy function obtained from the one-dimensional functional mentioned in 5 above.

The energies E_{MTF}, E_{DM}, E_{STF} and E_{HS} correspond to unique minimizers for the respective functionals. We denote the densities for these minimizers by ρ_{MTF}, ρ_{DM}, ρ_{STF} and $\bar{\rho}_{\text{HS}}$ respectively. We can prove that these densities approximate the quantum density ρ. However, to state these approximations we have to introduce different scalings in the different regions. In fact, the above approximating densities satisfy the following scaling relations

$$\rho_{\text{MTF}}(x; N, B, Z) = Z^2 \rho_{\text{MTF}}\left(Z^{1/3}x \, ; \frac{N}{Z}, \frac{B}{Z^{4/3}}, 1\right)$$

$$\rho_{\text{STF}}(x; N, B, Z) = Z^2 \left(\frac{B}{Z^{4/3}}\right)^{6/5} \rho_{\text{STF}}\left(\left(\frac{B}{Z^{4/3}}\right)^{2/5} Z^{1/3}x \, ; \frac{N}{Z}, 1, 1\right)$$

$$\rho_{\text{DM}}(x; N, B, Z) = Z^4 \rho_{\text{DM}}\left(Zx \, ; \frac{N}{Z}, \frac{B}{Z^3}, 1\right)$$

$$\bar{\rho}_{\text{HS}}(x_3; N, B, Z) = Z^2 \ln\left(\frac{B}{Z^3}\right) \bar{\rho}_{\text{HS}}\left(Z \ln\left(\frac{B}{Z^3}\right) x_3 \, ; \frac{N}{Z}, 1, 1\right)$$

THEOREM 3 (Convergence of the density) *In the five different regions the following relations hold as $Z \to \infty$. These limits are all in weak L^1_{loc} :*

(1–2) *If $B/Z^{4/3} \to \beta$, where $0 \leq \beta < \infty$ and if $N/Z = \lambda$ is fixed then*

$$Z^{-2}\rho(Z^{-1/3}x) \to \rho_{\text{MTF}}(x; \lambda, \beta, 1) \ .$$

(3) *If $B/Z^{4/3} \to \infty$ and $N/Z = \lambda$ is fixed then*

$$Z^{-2}\left(\frac{B}{Z^{4/3}}\right)^{-6/5} \rho\left(Z^{-1/3}\left(\frac{B}{Z^{4/3}}\right)^{-2/5}x\right) \to \rho_{\text{STF}}(x; \lambda, 1, 1) \ .$$

(4) *If $B/Z^{3} \to \eta$, where $0 < \eta < \infty$ and $N/Z = \lambda$ is fixed then*

$$Z^{-4}\rho_{\text{DM}}(Z^{-1}x) \to \rho_{\text{DM}}(x; \lambda, \eta, 1) \ .$$

(5) *If $B/Z^{3} \to \infty$ and $N/Z = \lambda$ is fixed then*

$$\frac{1}{Z^{2}\ln(B/Z^{3})} \overline{\rho}\left(\frac{x_{3}}{Z\ln(B/Z^{3})}\right) \to \overline{\rho}_{\text{HS}}(x_{3}; \lambda, 1, 1) \ .$$

3   The One-Body Hamiltonian

The spectrum of the one-body Hamiltonian $H_{\mathbf{A}}$ is described by the Landau bands $\varepsilon_{p\nu} = 2B\nu + p^{2}$, where p is the momentum along the field and $\nu = 0, 1, 2, \ldots$ is the index of the band. Owing to the spin degeneracy, the higher bands, $\nu \geq 1$, are twice as degenerate as the lowest band $\nu = 0$.

To calculate the energy of a large, complex atom one must first study the one-body Hamiltonian $H = H_{\mathbf{A}} + V(x)$, where V is an external potential. As usual, to calculate the ground state energy of a fermionic system we need to know the sum of the negative eigenvalues of the operator H (with $V \leq 0$ for simplicity).

In order to estimate accurately the sum of the negative eigenvalues of $H_{\mathbf{A}} + V(x)$ we need two things: (i) a lower bound for this quantity and (ii) an asymptotic (or semiclassical) limit formula for the quantity. These are provided by Theorems 4 and 5 below. The bound (i) is needed to control errors between the true answer and the semiclassical approximation. The semiclassical limit turns out to be relevant here (after some suitable scaling) because it is equivalent to the limit $Z \to \infty$.

There is an important difference between H_A and the operator $(\mathbf{p} - \mathbf{A})^2$ which has no spin dependence. While the spectrum of $(\mathbf{p} - \mathbf{A})^2$ is (B, ∞) the spectrum of H_A is $(0, \infty)$. Indeed, one can bound the sum of the negative eigenvalues of $(\mathbf{p} - \mathbf{A})^2 - V(x)$ by $-L \int |V(x)|^{5/2} dx$, (where L is some fixed constant) according to the standard Lieb-Thirring inequality (even with a magnetic field the proof of this inequality given in [10] is still correct if one appeals to the diamagnetic inequality). However, in the case of $H_A + V$ the question is somewhat more subtle. In fact, if $\int |V|^{3/2} < \infty$, the operator $(\mathbf{p} - \mathbf{A})^2 + V$ has a finite number of negative eigenvalues, while the operator $H_A + V$ can have infinitely many negative eigenvalues (compare [4]). We can, however, prove [9] the following bound which is important in our proofs.

THEOREM 4 *There exist universal constants $L_1, L_2 > 0$ such that if we let $e_j(B, V)$, $j = 1, 2, \ldots$ denote the negative eigenvalues of $H_A + V$ with $V \leq 0$ then*

$$\sum_j |e_j(B, V)| \leq L_1 B \int |V(x)|^{3/2} d^3 x + L_2 \int |V(x)|^{5/2} d^3 x . \quad (4)$$

We can choose L_1 as close to $2/3\pi$ as we please, compensating with L_2 large.

The first term on the right side is a contribution from the lowest band, $\nu = 0$. For large B this is the leading term.

We now ask the question of a semiclassical analog of (4). Thus, consider the operator

$$[(h\mathbf{p} - b\mathbf{a}(x)) \cdot \boldsymbol{\sigma}]^2 + v(x) , \quad (5)$$

where $\mathbf{a}(x) = \frac{1}{2}\hat{z} \times x$, $\hat{z} = (0, 0, 1)$ and $v \leq 0$.

If one computes the leading term in h^{-1} of the sum of the negative eigenvalues of (5) for fixed b one finds as in [3] that there is no b dependence. In our case, however, we shall not assume b fixed, or more precisely not assume that b is small compared with h^{-1}. The reason for this is that in the application to neutron stars it is not true, as we shall discuss below, that $b \ll h^{-1}$.

The interesting fact is, however, that we can prove ([9]) a semiclassical formula for the sum of the negative eigenvalues of the operator (5), which holds uniformly in b (even for large b).

THEOREM 5 *Let $e_j(h, b, v)$, $j = 1, \ldots$, denote the negative eigenvalues of the operator (5), with $v \leq 0$. Then*

$$\lim_{h \to 0} \left(\sum_j e_j(h, b, v) / E_{scl}(h, b, v) \right) = 1,$$

uniformly in b, where E_{scl} is the semiclassical approximation defined by

$$E_{scl}(h, b, v) = -\frac{1}{3\pi^2} h^{-2} b \int \left(|v(x)|^{3/2} + 2 \sum_{\nu=1}^{\infty} [|v(x)| - 2\nu bh]_+^{3/2} \right) d^3x .$$
(6)

Here $[t]_+ = t$ if $t > 0$, zero otherwise.

The formula (6) was already implicitly noted in [14]. The integrand in (6) looks peculiar, but it has the following simple physical interpretation. Take a cubic box of volume L^3 in $\mathbb{R}^3$ and let the number $\mu > 0$ be some fixed Fermi level (or chemical potential). Then add together all the negative eigenvalues of $H_A - \mu$. In the thermodynamic limit (large L) we can do this addition simply by using the known Landau levels, and the total energy per unit volume is the integrand in (6) in which $|v(x)|$ is set equal to μ.

For $bh \ll 1$, the right side of (6) reduces to the standard semiclassical formula from [3],

$$-\frac{2}{15\pi^2} h^{-3} \int |v(x)|^{5/2} d^3x .$$
(7)

(Recall that we are counting the spin which accounts for the 2 in front of the sum in (6).) For $bh \gg 1$, the sum in (6) is negligible, and we are left with the first term.

Formula (6) (with h replaced by 1) can be compared with the Lieb-Thirring inequality (4), which holds even outside the semiclassical regime. The two terms in (4) correspond to respectively the $b \to \infty$ (first term) and $b \to 0$ (last term) asymptotics of (6) .

As we know from elementary thermodynamics, the energy per unit volume as a function of the particle density ($\rho(x)$ in our case) is the Legendre transform of the pressure as a function of the chemical potential ($|v(x)|$). Thus, corresponding to $-(2/15\pi^2)|v(x)|^{5/2}$ in (7), there is the energy $(3/5)(3\pi^2)^{2/3}\rho(x)^{5/3}$, which is the usual kinetic energy expression in TF theory. Likewise, corresponding to (6) there is a kinetic energy which we call $w_B(\rho(x))$. It is no longer proportional to $\rho(x)^{5/3}$ but it is still a convex function of $\rho(x)$. It is proportional to $\rho(x)^3/B^2$ for small ρ, while it is asymptotically equal to $(3/5)(3\pi^2)^{2/3}\rho(x)^{5/3}$ as $\rho(x) \to \infty$.

4   The Many-Electron Atom

The essential ingredient in the study of the many-electron Hamiltonian H_N is to reduce it to a one-electron problem $H_\mathbf{A} + V_{\text{eff}}(x)$ with an effective mean field potential $V_{\text{eff}}(x) = -Z/|x| + \int |x-y|^{-1}\rho(y)d^3y$. This reduction involves approximating the repulsive energy

$$\int \|\psi(x_1,\ldots,x_N)\|^2 \sum_{1\leq i<j\leq N} |x_i - x_j|^{-1} d^3x_1 \ldots d^3x_N \quad ,$$

in the ground state ψ by

$$\tfrac{1}{2} \int\int \rho(x)\rho(y)|x - y|^{-1} d^3x d^3y \quad .$$

In standard TF theory the justification of this approximation is done by using the correlation inequality of Lieb and Oxford (see [6] and [7]). This very same argument (and inequality) work in the presence of a magnetic field. If B is not too large compared with Z it continues to be effective. However, in the hyper-strong case $B \gg Z^3$ the argument is no longer effective, the reason being that the correlation estimate is three dimensional in nature, while the atom is now effectively one-dimensional. The proof of a correlation estimate applicable in the hyper-strong case is difficult and will appear elsewhere ([9]).

The density ρ appearing in the mean field potential V_{eff} will not be taken to be the exact (unknown) density of the ground state, but rather an approximation to the exact density obtained from the density functionals that we shall now define.

Armed with the foregoing, we introduce a (magnetic field dependent) TF theory by means of the following functional of the unknown electron density $\rho(x)$:

$$\mathcal{E}_{\text{MTF}}(\rho) = \int w_B(\rho(x))d^3x - \int Z|x|^{-1}\rho(x)d^3x$$
$$+ \tfrac{1}{2} \iint \rho(x)|x-y|^{-1}\rho(y)d^3x d^3y. \tag{8}$$

It differs from the usual TF functional only in the replacement of $(\text{const.})\rho(x)^{5/3}$ by $w_B(\rho(x))$. We call this functional the **Magnetic Thomas-Fermi Functional**. It is studied in detail in [9]. The paper [13] seems to be the earliest reference that uses a Thomas-Fermi theory that takes *all* Landau levels into account. This theory was also studied in [2] and put on a rigorous basis in [14] for the regime $B \sim Z^{4/3}$.

We now choose our density ρ to be the unique minimizer for $\mathcal{E}_{\text{MTF}}$ constrained to the set $\int \rho \leq N$. We define the energy function that appears in Theorem 1 to be the infimum

$$E_{\text{MTF}}(N, B, Z) = \inf_{\int \rho \leq N} \mathcal{E}_{\text{MTF}}(\rho).$$

Theorems 4 and 5 play an essential role in the proof of Theorem 1. What makes the proof work when $B \ll Z^3$ is the fact that in the analysis of the mean-field, one-particle Hamiltonian, $H_A + V_{\text{eff}}(x)$, with $V_{\text{eff}}(x) = -Z/|x| + \int |x-y|^{-1}\rho(y)d^3y$, and with ρ being the density that minimizes the TF energy, we are in the semiclassical regime. The potential $V_{\text{eff}}(x)$ has the following behavior in Z and B

$$V_{\text{eff}}(x) = Z^{4/3}v(Z^{1/3}x) \quad \text{if } B \lesssim Z^{4/3}$$
$$V_{\text{eff}}(x) = Z^{4/5}B^{2/5}v(Z^{-1/5}B^{2/5}x) \quad \text{if } B \gtrsim Z^{4/3}, \tag{9}$$

where v is a function that does not depend significantly on B and Z.

Concentrating on the case $B \gtrsim Z^{4/3}$ we see, by a simple rescaling, that the Hamiltonian $H_A + V_{\text{eff}}(x)$ is unitarily equivalent to the operator

$$Z^{4/5}B^{2/5}[((h\mathbf{p} - b\mathbf{a}(x)) \cdot \boldsymbol{\sigma})^2 + v(x)], \tag{10}$$

where
$$h = (B/Z^3)^{1/5} \quad \text{and} \quad b = (B^2/Z)^{1/5}. \tag{11}$$

In the opposite case, when $B \lesssim Z^{4/3}$, we get $Z^{4/3}$ in place of $Z^{4/5}B^{2/5}$ in (10) and

$$h = Z^{-1/3} \quad \text{and} \quad b = B/Z. \tag{12}$$

When h is small we can study (10) by semiclassical methods.

If $B \gg Z^{4/3}$ we can replace $w_B(\rho)$ by its asymptotic form and we define the **Strong Thomas-Fermi** functional

$$\mathcal{E}_{\text{STF}}(\rho) = \tfrac{4}{3}\pi^4 B^{-2} \int \rho(x)^3 d^3x - \int Z|x|^{-1}\rho(x)d^3x$$
$$+ \tfrac{1}{2} \iint \rho(x)|x - y|^{-1}\rho(y)d^3x d^3y.$$

The analysis of E_{MTF} and E_{STF}, which is a separate story in itself, leads to the conclusions stated in 1, 2 and 3 of Sect. 2. Conclusions 1 and 2 were proved by Yngvason [14]; 3 is new. Since the TF energy functional has a unique minimizing $\rho(x)$ (because $\mathcal{E}_{MTF}$ is strictly convex in ρ) this ρ must be spherically symmetric. Thus we are led to the following remarkable conclusion:

If $B/Z^3 \to 0$ as $Z \to \infty$, *the atom is always spherical (to leading order) despite the fact that B has a leading order effect on the ground state energy.*

In region 2, $B \approx Z^{4/3}$, we cannot say that all the electrons are in the lowest Landau band, but if $B \gg Z^{4/3}$, they are – as the following theorem states precisely.

THEOREM 6 *If Π_0^N is the projection in the physical Hilbert space onto the subspace where all electrons are in the lowest Landau band, we can define the* **confined energy**

$$E_{\text{conf}}(N, B, Z) \equiv \text{ground state energy of } \Pi_0^N H_N \Pi_0^N. \tag{13}$$

Then, if $N < \lambda Z$ for some fixed $\lambda > 0$, we have that

$$E_{\text{conf}}(N, B, Z)/E(N, B, Z) \to 1 \quad \text{if} \quad B \to \infty \quad \text{and if} \quad Z^{4/3}/B \to 0. \tag{14}$$

What happens if $B \approx Z^3$? Semiclassical analysis breaks down (in the sense of being no longer asymptotically exact as $Z \to \infty$). The atom is no longer spherical. However, the atom is so non-semiclassical (one person called it post-modern) that another analysis becomes possible. This analysis, which we discuss next, is reminiscent of Hartree theory for bosons - even though it is relevant for fermionic electrons!

It is only the motion parallel to the magnetic field which can no longer be described semiclassically. The motion perpendicular to the field is still well approximated classically. To be more precise, the atom consists of a bundle of one dimensional quantum systems indexed by the position $x_\perp = (x_1, x_2)$ perpendicular to the field **B**. The state of one of these one-dimensional systems is described by a finite family of orthogonal functions $e_{x_\perp}^{(j)}$, $j = 1, 2 \ldots$ in $L^2(\mathbf{R})$ which are not normalized but satisfy $\|e_{x_\perp}^{(j)}\| \leq B/2\pi$. This condition follows from the Pauli principle and the fact that the two-dimensional density of states in the lowest Landau band is exactly $B/2\pi$.

We can combine the functions $e_{x_\perp}^{(j)}$, $j = 1, 2, \ldots$ into a density matrix

$$\gamma : x_\perp \mapsto \gamma_{x_\perp}(x_3, y_3) = \sum_j e_{x_\perp}^{(j)}(x_3)\overline{e_{x_\perp}^{(j)}(y_3)} \quad .$$

Then γ satisfies

(a) $0 \leq \gamma_{x_\perp} \leq (B/2\pi)I$ as an operator on $L^2(\mathbf{R})$

(b) $\int_{\mathbf{R}^2} \text{Tr}_{L^2(\mathbf{R})}[\gamma_{x_\perp}]d^2x_\perp = N$ = the total number of electrons

We can now approximate the energy by the functional

$$\mathcal{E}_{\text{DM}}(\gamma) = \int_{\mathbf{R}^2} \text{Tr}_{L^2(\mathbf{R})}[(-\partial_3^2 - Z|x|^{-1})\gamma_{x_\perp}]d^2x_\perp$$
$$+ \tfrac{1}{2} \iint \rho_\gamma(x)\rho_\gamma(y)|x - y|^{-1}d^3x d^3y \quad ,$$

where $\rho_\gamma(x) = \gamma_{x_\perp}(x_3, x_3)$.

We denote

$$E_{\text{DM}}(N, B, Z) = \inf\{\mathcal{E}(\gamma) : \gamma \text{ satisfies (a) and (b) above}\}.$$

This is the function appearing in Theorem 2. The Pauli principle comes into play in this theory only in condition (a). The proof of Theorem 2 is straightforward as soon as one has made the reduction to a one body problem and realized that condition (a) follows from the confinement to the lowest Landau band.

The Euler-Lagrange equation for the $\mathcal{E}_{\text{DM}}$ minimization problem implies that the functions $e_{x_\perp}^{(j)}$ are eigenfunctions of the one-dimensional Schrödinger operator $h_{x_\perp} = -\frac{d^2}{dx_3^2} - V_{\text{eff}}(x)$ where, as before, the effective potential is $V_{\text{eff}}(x) = -Z/|x| + \int |x-y|^{-1}\rho_\gamma(y)d^3y$ with ρ_γ being the density corresponding to the minimizer γ for $\mathcal{E}_{\text{DM}}$

5   The Super Strong Case $B \gg Z^3$

We shall present here the correct energy functional of the density when $B \gg Z^3$, and very briefly indicate what is involved in proving the correctness of the approximation.

The first step is to show that when B/Z^3 is larger than some critical value then the minimizing γ for $\mathcal{E}_{\text{DM}}$ is rank one for every $x_\perp$. Since the eigenfunction of $\gamma_{x_\perp}$ must be the ground state of $h_{x_\perp}$ we can conclude that it is a positive function. In this case we can write $\gamma_{x_\perp}(x_3, y_3) = \sqrt{\rho(x_\perp, x_3)}\sqrt{\rho(x_\perp, y_3)}$ where $\rho(x) = \rho_\gamma(x)$.

The functional $\mathcal{E}_{\text{DM}}$ thus becomes a density functional when B/Z^3 is large enough.

$$\mathcal{E}_{\text{DM}}(\gamma) = \mathcal{E}_{\text{SS}}(\rho) = \int \left(\frac{\partial}{\partial x_3}\sqrt{\rho(x)}\right)^2 d^3x - \int \frac{Z}{|x|}\rho(x)d^3x$$
$$+ \frac{1}{2}\iint \rho(x)|x-y|^{-1}\rho(y)d^3xd^3y, \quad (15)$$

with the condition that

$$\int \rho(x_1, x_2, x_3)dx_3 \leq \frac{B}{2\pi} \text{ for all } (x_1, x_2). \quad (16)$$

Then

$$E_{\text{DM}}(N, B, Z) = E_{\text{SS}}(N, B, Z))$$
$$= \inf\left\{\mathcal{E}_{\text{SS}}(\rho) : \int \rho \leq N, \rho \text{ satisfies (16)}\right\} \quad (17)$$

We can now ask for the limit of $\mathcal{E}_{SS}$ if $B/Z^3 \to \infty$, $Z \to \infty$ and N/Z is fixed. With some effort one can prove that $\mathcal{E}_{SS}$ then simplifies to another functional, which we call the **hyper-strong functional** of a *one-dimensional* density $\rho_1(x), x \in \mathbb{R}$. That is, the atom is now so thin compared to its length that only the average density and its variation along the direction parallel to B matter.

It is convenient, in defining this average density, to rescale the variables. Thus, setting $\eta \equiv B/(2\pi Z^3)$, and taking $(Z \ln \eta)^{-1}$ as the unit of length, we define

$$\rho_1(x) \equiv \frac{1}{Z^2 \ln \eta} \overline{\rho}\left(\frac{1}{Z \ln \eta}x\right) \equiv \frac{1}{Z^2 \ln \eta} \int \rho\left(x_1, x_2, \frac{1}{Z \ln \eta}x\right) dx_1 dx_2, \tag{18}$$

which has the normalization $\int \rho_1(x)dx = N/Z$. The hyper-strong functional is

$$\mathcal{E}_{HS}(\rho_1) = \int\limits_{-\infty}^{\infty} \left(\frac{d}{dx}\sqrt{\rho_1(x)}\right)^2 dx - \rho_1(0) + \frac{1}{2}\int\limits_{-\infty}^{\infty} \rho_1(x)^2 dx. \tag{19}$$

In other words, apart from some scalings, *the Coulomb potential is replaced by a Dirac delta function!* Using (19) we define a rescaled energy

$$E_{HS}(N/Z) \equiv \inf_{\int \rho_1 = N/Z} \mathcal{E}_{HS}(\rho_1). \tag{20}$$

We assert that under the conditions stated above,

$$Z^3 (\ln \eta)^2 E_{HS}(N/Z)/E(N, B, Z) \to 1$$

as $Z \to \infty, B/Z^3 \to \infty$ and N/Z is fixed.

A remarkable fact is that the minimizing ρ_1 can be evaluated exactly. The Euler-Lagrange equation is (with $\psi^2 \equiv \rho_1$ and Lagrange multiplier μ)

$$-\ddot{\psi}(x) - \psi(0)\delta(x) + \psi^3(x) = -\mu\psi(x). \tag{21}$$

With $\lambda \equiv N/Z$, there are solutions only for $\lambda \le 2$ (not $\lambda \le 1$ as in TF theory):

$$\begin{aligned}
\psi(x) &= \frac{\sqrt{2}(2-\lambda)}{2\sinh[\frac{1}{4}(2-\lambda)|x|+c]} &&\text{for } \lambda < 2 \\
\psi(x) &= \sqrt{2}(2+|x|)^{-1} &&\text{for } \lambda = 2,
\end{aligned} \tag{22}$$

with $\tanh c = (2 - \lambda)/2$. The energy is

$$E_{\mathrm{HS}}(\lambda) = \mathcal{E}_{\mathrm{HS}}(\psi^2) = -\frac{1}{4}\lambda + \frac{1}{8}\lambda^2 - \frac{1}{48}\lambda^3. \qquad (23)$$

Bibliography

[1] I. Fushiki, E. H. Gudmundsson, and C. J. Pethick, *Surface structure of neutron stars with high magnetic fields,* Astrophys. Jour. *342,* (1989), 958–975.

[2] I. Fushiki, E. H. Gudmundsson, C. J. Pethick, and J. Yngvason, *Matter in a magnetic field in the Thomas-Fermi and related theories,* Ann. of Phys. (in press).

[3] B. Helffer and D. Robert, *Calcul fonctionel par la transformeé de Mellin et operateurs admissible,* Jour. Func. Anal. *53,* (1983), 246–268.

[4] V. Ivrii, in preparation.

[5] B. B. Kadomtsev, *Heavy Atoms in an Ultrastrong Magnetic Field,* Soviet Phys. JETP *31,* (1970), 945–947.

[6] E. H. Lieb, *Thomas-Fermi and related theories of atoms and molecules,* Rev. Mod. Phys. *53,* (1981) 603–641. Errata, *54,* (1982) 311.

[7] E. H. Lieb and S. Oxford, *Improved lower bound on the indirect Coulomb energy,* Int. J. Quant. Chem. *19,* (1981), 427–439.

[8] E. H. Lieb and B. Simon, *The Thomas-Fermi theory of atoms, molecules and solids,* Adv. in Math. *23,* (1977), 22–116.

[9] E.H. Lieb, J. P. Solovej and J. Yngvason, in preparation.

[10] E. H. Lieb and W. E. Thirring, *A bound for the moments of the eigenvalues of the Schrödinger Hamiltonian and their relation to Sobolev inequalities,* in *Studies in Mathematical Physics: Essays in Honor of Valentine Bargmann,* ed. E. H. Lieb, B. Simon and A. S. Wightman, Princeton University Press (1977).

[11] R. O. Mueller, A. R. P. Rau and L. Spruch, *Statistical Model of Atoms in Intense Magnetic Fields*, Phys. Rev. Lett. *26*, (1971) 1136.

[12] M. Ruderman, *Matter in superstrong magnetic fields: The surface of a neutron star*, Phys. Rev. Lett. *27*, (1971), 1306–1308.

[13] Y. Tomishima and K. Yonei, *Thomas Fermi theory for atoms in a strong magnetic field*, Progr. Theor. Phys., *59*, (1978), 683–696.

[14] J. Yngvason, *Thomas-Fermi theory for matter in a magnetic field as a limit of quantum mechanics*, Lett. Math. Phys. *22*, (1991) 107–117.

Algebraic Riccati Equations Arising in Game Theory and in H^∞-Control Problems for a Class of Abstract Systems

C. McMillan and R. Triggiani
University of Virginia

Abstract

We consider the abstract framework of [4, class (H.2)] which models a variety of mixed partial differential equation problems in a smooth bounded domain $\Omega \subset \mathbb{R}^n$, arbitrary n, with boundary $L_2(0, T; \partial\Omega)$-control. These include second-order hyperbolic equations, first-order hyperbolic systems, Euler-Bernoulli, Kirchhoff, Schroedinger equations, etc. For these dynamics we set and solve a min-max game theory problem – a fortiori the H^∞-robust stabilization problem – in terms of an algebraic Riccati equation to express the optimal quantities in pointwise feedback form.

1 Introduction

1.1 Problem Setting

Let U (control) and Y (state) be separable Hilbert spaces. We introduce the following abstract state equation

$$\dot{y}(t) = Ay(t) + Bu(t) + Gw(t) \quad \text{in } [D(A^*)]'; \quad y(0) = y_0 \in Y \quad (1)$$

Differential Equations with Applications to Mathematical Physics

Here, the function $u \in L_2(0, \infty; U)$ is the control and $w \in L_2(0, \infty; Y)$ is a deterministic disturbance. The dynamics (1) is subject to the following assumptions, which will be maintained throughout the paper:

(H.1) $A : Y \subset D(A) \longrightarrow Y$ is the infinitesimal generator of a strongly continuous (s.c.) semigroup e^{At} on the Hilbert space Y;

(H.2) B: continuous $U \longrightarrow [D(A^*)]'$; or, equivalently, $A^{-1}B \in \mathcal{L}(U, Y)$, where $[D(A^*)]'$ denotes the dual of $D(A^*)$ with respect to the Y-topology, and A^* is the Y-adjoint of A;

(H.3) the following abstract trace regularity holds (see Remark 1.1): the operator $B^* e^{A^* t}$ admits a continuous extension, denoted by the same symbol, from $Y \longrightarrow L_2(0, T; U)$:

$$\int_0^T \| B^* e^{A^* t} x \|_Y^2 \, dt \leq c_T \| x \|_Y^2 \quad \forall \, T < \infty; \qquad (2)$$

where B^* is the dual of B, satisfies $B^* \in \mathcal{L}(D(A^*), U)$ after identifying $[D(A^*)]''$ with $D(A^*)$.

(H.4) G and R are bounded operators on Y, i.e. $G, R \in \mathcal{L}(Y)$;

The solution to the state equation (1) is given explicitly by

$$y(t) = y(t; y_0) = e^{At} y_0 + (Lu)(t) + (Ww)(t) \qquad (3)$$

where, for the problem (1) defined on the interval $[0, T]$,

$$(Lu)(t) \; = \; \int_0^t e^{A(t-\tau)} Bu(\tau) d\tau \qquad (4)$$
$$: \; continuous \; L_2(0, \infty; U) \longrightarrow C([0, T]; Y)$$

by duality on (2), and

$$(Ww)(t) \; = \; \int_0^t e^{A(t-\tau)} Gw(\tau) d\tau \qquad (5)$$
$$: \; continuous \; L_2(0, \infty; Y) \longrightarrow C([0, T]; Y)$$

Remark 1.1: Assumption (H.3) = (2) is an abstract trace theory property. Over the past ten years, this property has been proved to hold true for many classes of partial differential equations by purely P.D.E.'s methods (energy methods either in differential or in pseudo-differential form), including: second order hyperbolic equations; Euler-Bernoulli, Kirchhoff, and Schroedinger equations; first order hyperbolic systems, etc., all in arbitrary space dimensions and on explicitly identified spaces; see e.g. [4, class (H.2)].

1.2 Game Theory Problem

For a fixed $\gamma > 0$, we associate with (1) the cost functional

$$
\begin{aligned}
J(u,w) &= J(u,w,y(u,w)) \\
&= \int_0^\infty [\|Ry(t)\|_Y^2 + \|u(t)\|_U^2 - \gamma^2 \|w(t)\|_Y^2] dt \quad (6)
\end{aligned}
$$

where $y(t) = y(t; y_0)$ is given by (3). The aim of this paper is to study the following game-theory problem:

$$
\sup_w \inf_u J(u,w) \quad (7)
$$

where the infimum is taken over all $u \in L_2(0,\infty; U)$, for w fixed, and the supremum is taken over all $w \in L_2(0,\infty; Y)$. This problem is known to be equivalent to the so-called H^∞-robust stabilization problem [1].

In our approach here to problem (7), we shall critically rely on the treatment of [1; sect 5], [2] in the case $w \equiv 0$. The case where e^{At} exponentially stable (add damping to the mixed partial differential equation problems of Remark 1.1) admits a much simpler, fully explicit, and more informative treatment [5]. Here, we shall consider the general case [6]. Another treatment is in [1], which also critically falls into [2].

1.3 Main Results

Theorem 1.3.1 *Assume (H.1) - (H.4) as well as the "Finite Cost Condition" and the "Detectability Condition" (see [4]: <u>all</u> these assumptions are automatically satisfied for mixed partial differential equation problems of Remark 1.1). There exists an intrinsic value* (critical) $\gamma_c > 0$ of γ, explicitly defined in terms of the problem data in Eq. (29) below, such that:

(a) if $0 < \gamma < \gamma_c$, then the supremum in w in (7) leads to $+\infty$ and the min-max problem has no finite solution.

(b) if $\gamma > \gamma_c$, then:

(i) there exists a unique optimal solution $\{u^*(\cdot \ ; y_0), w^*(\cdot \ ; y_0);$ $y^*(\cdot \ ; y_0)\}$ of problem (7);

$$(ii) \qquad u^*(t; y_0) = -B^* P y^*(t; y_0) \in L_2(0, \infty; U) \qquad (8)$$

where P is the unique bounded, nonnegative self-adjoint operator which satisfies the following Algebraic Riccati Equation, ARE_γ for all $x, z \in D(A)$:

$$(PAx, z)_Y + (Px, Az)_Y + (Rx, Rz)_Y = (B^* Px, B^* Pz)_Y$$
$$-\gamma^{-2}(G^* Px, G^* Pz)_Y \quad (9)$$

with the property

$$B^* P \in \mathcal{L}(D(A); Y); \qquad (10)$$

(iii) the operator (F stands for "feedback")

$$A_F = A - BB^* P + \gamma^{-2} GG^* P \qquad (11)$$

is the generator of a s.c. semigroup on Y and, in fact, for $y_0 \in Y$:

$$y^*(t; y_0) = e^{(A - BB^* P + \gamma^{-2} GG^* P)t} y_0, \quad t \geq 0 \qquad (12)$$

and, moreover, the semigroup is uniformly (exponentially) stable on Y.

$$(iv) \qquad \gamma^2 w^*(t; y_0) = G^* P y^*(t; y_0); \qquad (13)$$

(v) for $y_0 \in Y$

$$(Py_0, y_0) = J^*(y_0) = J(u^*, w^*, y^*) = \sup_w \inf_u J(u, w, y) \qquad (14)$$

Other properties are given in [6]. □

2 Scheme of Proof of Theorem 1.3.1

Naturally, the proof proceeds along two main steps, (i) and (ii) below.

(i) First, one studies the minimization problem $\inf_u J$ holding $w \in L_2(0, \infty; Y)$ fixed. This is a standard, quadratic (strictly convex) problem, which, for the present abstract class, can be studied following the methods of [2, sects 4, 5]: one first studies the minimization in u over a finite time interval $[0, T]$, characterizes here the optimal solution, and then considers the limit process as $T \uparrow \infty$, as in [2, sect 4]. Now, however, due to the presence of w, it is technically important to adapt to present circumstances, the idea of "decoupling" (expressed by Eq. (15) below) between the known case $w \equiv 0$, and a convenient formulation of the case $w \neq 0$, see (19). As a result of a technical treatment [6], one culminates the limit process $T \uparrow \infty$ with the following formulas

$$u^0_{w,\infty}(\,\cdot\,; y_0) = -B^* p_{w,\infty}(\,\cdot\,; y_0) \tag{15}$$

$$p_{w,\infty}(t; y_0) = P_{0,\infty} y^0_{w,\infty}(\,\cdot\,; y_0) + r_{w,\infty}(t), \quad t > 0, \quad in \;\; Y \tag{16}$$

where $\{u^0_{w,\infty}(\,\cdot\,; y_0), y^0_{w,\infty}(\,\cdot\,; y_0)\}$ is the unique optimal pair of the inf J-process in u, holding w fixed. In Eq. (16), $P_{0,\infty}$ is the unique Algebraic Riccati operator corresponding to the case $w \equiv 0$, and guaranteed by [2] via the Finite Cost Condition and the Detectability Condition. Thus, $P_{0,\infty}$ is a bounded, nonnegative self-adjoint operator in Y, and

$$A_{P_{0,\infty}} = A - BB^* P_{0,\infty} \tag{17}$$

is the generator of a s.c. uniformly stable semigroup: there exist $M \geq 1$ and $\delta > 0$ such that

$$\|e^{A_{P_{0,\infty}} t}\|_{\mathcal{L}(Y)} \leq M e^{-\delta t}, \quad t \geq 0 \tag{18}$$

A key feature of the function $r_{w,\infty}(t)$, which unlike $P_{0,\infty}$, refers now to the $0 \neq w$ fixed, is that $r_{w,\infty}(t)$ satisfies a differential equation

$$\dot{r}_{w,\infty} = -A^*_{P_{0,\infty}} r_{w,\infty}(t) - P_{0,\infty} Gw \tag{19}$$

$$r_{w,\infty}(t) = \int_t^\infty e^{A_{P_0,\infty}(\tau-t)} P_{0,\infty} G w(\tau) d\tau \tag{20}$$

with *stable* generator $A^*_{P_0,\infty}$. Similarly, $p_{w,\infty}$ and the optimal dynamics $y^0_{w,\infty}$ can be rewritten as to satisfy differential equations by *stable* semigroups:

$$\dot{p}_{w,\infty}(t; y_0) = -A^*_{P_0,\infty} p_{w,\infty}(t; y_0) \quad + P_{0,\infty} B u_{w,\infty}(t; y_0)$$
$$- R^* R y_{w,\infty}(t; y_0) \tag{21}$$

$$\dot{y}_{w,\infty}(t; y_0) = A_{P_0,\infty} y_{w,\infty}(t; y_0) - B B^* r_{w,\infty}(t) + G w(t) \tag{22}$$

where, in addition, one can prove the technical result that

$$B^* r_{w,\infty}(t) \in L_2(0, \infty; U) \tag{23}$$

A related technical issue [6], which employs the Algebraic Riccati Equation satisfied by $P_{0,\infty}$ (case $w \equiv 0$) [2] is that:

$$(\mathcal{L}_{P_0,\infty} u)(t) \quad \equiv \int_0^t e^{A_{P_0,\infty}(t-\tau)} B u(\tau) d\tau \tag{24}$$
$$: \text{ continuous } L_2(0, \infty; U) \longrightarrow L_2(0, \infty; Y)$$

with L_2-adjoint continuous $L_2(0, \infty; Y) \longrightarrow L_2(0, \infty; U)$:

$$(\mathcal{L}^*_{P_0,\infty} v)(t) \equiv B^* \int_t^\infty e^{A^*_{P_0,\infty}(\tau-t)} v(\tau) d\tau \tag{25}$$

We similarly introduce the operator

$$(\mathcal{W}_{P_0,\infty} w)(t) \equiv \int_0^t e^{A_{P_0,\infty}(t-\tau)} G u(\tau) d\tau \tag{26}$$

and its L_2-adjoint

$$(\mathcal{W}^*_{P_0,\infty} f)(t) \equiv G^* \int_t^\infty e^{A^*_{P_0,\infty}(\tau-t)} v(\tau) d\tau \tag{27}$$

both continuous $L_2(0, \infty; Y) \longrightarrow$ itself. With these preliminaries, we now introduce the self-adjoint operator in $\mathcal{L}(L_2(0, \infty; Y))$ by:

$$S = G^* P_{0,\infty} \mathcal{L}_{P_0,\infty} \mathcal{L}^*_{P_0,\infty} P_{0,\infty} G - [G^* P_{0,\infty} \mathcal{W}_{P_0,\infty} + \mathcal{W}^*_{P_0,\infty} P_{0,\infty} G] \tag{28}$$

We now define the critical value of γ, $\gamma_c > 0$, by

$$\gamma_c^2 \equiv max\{0, - \inf_{\|w\|=1} (Sw, w)_{L_2(0,\infty;Y)}\} \tag{29}$$

A critical technical result [6], is that the optimal cost $J_w^0(y_0)$ corresponding to the infimum in u can be expressed as

$$-J_w^0(y_0) = (E_\gamma w, w)_{L_2(0,\infty;Y)} \quad +(w, a_{y_0})_{L_2(0,\infty;Y)}$$
$$+(P_{0,\infty}y_0, y_0)_Y \tag{30}$$

where

$$E_\gamma = \gamma^2 I - S \tag{31}$$

and a_{y_0} is a vector in $L_2(0,\infty;Y)$ depending on y_0.

(ii) The second step is to study the infimum of $-J_w^0(y_0)$ over $w \in L_2(0,\infty;Y)$. Eqs. (29) - (31) reveal that for $\gamma > \gamma_c$, the dominant quadratic term in (30) is coercive and then a unique optimal $w^*(\cdot\,;y_0)$ can be asserted. Because of the stability property of the generator in (19) (or (20)), (21), and (22), one can then characterize directly such optimal w^* over the infinite time interval $[0,\infty]$, via, say, Lagrange Multiplier Theory (Liusternik's Theorem). The result is

$$\gamma^2 w^*(t; y_0) = G^* p^*(t; y_0), \quad \gamma > \gamma_c \tag{32}$$

where $p^*(t; y_0) = p_{w=w^*,\infty}(t; y_0)$. Moreover, it is possible to express w^* explicitly in terms of the problem data via E_γ^{-1}:

$$w^*(\cdot\,;y_0) = E_\gamma^{-1}[G^* P_{0,\infty} e^{A_{P_{0,\infty}} \cdot} y_0], \quad \gamma > \gamma_c \tag{33}$$

From here, then, one first finds (as in [3]) the transition property for w^* when $\gamma > \gamma_c$:

$$w^*(t + \sigma; y_0) = w^*(\sigma; y^*(t; y_0)) \tag{34}$$

for t fixed, the equality being intended in $L_2(0,\infty;Y)$ in σ. Next, using (34) and (20), one finds a transition property for $\gamma > \gamma_c$:

$$r^*(t + \sigma; y_0) = r^*(\sigma; y^*(t; y_0)) \underset{(in\ \sigma)}{\in} C_b([0,\infty];Y) \tag{35}$$

Finally, using (34) and (35), one finds the semigroup property for $\gamma > \gamma_c$:

$$y^*(t + \sigma; y_0) = y^*(\sigma; y^*(t; y_0)) \tag{36}$$

for $y^*(t; y_0) = \Phi(t)y_0$, whereby $\Phi(t)$ is then a s.c. semigroup on Y. An analogous transition property for p^* is likewise valid for $\gamma > \gamma_c$:

$$p^*(t + \sigma; y_0) = p^*(\sigma; y^*(t; y_0)) \tag{37}$$

via (35), (36), and (19). The operator P is defined by

$$Px \equiv p^*(0; x) \tag{38}$$

and the semigroup $\Phi(t)$ is then, in fact,

$$\Phi(t)y_0 = e^{(A - BB^*P + \gamma^{-2}GG^*P)t}y_0, \quad t \geq 0 \tag{39}$$

One then can fall into the technical treatment of [2, sect 4], replacing the operator A_F there with the operator

$$A_F = A - BB^*P + \gamma^{-2}GG^*P \tag{40}$$

here, to complete the proof of Theorem 1.3.1.

Bibliography

[1] V. Barbu. H^∞-boundary control with state feedback; the hyperbolic case, preprint 1992.

[2] F. Flandoli, I. Lasiecka, and R. Triggiani. Algebraic Riccati equations with non-smoothing observation arising in hyperbolic and Euler-Bernoulli equations, Annali di Matematica Pura et Applicata, IV Vol. CLIII (1989), pp. 307-382.

[3] I. Lasiecka and R. Triggiani. Riccati equations for hyperbolic partial differential equations with $L_2(0, T; L_2(0, \infty; Y))$-Dirichlet boundary terms, SIAM J. Control Optimiz., 24 (1986), pp. 884-924.

[4] I. Lasiecka and R. Triggiani. *Differential and algebraic Riccati equations with application to boundary/point control problems: continuous theory and approximation theory*, Springer-Verlag Lectures Notes LNCIS series Vol. # 164 (1991), pp. 160.

[5] C. McMillan and R. Triggiani. *Min-Max Game theory and algebraic Riccati equations for boundary control problems with continuous input-solution map. Part I: the stable case*, to appear in Proceedings of the 3rd International Workshop-Conference on evolution Equations, Control Theory, and Biomathematics, held at Han-sur-Lesse, Belgium, October 1991, to be published by Marcel Dekker.

[6] C. McMillan and R. Triggiani. *Min-Max Game theory and algebraic Riccati equations for boundary control problems with continuous input-solution map. Part II: the general case*, to appear.

Symmetries and Symbolic Computation

M. C. Nucci
Dipartimento di Matematica
Università di Perugia,
06100 Perugia, Italy

Abstract

In this paper we show some applications of our interactive RE-DUCE programs for calculating classical, non-classical, and Lie-Bäcklund symmetries of differential equations. These programs are easy to use and do not require an in-depth knowledge of LISP or REDUCE. They were designed for the unsophisticated user who is knowledgeable in the area of symmetries of differential equations.

1 Introduction

It is well-known that the main obstacle to the application of the Lie group theories [23], [1], [4], [22], [6], [24] is the extensive calculations they involve. At present many computer algebra softwares are available, such as MAPLE by B. Char at Waterloo, MACSYMA by the Mathlab Group at MIT, REDUCE by A.C. Hearn at the Rand Corporation, SMP by S. Wolfram, and SCRATCHPAD II by R.D. Jenks and D. Yun at IBM. Also ad-hoc programs were developed to find the classical symmetries of differential equations*, i.e. perform the so-called "group analysis" [23].

These programs may be divided into the following two groups:

---

*A comprehensive list is given in [9].

1. automatic packages where you put in your equation and get the answer without any interaction with the computer. Perhaps the best example is SPDE by F. Schwarz [25] which runs with REDUCE version 3.3 and higher [12];

2. interactive programs which require the user to make specific choices at different stages of the computation. An example is [13] which also runs with REDUCE.

In the following we emphasize the differences between these two methods. For this purpose we make a comparison of SPDE and [13]. The former is able to find classical symmetries of many type of differential equations with the following exceptions:

- equations with arbitrary functions of the unknown and its derivatives, as $u_{tt} = [f(u)u_x]_x$ [2];

- overdetermined system of equations.

Unfortunately, because SPDE is not reliable all the time, interactive programs must be used. In [13] such programs were developed. They should be able to handle the above equations and also find their Lie-Bäcklund symmetries [4], [22], [6], but:

- they require a good knowledge of LISP [11];

- the non-expert user cannot modify them for his own needs (e.g. finding the non-classical symmetries [5], [16]).

In an effort to overcome these problems, we have developed [17] easy to use interactive programs which do not require an in-depth knowledge of LISP or REDUCE. In fact, to use them one only needs to know a single LISP command and have a very basic familiarity with REDUCE (version 3.3 or higher). The programs labelled GA automatically construct the determining equations for the classical symmetries; with only minor modifications, all the other programs are derived. They calculate the non-classical (SGA), and the Lie-Bäcklund (GS) symmetries of any differential equation. When any of these programs is loaded, REDUCE will automatically run it and construct the determining equations. At this point the user will begin to interact with the computer. In [17] we presented several computer sessions for calculating classical, non-classical, and Lie-Bäcklund symmetries of known equations, including the classical symmetries of $u_{tt} = [f(u)u_x]_x$. Further applications can be found in various articles [3], [18], [19], [20], [21]. Here we show some other results obtained by using our interactive REDUCE programs: the classical symmetries of an overdetermined system in fluid mechanics, the nonclassical symmetries of Burgers' equation, and the third order Lie-Bäcklund symmetries of sine-Gordon equation and nonlinear reaction-diffusion equation [7]. Although our programs were originally developed for a mainframe IBM 4381, the following outputs were obtained by running our programs on a SUN SPARCstation I of the School of Mathematics at Georgia Institute of Technology, Atlanta (U.S.A). We underline the importance of a REDUCE-LaTeX translator [10] in reporting the results.

2   Classical Symmetries of an Overdetermined System

The following system of 7 first order differential equations with 3 independent (t, x, y) and 6 dependent variables (u, v, h, u', v', h') models the motion of two shallow immiscible inviscid incompressible fluids subject to the force of gravity and contained in a rigid basin rotating with the Earth. More details can be found in U. Ramgulam's Ph.D. thesis supervised by C. Rogers at Loughborough University of Technology, Loughborough (U.K.).

$$u_t + uu_x + vu_y - fv + \frac{\rho'}{\rho}\left(Z + h'\right)_x + \left(1 - \frac{\rho'}{\rho}\right)\left(Z + h\right)_x = 0 \quad (1)$$

$$v_t + uv_x + vv_y + fu + \frac{\rho'}{\rho}\left(Z + h'\right)_y + \left(1 - \frac{\rho'}{\rho}\right)\left(Z + h\right)_y = 0 \quad (2)$$

$$h_t + \left(uh\right)_x + \left(vh\right)_y = 0 \quad (3)$$

$$u'_t + u'u'_x + v'u'_y - fv' + \left(Z + h'\right)_x = 0 \quad (4)$$

$$v'_t + u'v'_x + v'v'_y + fu' + \left(Z + h'\right)_y = 0 \tag{5}$$

$$h'_t + \left(u'h'\right)_x + \left(v'h'\right)_y = 0 \tag{6}$$

$$\left(h' - h\right)_t + \left(u'(h' - h)\right)_x + \left(v'(h' - h)\right)_y = 0 \tag{7}$$

Note that $z = Z(x, y)$ is the equation of the basin surface, f is the constant Coriolis parameter, (u, v) and (u', v') the x and y components of the velocity of the first and second fluid, respectively, ρ and ρ' the density of the first and second fluid, respectively, h and h' the vertical distance from the free surface to the basin of the first and second fluid, respectively. The classical symmetry analysis of the partial differential equations which model the motion of a rotating shallow liquid in a rigid basin was performed in [15] by using a MACSYMA program [8]. Here we perform the classical symmetry analysis of (1-7) by using our interactive REDUCE program, and after having solved the system (1-7) algebraically for u_t, v_t, h_t, u'_t, v'_y, v'_t, and h'_t, i.e.:

$$u_t = -\left(uu_x + vu_y - fv + \frac{\rho'}{\rho}\left(\frac{\partial Z}{\partial x} + h'_x\right) + \left(1 - \frac{\rho'}{\rho}\right)\left(\frac{\partial Z}{\partial x} + h_x\right)\right) \tag{8}$$

$$v_t = -\left(uv_x + vv_y + fu + \frac{\rho'}{\rho}\left(\frac{\partial Z}{\partial y} + h'_y\right) + \left(1 - \frac{\rho'}{\rho}\right)\left(\frac{\partial Z}{\partial y} + h_y\right)\right) \tag{9}$$

$$h_t = -\left(u_x h + uh_x + v_y h + vh_y\right) \tag{10}$$

$$u'_t = -\left(u'u'_x + v'u'_y - fv' + \frac{\partial Z}{\partial x} + h'_x\right) \tag{11}$$

$$v'_y = \left(- h_t - u'_x h - u'h_x - v'h_y\right)/h \tag{12}$$

$$v'_t = -\left(u'v'_x + v'v'_y + fu' + \frac{\partial Z}{\partial y} + h'_y\right) \tag{13}$$

$$h'_t = h_t - u'_x h' + u'_x h - h'v'_y + hv'_y - u'h'_x + u'h_x - v'h'_y + v'h_y \tag{14}$$

The classical symmetry analysis consists of looking for the Lie group of infinitesimal transformation which leaves (1-7) invariant. We find that for a basin of general form $z = Z(x, y)$ the generator of the Lie group is given by the following operator:

$$Q = V_1 \partial_t \quad + \quad V_2 \partial_x + V_3 \partial_y + G_1 \partial_u$$
$$+ \quad G_2 \partial_v + G_3 \partial_h + G_4 \partial_{u'} + G_5 \partial_{v'} + G_6 \partial_{h'} \quad (15)$$

with:

$$V_1 = \alpha \tag{16}$$

$$V_2 = \left(\left(\frac{d\alpha}{dt} + 2c_1\right)x + \left(\alpha f + 2c_2\right)y + 2\beta\right)/2 \tag{17}$$

$$V_3 = \left(\left(\frac{d\alpha}{dt} + 2c_1\right)y - \left(\alpha f + 2c_2\right)x + 2\gamma\right)/2 \tag{18}$$

$$G_1 = \left(\frac{d^2\alpha}{dt^2}x - \frac{d\alpha}{dt}u + \frac{d\alpha}{dt}yf + 2\frac{d\beta}{dt} + 2uc_1 + v\alpha f + 2vc_2\right)/2 \tag{19}$$

$$G_2 = \left(\frac{d^2\alpha}{dt^2}y - \frac{d\alpha}{dt}v - \frac{d\alpha}{dt}xf + 2\frac{d\gamma}{dt} - u\alpha f - 2uc_2 + 2vc_1\right)/2 \tag{20}$$

$$G_3 = -\left(\frac{d\alpha}{dt} - 2c_1\right)h \tag{21}$$

$$G_4 = \left(\frac{d^2\alpha}{dt^2}x - \frac{d\alpha}{dt}u' + \frac{d\alpha}{dt}yf + 2\frac{d\beta}{dt} + 2u'c_1 + v'\alpha f + 2v'c_2\right)/2 \tag{22}$$

$$G_5 = -\left(\left(v' + xf\right)\frac{d\alpha}{dt} + \alpha f + 2c_2 u' - \frac{d^2\alpha}{dt^2}y - 2\frac{d\gamma}{dt} - 2v'c_1\right)/2 \tag{23}$$

$$G_6 = -\left(\frac{d\alpha}{dt} - 2c_1\right)h' \tag{24}$$

where c_1 and c_2 are constants, whereas $\alpha = \alpha(t)$, $\beta = \beta(t)$, and $\gamma = \gamma(t)$ are functions of time, subject to the constraints:

$$\psi_x = 0 \qquad \psi_y = 0 \tag{25}$$

with:

$$\psi = 2\left[\left(\alpha f + 2c_2\right)x - \frac{d\alpha}{dt}y - 2yc_1 - 2\gamma\right]\frac{\partial Z}{\partial y}$$
$$-2\left[\left(\alpha f + 2c_2\right)y + \frac{d\alpha}{dt}x + 2xc_1 + 2\beta\right]\frac{\partial Z}{\partial x}$$
$$-\left(x^2 f^2 + y^2 f^2 + 4Z\right)\frac{d\alpha}{dt} - \left(x^2 + y^2\right)\frac{d^3\alpha}{dt^3}$$
$$-4\frac{d^2\gamma}{dt^2}y + 4\frac{d\gamma}{dt}xf - 4\frac{d^2\beta}{dt^2}x - 4\frac{d\beta}{dt}yf + 8Zc_1 \tag{26}$$

3   Non-Classical Symmetries of Burgers' Equation

The non-classical symmetry analysis of the Burgers' equation [1]:

$$u_t = u_{xx} + uu_x \tag{27}$$

consists in adding another "restriction" on u given by the invariant surface condition:

$$V_1(t, x, u)u_t + V_2(t, x, u)u_x - G(t, x, u) = 0 \tag{28}$$

which is associated with the symmetry generator, i.e.:

$$V_1 \partial_t + V_2 \partial_x + G \partial_u \tag{29}$$

By using our interactive REDUCE programs, we obtain three cases in addition to the classical one:

CASE 1

$$V_1 = 1 \qquad V_2 = -u \qquad G = 0 \tag{30}$$

CASE 2

$$V_1 = 1; \quad V_2 = (u + 2\alpha)/2; \quad G = -\left(u^3 + 2u^2\alpha - 4u\beta - 4\gamma\right)/4 \tag{31}$$

where α, β, γ are functions of (t, x) which must satisfy the Burgers-heat system [5]:

$$\frac{\partial \alpha}{\partial t} - \frac{\partial^2 \alpha}{\partial x^2} + 2\frac{\partial \alpha}{\partial x}\alpha + 2\frac{\partial \beta}{\partial x} = 0 \tag{32}$$

$$2\frac{\partial \alpha}{\partial x}\beta + \frac{\partial \beta}{\partial t} - \frac{\partial^2 \beta}{\partial x^2} - \frac{\partial \gamma}{\partial x} = 0 \tag{33}$$

$$2\frac{\partial \alpha}{\partial x}\gamma + \frac{\partial \gamma}{\partial t} - \frac{\partial^2 \gamma}{\partial x^2} = 0 \tag{34}$$

CASE 3

$$V_1 = 0 \qquad V_2 = 1 \qquad G = G(t, x, u) \tag{35}$$

where G must satisfy:

$$2GG_{xu} + G^2 G_{uu} + G_{xx} + uG_x - G_t + G^2 = 0 \tag{36}$$

4 Third Order Lie-Bäcklund Symmetries

4.1 Sine-Gordon Equation

Here we compute the third order Lie-Bäcklund symmetries of the sine-Gordon equation:

$$u_{xt} = \sin(u) \tag{37}$$

The corresponding infinitesimal generator is given by:

$$\Omega(x, t, u, u_t, u_x, u_{tt}, u_{xx}, u_{ttt}, u_{xxx})\partial_u \tag{38}$$

By using [17] we obtain:

$$\Omega = \Big(2tu_t c_4 - 2x u_x c_4 + u_t^3 c_2 + 2u_t c_5 + u_x^3 c_1$$
$$+ 2u_x c_3 + 2u_{ttt} c_2 + 2u_{xxx} c_1 \Big)/2 \tag{39}$$

where c_i ($i = 1, 2, 3, 4$) are arbitrary constants. This Lie-Bäcklund symmetry generator contains as particular cases those found by Kumei [14]. In [22] (Exercise 5.21, pag. 372), Olver noticed that the explicit dependence on the independent variables is required if one wants to generate a non-trivial conservation law of the sine-Gordon equation.

4.2 Nonlinear Reaction-Diffusion Equation

Here we find the conditions under which third order Lie-Bäcklund symmetries exist for nonlinear reaction-diffusion equation of the type:

$$u_t = [H(u)u_x]_x + J(u)u_x + K(x, t, u) \tag{40}$$

The symmetry generator is given by:

$$\Omega(x, t, u, u_x, u_{xx}, u_{xxx})\partial_u \tag{41}$$

The third order Lie-Bäcklund symmetries of (39) were found in [7] in the case where $J' = 0$. By using [17] we find that third order Lie-Bäcklund symmetries of (39) exist if H, J, K are of the following form:

$$H = 1/\Big(uA_1 + A_2\Big)^2 \tag{42}$$

$$J = \left(\left(uA_1 + A_2\right)^2 A_4 + A_3\right)/\left(uA_1 + A_2\right)^2 \tag{43}$$

$$K = -\left(2\alpha^{2/3}e^{A_3(A_4t+x)}A_3c_2 - \frac{d\,\alpha}{d\,t}uA_1 - \frac{d\,\alpha}{d\,t}A_2\right)/\left(3\alpha A_1\right) \tag{44}$$

where A_i $(i = 1,2,3,4)$ and c_2 are constants, and $\alpha = \alpha(t)$ is a function of time. The symmetry operator (40) is given by:

$$\Omega = -\Big(4\alpha^{2/3}e^{2A_3(A_4t+x)}u^5\alpha A_3 A_1^5 c_2^2 + 20\alpha^{2/3}e^{2A_3(A_4t+x)}u^4\alpha A_3 A_1^4 A_2 c_2^2$$

$$+40\alpha^{2/3}e^{2A_3(A_4t+x)}u^3\alpha A_3 A_1^3 A_2^2 c_2^2$$

$$+40\alpha^{2/3}e^{2A_3(A_4t+x)}u^2\alpha A_3 A_1^2 A_2^3 c_2^2 + 20\alpha^{2/3}e^{2tA_3 A_4 + 2xA_3}u\alpha A_3 A_1 A_2^4 c_2^2$$

$$+4\alpha^{2/3}e^{2A_3(A_4t+x)}\alpha A_3 A_2^5 c_2^2$$

$$-2\alpha e^{A_3(A_4t+x)}\frac{d\,\alpha}{d\,t}u^6 A_1^6 c_2 - 12\alpha e^{A_3(A_4t+x)}\frac{d\,\alpha}{d\,t}u^5 A_1^5 A_2 c_2$$

$$-30\alpha e^{A_3(A_4t+x)}\frac{d\,\alpha}{d\,t}u^4 A_1^4 A_2^2 c_2 - 40\alpha e^{A_3(A_4t+x)}\frac{d\,\alpha}{d\,t}u^3 A_1^3 A_2^3 c_2$$

$$-30\alpha e^{tA_3 A_4 + xA_3}\frac{d\,\alpha}{d\,t}u^2 A_1^2 A_2^4 c_2$$

$$-12\alpha e^{A_3(A_4t+x)}\frac{d\,\alpha}{d\,t}uA_1 A_2^5 c_2 - 2\alpha e^{A_3(A_4t+x)}\frac{d\,\alpha}{d\,t}A_2^6 c_2$$

$$-18\alpha e^{A_3(A_4t+x)}u^3 u_x\alpha A_3 A_1^4 c_2 - 6\alpha e^{A_3(A_4t+x)}u^3 u_{xx}\alpha A_1^4 c_2$$

$$+12\alpha e^{tA_3 A_4 + xA_3}u^2 u_x^2\alpha A_1^4 c_2 - 54\alpha e^{A_3(A_4t+x)}u^2 u_x\alpha A_3 A_1^3 A_2 c_2$$

$$-18\alpha e^{A_3(A_4t+x)}u^2 u_{xx}\alpha A_1^3 A_2 c_2 + 24\alpha e^{A_3(A_4t+x)}uu_x^2\alpha A_1^3 A_2 c_2$$

$$-54\alpha e^{A_3(A_4t+x)}uu_x\alpha A_3 A_1^2 A_2^2 c_2 - 18\alpha e^{tA_3 A_4 + xA_3}uu_{xx}\alpha A_1^2 A_2^2 c_2$$

$$+12\alpha e^{A_3(A_4t+x)}u_x^2\alpha A_1^2 A_2^2 c_2$$

$$-18\alpha e^{A_3(A_4t+x)}u_x\alpha A_3 A_1 A_2^3 c_2 - 6\alpha e^{A_3(A_4t+x)}u_{xx}\alpha A_1 A_2^3 c_2$$

$$+2e^{tA_3 A_4 + xA_3}\frac{d\,\alpha}{d\,t}u^6\alpha A_1^6 c_2 + 12e^{tA_3 A_4 + xA_3}\frac{d\,\alpha}{d\,t}u^5\alpha A_1^5 A_2 c_2$$

$$+30e^{tA_3 A_4 + xA_3}\frac{d\,\alpha}{d\,t}u^4\alpha A_1^4 A_2^2 c_2$$

$$+40e^{A_3(A_4t+x)}\frac{d\,\alpha}{d\,t}u^3\alpha A_1^3 A_2^3 c_2 + 30e^{A_3(A_4t+x)}\frac{d\,\alpha}{d\,t}u^2\alpha A_1^2 A_2^4 c_2$$

$$+12e^{A_3(A_4t+x)}\frac{\mathrm{d}\,\alpha}{\mathrm{d}\,t}u\alpha A_1A_2^5c_2 + 2e^{(A_4t+x)A_3}\frac{\mathrm{d}\,\alpha}{\mathrm{d}\,t}\alpha A_2^6c_2$$

$$+4e^{A_3(A_4t+x)}u^5\alpha\beta A_3A_1^5c_2 + 12e^{tA_3A_4+xA_3}u^4\alpha^2 A_3^2A_1^4c_2$$

$$+20e^{A_3(A_4t+x)}u^4\alpha\beta A_3A_1^4A_2c_2 + 48e^{A_3(A_4t+x)}u^3\alpha^2 A_3^2A_1^3A_2c_2$$

$$+40e^{A_3(A_4t+x)}u^3\alpha\beta A_3A_1^3A_2^2c_2 + 72e^{tA_3A_4+xA_3}u^2\alpha^2 A_3^2A_1^2A_2^2c_2$$

$$+40e^{A_3(A_4t+x)}u^2\alpha\beta A_3A_1^2A_2^3c_2 + 48e^{A_3(A_4t+x)}u\alpha^2 A_3^2A_1A_2^3c_2$$

$$+20e^{A_3(A_4t+x)}u\alpha\beta A_3A_1A_2^4c_2 + 12e^{A_3(A_4t+x)}\alpha^2 A_3^2A_2^4c_2$$

$$+4e^{tA_3A_4+xA_3}\alpha\beta A_3A_2^5c_2 - 2\sqrt[3]{\alpha}\frac{\mathrm{d}\,\alpha}{\mathrm{d}\,t}u^6\beta A_1^6$$

$$-12\sqrt[3]{\alpha}\frac{\mathrm{d}\,\alpha}{\mathrm{d}\,t}u^5\beta A_1^5A_2 - 30\sqrt[3]{\alpha}\frac{\mathrm{d}\,\alpha}{\mathrm{d}\,t}u^4\beta A_1^4A_2^2$$

$$-40\sqrt[3]{\alpha}\frac{\mathrm{d}\,\alpha}{\mathrm{d}\,t}u^3\beta A_1^3A_2^3 - 30\sqrt[3]{\alpha}\frac{\mathrm{d}\,\alpha}{\mathrm{d}\,t}u^2\beta A_1^2A_2^4$$

$$-12\sqrt[3]{\alpha}\frac{\mathrm{d}\,\alpha}{\mathrm{d}\,t}u\beta A_1A_2^5 - 2\sqrt[3]{\alpha}\frac{\mathrm{d}\,\alpha}{\mathrm{d}\,t}\beta A_2^6 + 3\sqrt[3]{\alpha}\frac{\mathrm{d}\,\beta}{\mathrm{d}\,t}u^6\alpha A_1^6$$

$$+18\sqrt[3]{\alpha}\frac{\mathrm{d}\,\beta}{\mathrm{d}\,t}u^5\alpha A_1^5A_2 + 45\sqrt[3]{\alpha}\frac{\mathrm{d}\,\beta}{\mathrm{d}\,t}u^4\alpha A_1^4A_2^2 + 60\sqrt[3]{\alpha}\frac{\mathrm{d}\,\beta}{\mathrm{d}\,t}u^3\alpha A_1^3A_2^3$$

$$+45\sqrt[3]{\alpha}\frac{\mathrm{d}\,\beta}{\mathrm{d}\,t}u^2\alpha A_1^2A_2^4 + 18\sqrt[3]{\alpha}\frac{\mathrm{d}\,\beta}{\mathrm{d}\,t}u\alpha A_1A_2^5$$

$$+3\sqrt[3]{\alpha}\frac{\mathrm{d}\,\beta}{\mathrm{d}\,t}\alpha A_2^6 - 6\sqrt[3]{\alpha}u^5u_x\alpha A_1^6c_1 - 30\sqrt[3]{\alpha}u^4u_x\alpha A_1^5A_2c_1$$

$$-6\sqrt[3]{\alpha}u^3u_x\alpha\beta A_3A_1^4 - 60\sqrt[3]{\alpha}u^3u_x\alpha A_1^4A_2^2c_1 - 6\sqrt[3]{\alpha}u^3u_{xx}\alpha\beta A_1^4$$

$$+12\sqrt[3]{\alpha}u^2u_x^2\alpha\beta A_1^4 - 12\sqrt[3]{\alpha}u^2u_x\alpha^2 A_3^2A_1^3 - 18\sqrt[3]{\alpha}u^2u_x\alpha\beta A_3A_1^3A_2$$

$$-60\sqrt[3]{\alpha}u^2u_x\alpha A_1^3A_2^3c_1 - 18\sqrt[3]{\alpha}u^2u_{xx}\alpha^2 A_3A_1^3 - 18\sqrt[3]{\alpha}u^2u_{xx}\alpha\beta A_1^3A_2$$

$$-6\sqrt[3]{\alpha}u^2u_{xxx}\alpha^2 A_1^3 + 54\sqrt[3]{\alpha}uu_x^2\alpha^2 A_3A_1^3 + 24\sqrt[3]{\alpha}uu_x^2\alpha\beta A_1^3A_2$$

$$+54\sqrt[3]{\alpha}uu_xu_{xx}\alpha^2 A_1^3 - 24\sqrt[3]{\alpha}uu_x\alpha^2 A_3^2A_1^2A_2$$

$$-18\sqrt[3]{\alpha}uu_x\alpha\beta A_3A_1^2A_2^2 - 30\sqrt[3]{\alpha}uu_x\alpha A_1^2A_2^4c_1 - 36\sqrt[3]{\alpha}uu_{xx}\alpha^2 A_3A_1^2A_2$$

$$-18\sqrt[3]{\alpha}uu_{xx}\alpha\beta A_1^2A_2^2 - 12\sqrt[3]{\alpha}uu_{xxx}\alpha^2 A_1^2A_2$$

$$-72\sqrt[3]{\alpha}u_x^3\alpha^2 A_1^3 + 54\sqrt[3]{\alpha}u_x^2\alpha^2 A_3A_1^2A_2 + 12\sqrt[3]{\alpha}u_x^2\alpha\beta A_1^2A_2^2$$

$$+54\sqrt[3]{\alpha}u_x u_{xx}\alpha^2 A_1^2 A_2 - 12\sqrt[3]{\alpha}u_x\alpha^2 A_3^2 A_1 A_2^2$$

$$-6\sqrt[3]{\alpha}u_x\alpha\beta A_3 A_1 A_2^3 - 6\sqrt[3]{\alpha}u_x\alpha A_1 A_2^5 c_1$$

$$-18\sqrt[3]{\alpha}u_{xx}\alpha^2 A_3 A_1 A_2^2 - 6\sqrt[3]{\alpha}u_{xx}\alpha\beta A_1 A_2^3$$

$$-6\sqrt[3]{\alpha}u_{xxx}\alpha^2 A_1 A_2^2\Big)/\Big(6\sqrt[3]{\alpha}\big(uA_1 + A_2\big)^5\alpha A_1\Big) \qquad (45)$$

where c_1 is a constant and $\beta = \beta(t)$ is a function of time which must satisfy:

$$2\frac{\mathrm{d}\,\alpha}{\mathrm{d}\,t}\beta - 3\frac{\mathrm{d}\,\beta}{\mathrm{d}\,t}\alpha - 6\alpha A_3 c_1 = 0 \qquad (46)$$

Bibliography

[1] W. F. Ames, *Nonlinear Partial Differential Equations in Engineering*,Vol. 2 Academic Press, New York (1972).

[2] W. F. Ames, E. Adams and R. J. Lohner, *New classes of symmetries for partial differential equations*, Int. J. Non-Linear Mech. **16** (1981) 439.

[3] W. F. Ames and M. C. Nucci, *Waves in hole enlargement: symmetry analysis*, in IMACS '91, Proceedings of the 13th World Congress on Computation and Applied Mathematics, Vol.1, Clarion Press, Dublin (1991) 343.

[4] R. L. Anderson and N. H. Ibraghimov, *Lie-Bäcklund Transformations in Applications*, SIAM, Philadelphia (1979).

[5] G. W. Bluman and J. D. Cole, *The general similarity solution of the heat equation*, J. Math. Mech. **18** (1969) 1025.

[6] G. W. Bluman and S. Kumei, *Symmetries and Differential Equations*, Springer-Verlag, Berlin (1989).

[7] P. Broadbridge and C. Rogers, *On a nonlinear reaction-diffusion boundary-value problem: application of a Lie-Bäcklund symmetry*, Preprint (1991).

[8] B. Champagne and P. Winternitz, *A MACSYMA program for calculating the symmetry group of a system of differential equations*, Report CRM-1278 (1985).

[9] B. Champagne, W. Hereman and P. Winternitz, *The computer calculation of Lie point symmetries of large systems of differential equations*, Comput. Phys. Comm. **66** (1991) 319.

[10] L. Drska, R. Liska and M. Sinor, *Two practical packages for computational physics- GCPM, RLFI*, Comput. Phys. Comm. **61** (1990) 225.

[11] J. Fitch, *Manual for Standard LISP on IBM System 360 and 370*, University of Utah, Salt Lake City (1978).

[12] A.C. Hearn, *REDUCE 3.4 User's Manual*, Rand Corp., Santa Monica (1991).

[13] P. H. M. Kersten, *Infinitesimal Symmetries: a Computational Approach*, CWI, Amsterdam (1987).

[14] S. Kumei, *Invariance transformations, invariance group transformations, and invariance groups of the sine-Gordon equation*, J. Math. Phys. **16** (1975) 2461.

[15] D. Levi, M. C. Nucci, C. Rogers, and P. Winternitz, *Group theoretical analysis of a rotating shallow liquid in a rigid container*, J. Phys. A: Math. Gen. **22** (1989) 4743.

[16] D. Levi and P. Winternitz, *Non-Classical Symmetry Reduction: Example of the Boussinesq Equation*, J. Phys. A: Math. Gen. **22** (1989) 2915.

[17] M. C. Nucci, *Interactive REDUCE programs for calculating classical, non-classical and Lie-Bäcklund symmetries of differential equations*, ©1990 M. C. Nucci, Preprint GT Math: 062090-051 (1990).

[18] M. C. Nucci, *Interactive REDUCE programs for calculating classical, non-classical, and approximate symmetries of differential equations*, in IMACS '91, Proceedings of the 13th World

Congress on Computation and Applied Mathematics, Vol.1, Clarion Press, Dublin (1991) 349.

[19] M. C. Nucci, *Symmetries of linear, C-integrable, S-integrable, and non-integrable equations,* in Proceedings of NEEDS '91, World Scientific, Singapore (1992) to appear.

[20] M. C. Nucci and W. F. Ames, *Classical and nonclassical symmetries of the Helmholtz equation,* J. Math. Anal. Appl. (1992) to appear.

[21] M. C. Nucci and P. A. Clarkson, *The nonclassical method is more general than the direct method for symmetry reductions:an example of the Fitzhugh-Nagumo equation,* Phys. Lett. A **164** (1992) 49.

[22] P. J. Olver, *Applications of Lie Groups to Differential Equations,* Springer-Verlag, Berlin (1986).

[23] L. V. Ovsjannikov, *Group Analysis of Differential Equations,* Academic Press, New York (1982).

[24] C. Rogers and W. F. Ames, *Nonlinear Boundary Value Problems in Science and Engineering,* Academic Press, New York (1989).

[25] F. Schwarz, *The Package SPDE for Determining Symmetries of Partial Differential Equations User's Manual,* Rand Corp., Santa Monica (1991).

On Stabilizing Ill–Posed Cauchy Problems for the Navier–Stokes Equations

L. E. Payne
Department of Mathematics
Cornell University
Ithaca, NY 14853

Abstract

In this paper we discuss a number of ill–posed problems that arise in attempts to solve the Navier–Stokes equations backward in time. In particular we provide criteria which are sufficient to stabilize solutions against errors in the "final" data, in the "final" time geometry, and in the spatial geometry. Other continuous dependence results will appear in a forthcoming paper.

1 Introduction

It is well known that attempts to solve a system of equations modeling an evolutionary process backward in time usually lead to mathematical problems that are not well posed. Solutions of such problems typically do not exist, and when they do these solutions do not depend continuously on the data, coefficients, or geometry; in fact they typically fail to depend continuously on any quantities which are subject to error in setting up the system of equations which model the physical process.

Differential Equations with
Applications to Mathematical
Physics

In studying such problems mathematicians usually do not concern themselves with the lack of existence. Rather they are willing to accept as a "solution", a function in an appropriately constrained subspace which sufficiently closely approximates the data and which is a "near" solution of the governing equations. If the mathematical problem is an ill–posed Cauchy problem for an evolutionary system, then the main concern in the literature has been with the question of stabilizing such an inherently unstable system against errors in the Cauchy data. Until quite recently little attention has been given to the question of stabilizing the system against errors in coefficients, geometry, etc.

To stabilize such problems against errors in the Cauchy data, it has been the custom to require not only that the socalled "solution" approximate the data well but also that it belong to some appropriately defined constraint set (see e.g. Payne [4]). It is this constraint set restriction which stabilizes the problem against errors in the Cauchy data. Any constraint set restriction should of course be realizable and as weak as practically possible. Unfortunately, this constraint restriction has the effect of making otherwise linear problems, nonlinear — a fact which complicates the total problem. To be of any practical use the constraint restriction should simultaneously stabilize the problem against all possible sources of error, and since the constrained problem is nonlinear we cannot automatically decompose the problem and treat the various sources of error separately. Nevertheless, this is usually what we do for two reasons. In the first place we cannot even characterize the errors made in setting up the model system — errors due to use of inexact physical laws, treating a fluid as a continuum, etc. Secondly, the problem itself would become so messy and complicated that it is unlikely that it could be treated even if we were able to characterize the modeling errors.

The simplest example of the type of problem we have been discussing is that of solving the heat equation backward in time. Many methods have been proposed for stabilizing this problem against errors in the Cauchy data (see e.g. the references cited in [4]). The question of continuous dependence on the spatial geometry was investigated by Crooke and Payne [1]. Another system, whose past history has been studied, is the Navier–Stokes system — the first results being those of Knops and Payne [3] who succeeded in stabilizing the problem against errors in the Cauchy data.

Since many of the important evolutionary systems we encounter in continuum physics involve the Navier–Stokes equations coupled with other equations, it is clear that if we wish to study the past history of one of these complicated systems we must first know how to stabilize the Navier–Stokes equations themselves. Thus in this paper we concentrate on some recent results on the stabilization of solutions of the Navier–Stokes equations backward in time.

As indicated earlier the first attempt at stabilizing solutions of the Navier–Stokes equations backward in time, against errors in the "final" data, was made by Knops and Payne [3], who showed that solutions of the Cauchy problem defined on a bounded region of $\mathbf{R}^3$ and appropriately constrained do depend continuously on the data (in L_2). Using a slightly different measure Payne [5] was able to relax somewhat the constraint restriction. The equivalent problem for an exterior region has been dealt with by Straughan [9] and by Galdi and Straughan [2]. In [5], Payne was able to stabilize the "backward Cauchy problem" for the Navier–Stokes equations against errors in the initial time geometry, and in [6] he succeeded in stabilizing the same problem against errors in the spatial geometry. We mention also that a constraint restriction which stabilizes this ill-posed Cauchy problem against a certain type of modeling error was found by Payne and Straughan [7], and the question of stabilizing against errors in body force, the viscosity coefficient, boundary data and another type of modeling error will be discussed by Ames and Payne in a forthcoming work.

In this paper, instead of investigating solutions of the forward Navier–Stokes equations backward in time we change the time variable t to $-t$ and study the backward Navier–Stokes equations forward in time. Our "final" value problem thus becomes an initial value problem for the backward Navier–Stokes equations. For simplicity we assume that the data and geometry are such that classical solutions exist on the indicated space–time regions, although, as pointed out in the cited references, the results which we shall state actually hold for appropriately defined weak solutions.

The specific problem we consider is the following: we are concerned with solutions $u_i(x,t)$ of

$$\left. \begin{array}{rcl} u_{i,t} - u_j u_{i,j} + \nu u_{i,jj} & = & p_{,i} \\ u_{j,j} & = & 0 \end{array} \right\} \quad \text{in } D \times (0,T) \qquad (1.1)$$

with

$$u_i(x,0) = f_i(x) \qquad (1.2)$$

and

$$u_i(x,t) = 0 \text{ on } \partial D \times [0,T]. \qquad (1.3)$$

In (1.1) and in what follows a comma denotes partial differentiation and the convention of summing over repeated indices (from 1 to 3) in any term, is adopted. In (1.1), ν is the coefficient of kinematic viscosity, p is the unknown pressure term (divided by the constant density) and D is a bounded region in $\mathbf{R}^3$ with sufficiently smooth boundary ∂D.

Let Ω designate a general domain in (x,t) space, where $x = (x_1, x_2, x_3)$. We shall define three different sets of functions $\mathcal{M}_1(\Omega)$, $\mathcal{M}_2(\Omega)$, $\mathcal{M}_3(\Omega)$.

1) A function $\psi_i(x,t)$ will be said to belong to $\mathcal{M}_1(\Omega)$ if

$$\sup_{(x,t)\in\Omega} [\psi_i \psi_i] \leq M_1^2; \qquad (1.4)$$

2) A function $\psi_i(x,t)$ will be said to belong to $\mathcal{M}_2(\Omega)$ if

$$\sup_{(x,t)\in\Omega} [\psi_i \psi_i + \psi_{i,j} \psi_{i,j}] \leq M_2^2; \qquad (1.5)$$

and

3) A function $\psi_i(x,t)$ will be said to belong to $\mathcal{M}_3(\Omega)$ if

$$\sup_{(x,t)\in\Omega} [\psi_i u_i + \psi_{i,j} \psi_{i,j} + \psi_{i,t} \psi_{i,t}] \leq M_3^2. \qquad (1.6)$$

Here M_1, M_2 and M_3 are constants which will in general depend on Ω. These sets $\mathcal{M}_1$, $\mathcal{M}_2$ and $\mathcal{M}_3$ will be used as constraint restrictions, which solutions in various cases will be required to satisfy. The appropriate restrictions will lead to different types of continuous dependence results.

2   Continuous Dependence Results

In this section we reproduce a number of continuous dependence results that have been derived for solutions of (1.1)–(1.3). It should be emphasized that the constraint restrictions imposed in order to derive these results are sufficient conditions, but they may be more stringent than necessary. It will certainly be worthwhile to try to relax these requirements.

We present first a result that was derived in [5].

2.1   Continuous Dependence on the Initial Data

Let $u_i(x,t)$ be a solution of (1.1)–(1.3) corresponding to pressure p and initial data $f_i(x)$ and $v_i(x,t)$ be a solution with pressure g and initial data $\tilde{f}_i(x)$. Then if we set

$$w_i = v_i - v_i \tag{2.1}$$

we have the following result.

Theorem 1 *Let* $u_i \in \mathcal{M}_1$ *and* $v_i \in \mathcal{M}_2$ *in* $D \times (0,T)$, *then it is possible to compute an explicit constant* K *and a function* $\delta(t)$ $(0 < \delta(t) \leq 1)$ *independent of* u_i *and* v_i *such that for* $0 \leq t < T$

$$\int_0^t \|w\|_D^2 d\eta \leq K\sigma^{2\delta(t)}, \ \delta(T) = 0 \tag{2.2}$$

where

$$\sigma^2 = \|f - \tilde{f}\|_D^2. \tag{2.3}$$

In (2.2) and (2.3) $\|\cdot\|_D$ denotes the ordinary L_2 norm in D. This clearly implies Hölder continuous dependence on the initial data. In fact, if $v_i \in \mathcal{M}_2$ denotes a smooth base flow and $u_i \in \mathcal{M}_1$ is a perturbed flow then provided f_i and $\tilde{f}_i$ are close, $u_i(x,t)$ and $v_i(x,t)$ will be close for $0 \leq t \leq t_1 < T$ in the sense indicated by (2.2). This theorem represents a slight improvement over the earlier result in [3], but numerical evidence seems to indicate that the results are very conservative, in that weaker constraint restrictions should be possible; also the exponent $\delta(t)$ seems to be smaller than necessary.

The proof of Theorem 1 makes use of the fact that the quantity $\Phi(t)$ defined by

$$\Phi(t) = \int_0^t \|w\|_0^2 d\eta + (T - t + k)\|f - \tilde{f}\|_0^2 \tag{2.4}$$

satisfies, for an appropriately chosen constant k, a logarithmic convexity inequality. The proofs of the next two theorems employ similar arguments but are considerably more complicated. A brief sketch of the proofs is given at the end of the section.

2.2   Continuous Dependence on the Initial–Time Geometry

We now wish to compare the solution of (1.1)–(1.3) with the solution $V_i(x,t)$ of

$$v_{i,t} - v_j v_{i,j} + \nu v_{i,jj} = q_{,i} \tag{2.5}$$

$$v_{j,j} = 0 \tag{2.6}$$

in the region $\Omega(F)$ defined by

$$\Omega(F) = \{(x,t); \ F(x) < t < T, \ x \in D\} \tag{2.7}$$

and satisfying

$$v_i(x, F(x)) = f_i(x) \tag{2.8}$$

$$v_i(x,t) = 0, \ F(x) \le t \le T, \ x \in D. \tag{2.9}$$

The problem (2.5)–(2.9) might arise if the initial data were measured on some surfaces $t = F(x)$ rather than at time $t = 0$. These data are, however, assigned at $t = 0$ thus leading to problem (1.1)–(1.3). Then if in particular

$$|F(x)| < \varepsilon \tag{2.10}$$

and the solutions u_i and v_i are appropriately constrained we would like to determine whether w_i given by

$$w_i = u_i - v_i \tag{2.11}$$

is small on the interval $[\varepsilon, T)$. We state now the following theorem which was proved in [5].

Theorem 2 *If $u_i(x,t) \in \mathcal{M}_1$ in $D \times (0,T)$ and $v_i(x,t) \in \mathcal{M}_2$ in $\Omega(F)$, then it is possible to compute an explicit constant $\tilde{K}$ and a function $\tilde{\delta}(t)$ $(0 < \tilde{\delta}(t) \leq 1)$ independent of u_i and v_i such that for $\varepsilon \leq t < T$*

$$\int_\varepsilon^t \|w\|_D^2 d\eta \leq \tilde{K}\varepsilon^{\tilde{\delta}(t)}. \tag{2.12}$$

This is the desired continuous dependence result.

2.3   Continuous Dependence on Spatial Geometry

Although continuous dependence on spatial geometry has received little attention in the literature it is nevertheless vitally important. In the first place when modeling a physical problem the geometry of the domain can seldom be prescribed with absolute precision. Secondly, and more importantly perhaps, if we are to have any hope of solving the problem numerically we must be able to stabilize the problem against errors in geometry since elements or meshes will seldom fit the domain exactly.

The first paper which dealt with the question of stabilizing ill posed problems against errors in spatial geometry was that of Crooke and Payne [1] who developed criteria for stabilizing the backward heat equation against geometric errors. Little else on this question has appeared in the literature (see [8]).

In this case we wish to compare $u_i^1(x,t)$ and $u_i^2(x,t)$ where $u_i^\alpha(x,t)$ satisfies, for $\alpha = 1, 2$,

$$\left. \begin{array}{r} u_{i,t}^\alpha - u_j^\alpha u_{i,j}^\alpha + \nu u_{i,jj}^\alpha = p_{,i}^\alpha \\ u_{j,j}^\alpha = 0 \end{array} \right\} \quad \text{in } D_\alpha \times (0,T) \tag{2.13}$$

with

$$u_i^\alpha = 0 \text{ on } \partial D_\alpha \times [0,T] \tag{2.14}$$

and

$$u_i^\alpha(x,0) = f_i^\alpha(x) \quad x \in D^\alpha. \tag{2.15}$$

For simplicity we assume that $f_i^1 = f_i^2$ in $D_1 \cap D_2$. The first question we are to ask is how do we compare u_i^1 and u_i^2? We could for instance map D_1 (with its corresponding problem) onto D_2 and compare the problems on D_2. Alternatively we could compare u_i^1 and u_i^2 over $D_1 \cap D_2$. A third possibility would be to extend u_i^1 and u_i^2 as zero outside their respective domains of definition and compare the extended functions over $\mathbf{R}^3$. We shall here compare u_i^1 and u_i^2 over D, where

$$D = D_1 \cap D_2. \tag{2.16}$$

In this case we have the following theorem (see [6]) valid for regions D_1 and D_2 starshaped with respect to the origin.

Theorem 3 *If $u_i^1(x,t) \in \mathcal{M}_1$ in $D_1 \times (0,T)$ and $u_i^2(x,t) \in \mathcal{M}_3$ in $D_2 \times (0,T)$ then it is possible to compute an explicit $\hat{K}$ and a function $\hat{\delta}(t)$ $(0 < \hat{\delta}(t) \leq 1)$ independent of u_i^1 and u_i^2 such that for $0 \leq t < T$*

$$\int_0^t (t - \eta)\|w\|_D^2 d\eta \leq \hat{K}\tau^{\hat{\delta}(t)} \tag{2.17}$$

where τ is the maximum distance along a ray between ∂D_1 and ∂D_2. In (2.17)

$$w_i = u_i^1 - u_i^2. \tag{2.18}$$

This is the desired continuous dependence result when D_1 and D_2 are starshaped. If they are not both starshaped with respect to a single interior point of D, but can be decomposed into starshaped subregions it is possible to derive a result similar to (2.17), but of course τ has a somewhat different interpretation.

The proof of the first theorem involves showing that the $\Phi(t)$ of (2.4) satisfies an inequality of the form

$$\Phi\Phi'' - (\Phi')^2 \geq -C_1\Phi\Phi^1 - C_1\Phi^2 \tag{2.19}$$

for explicit constants C_1 and C_2. Setting

$$\tau = e^{-C_1 t} \tag{2.20}$$

we may then rewrite (2.18) as

$$\frac{d^2}{d\tau^2}\{\ln[\Phi\tau^{C_2/C_1^2}]\} \geq 0. \tag{2.21}$$

The convexity of the term in braces leads directly to (2.2) where $\delta(t)$ is given by

$$\delta(t) = [e^{-C_1 t} - e^{-C_1 T}] / [1 - e^{-C_1 T}]. \tag{2.22}$$

The proof of Theorem 2 is obtained in two steps. Replacing $\Phi(t)$ by

$$\Psi(T) = \int_\varepsilon^t \|w\|_D^2 d\eta + (T - t + k)\|w(\varepsilon)\|_D^2 \tag{2.23}$$

and following the above procedure we conclude that

$$\int_\varepsilon^t \|w\|_D^2 d\eta \leq K\|w(\varepsilon)\|_D^{2\delta^*(t)} \tag{2.24}$$

where $\delta^*(t)$ is the appropriate modification of (2.22). We next bound $\|w(\varepsilon)\|_D^2$ by continuing u_i as f_i for $t < 0$ and v_i as f_i for $t < F(x)$. Making use of a Poincarè inequality and bounding L_2 integrals of the extended functions u_i and v_i over the time interval $(-\varepsilon, \varepsilon)$, in terms of M_1, M_2 and data we are able to bound the right hand side of (2.24) by the right hand side of (2.12). The proof of Theorem 3 is more complicated due to the fact that in this case w_i does not vanish on ∂D. We therefore subtract from w_i an appropriate auxiliary function H_i which takes the same boundary values as w_i. We then apply the convexity arguments to $w_i - H_i$. To derive (2.17) we must of course derive bounds for various norms of the auxiliary function H_i.

3 Concluding Remarks

It is of course of interest to know whether in our continuous dependence results the constant T can be taken arbitrarily large. It is easily seen that if we take T to be infinite then no solution in $\mathcal{M}_1$ (and hence in $\mathcal{M}_2$ and $\mathcal{M}_3$) can exist. This follows immediately from the fact that if u_i is such a solution then

$$\begin{aligned} \frac{d}{dt}\|u\|_D^2 &= 2\nu \int_D u_{i,j} u_{i,j} dx \\ &\geq 2\nu\lambda\|u\|_D^2 \end{aligned} \tag{3.1}$$

where λ is the appropriate Poincarè constant. But (3.1) implies that

$$\|u(t)\|_D^2 \geq \|u(0)\|_D^2 \exp(2\nu\lambda t) \tag{3.2}$$

which leads to a contradiction since $u_i u_i$ cannot remain bounded as $t \to \infty$. Thus for $T = \infty$ our theorem is vacuous. In fact, for arbitrary T we have, letting $t \to T$ in (3.2)

$$M_1^2|D| \geq \left\{ \int_D f_i f_i dx \right\} \exp(2\nu\lambda T), \tag{3.3}$$

when $|D|$ is the volume measure of D. This means that for given f_i, T, ν and D, M_1 cannot be arbitrary but, in fact, must satisfy

$$M_1^2 \geq |D|^{-1}\left\{ \int_D f_i f_i dx \right\} \exp(2\nu\lambda T). \tag{3.4}$$

The same inequality must clearly be satisfied by M_2 with f_i replaced by $\tilde{f}_i$. Another way of looking at this is that given any f_i, ν, D and M_1 with

$$M_1^2 > \max_D f_i f_i \tag{3.5}$$

then our Theorem 1 will hold only if

$$T < \frac{1}{2\nu\lambda} \ln\left\{ \frac{M_1^2|D|}{\int_D f_i f_i dx} \right\}. \tag{3.6}$$

This is a necessary but not a sufficient condition. The point to be made is that given ν, D, and f_i, then the bound M_1 and the time T cannot be chosen independently. Thus if we wish to develop a numerical scheme for finding an approximate solution of (1.1)–(1.3) we must be careful to choose the M_i's sufficiently large. There is of course a trade off since the larger the M_i the larger the constants K in the theorem and the smaller the exponent δ.

As was pointed out in [3], [5], and [6] the continuous dependence results stated in Theorems 1–3 could actually have been derived under somewhat less restrictive constraint set restrictions. The L_∞ constraints could have in some cases been replaced by L_p constraints for suitable values of p. In this case, however, it becomes somewhat more difficult to make the inequalities explicit.

Bibliography

[1] P. S. Crooke and L. E. Payne, *Continuous dependence on geometry for the backward heat equation*, Math. Meth. in Appl. Sci., **6** 1984, pp. 433–438.

[2] G. P. Galdi and B. Straughan, *Stability of solutions to the Navier–Stokes equations backward in time*, Arch. Rat. Mech. Anal., **101** 1988, pp. 107–114.

[3] R. J. Knops and L. E. Payne, *On the stability of solutions of the Navier–Stokes equations backward in time*, Arch. Rat. Mech. Anal., **29** 1968, pp. 331–335.

[4] L. E. Payne, *Improperly posed problems in partial differential equations*, Regional Conference Series in Applied Mathematics #22, SIAM, Philadelphia, 1975.

[5] L. E. Payne, *Some remarks on ill–posed problems for viscous fluids*, Int. J. Eng. Sci., to appear, 1992.

[6] L. E. Payne, *Continuous dependence on spatial geometry for solutions of the Navier–Stokes equations backward in time*, to appear.

[7] L. E. Payne and B. Straughan, *Comparison of viscous flows backward in time with small data*, Int. J. Non–Linear Mech., **24** 1989, pp. 209–214.

[8] J. Persens, *On stabilizing ill–posed problems for partial differential equations under perturbations of the geometry of the domain*, Ph.D. thesis, Cornell University, 1986.

[9] B. Straughan, *Backward uniqueness and unique continuation for solutions to the Navier–Stokes equations on an exterior domain*, J. Math. Pures et Appl., **62** 1983, pp. 49–62.

Evans' Function, Melnikov's Integral, and Solitary Wave Instabilities

Robert L. Pego [*]
Department of Mathematics
University of Maryland
College Park, MD 20742

Michael I. Weinstein [*]
Department of Mathematics
University of Michigan
Ann Arbor, MI 48109

1 Introduction

In this note we discuss recent results on:

(i) a method for detecting the eigenvalues of systems of ordinary differential equations with asymptotically constant coefficients, and

(ii) applications of this method to the detection of instabilities and transitions to instability of solitary wave solutions to equations which model long wave propagation in dispersive media.

We shall illustrate and apply this theory for solitary waves of a generalization of the Korteweg-de Vries equation (gKdV):

$$\partial_t u + \partial_x f(u) + \partial_x^3 u = 0 \;, \tag{1}$$

where $f(u) = u^{p+1}/(p+1)$, and $p \geq 1$ is real.

A more detailed discussion including applications to generalizations of a Boussinesq equation (Bou), a regularized long wave equation of Benjamin, Bona & Mahoney (BBM), and a KdV-Burgers equation can be found in [15, 14]. The technique developed to study the spectrum of the linearized operator about the solitary wave plays

---

[*]Partially supported by grants from the National Science Foundation.

Differential Equations with
Applications to Mathematical
Physics

an important role in the proof of *asymptotic* stability of solitary waves of gKdV [16, 17].

The method we use is based on the study of Evans' function $D(\lambda)$, and in particular, new formulas for the derivatives of $D(\lambda)$. $D(\lambda)$ was introduced by J. W. Evans in his study of the stability of traveling wave solutions of reaction-diffusion systems that model nerve impulse propagation [8]. In addition to discussing how unstable eigenvalues are detected using $D(\lambda)$, we discuss how:

(iii) $D(\lambda)$ detects "resonance poles". These are pole singularities of a suitably defined resolvent operator, which play a role in the mechanism of transition to instability. This mechanism is quite different from that seen in transitions to instability in finite dimensional Hamiltonian systems. Resonance poles arise in quantum scattering theory (e.g. Augur states for the helium atom [18]) and in plasma physics (Landau damping for the Vlasov-Poisson system [5, 6]).

Finally, we point out:

(iv) a connection between our expression for $D'(\lambda)$ and the *Melnikov integral* (see [10, 13]), which was introduced to study the order of splitting, under perturbation, of the stable and unstable manifolds of a homoclinic point of an autonomous system of ODE's.

2   Solitary Waves and Linearized Stability

The generalized KdV equation admits solitary wave solutions for any $c > 0$, of the form $u(x,t) = u_c(x - ct)$ where $u_c(x) = \alpha \operatorname{sech}^{2/p}(\gamma x)$ with $\alpha = (c(p+1)(p+2)/2)^{1/p}$, $\gamma = pc^{1/2}/2$. The wave profile $u_c(x)$ decays to zero exponentially as $|x| \to \infty$.

To consider the stability of such waves, we study the evolution of small perturbations of such waves, writing $u(x,t) = u_c(x - ct) + v(x - ct, t)$. Neglecting terms nonlinear in the perturbation v, the linearized evolution equation for the perturbation of the wave is

$$\partial_t v + \partial_x((u_c^p - c)v) + \partial_x^3 v = 0 . \tag{2}$$

We look for solutions of the form $v = e^{\lambda t} Y(x)$, where $\lambda \in \mathbf{C}$ and Y satisfies

$$\partial_x L_c Y = \lambda Y , \tag{3}$$

where $L_c = -\partial_x^2 + c - u_c^p$. This eigenvalue problem takes the form of an eigenvalue problem commonly associated with linear Hamiltonian systems:

$$JLY = \lambda Y ,$$

where J is skew symmetric and L is self-adjoint. If (3) admits a square integrable solution for some λ with $\text{Re } \lambda \neq 0$, we call λ an unstable eigenvalue for (3) and Y the associated eigenfunction. (By reflection symmetry, $-\lambda$ is an eigenvalue if λ is.)

Previous work ([2, 3, 4, 12, 20, 21, 22], see also [9]) has shown that u_c is nonlinearly stable in H^1 (modulo spatial translations) if

$$\frac{d}{dc}\mathcal{N}[u_c] > 0 , \qquad (4)$$

and unstable if

$$\frac{d}{dc}\mathcal{N}[u_c] < 0 . \qquad (5)$$

Here the functional $\mathcal{N}[u] = \frac{1}{2}\int_{-\infty}^{\infty} u^2 \, dx$ is a generalized momentum associated with the Hamiltonian structure of (gKdV), and is independent of time for solutions. For the particular example at hand, (4) holds if and only if $p < 4$. The stability proofs rely on establishing that u_c is a local minimizer of a conserved energy functional, subject to the constraint of fixed momentum.

Here we discuss how: *If the instability condition (5) holds, a real unstable eigenvalue exists with $\lambda > 0$. This gives rise to a non-oscillatory and exponentially growing solution of the linearized evolution equation.* The same was proved in [15], for generalizations of the BBM and Boussinesq equations. These results clarify the mechanisms for the instability proved for gKdV and gBBM in [4] and [19]; see [12] for an alternative approach to studying linear exponential instability. Our result concerning gBou seems to be the first regarding the stability or instability of the solitary waves of this equation. The methods used in the works mentioned above apparently fail to decide stability in this equation.

3 Evans' Function and Unstable Eigenvalues

The method we use to study the existence of eigenvalues for (3)
is related to the study of eigenvalues in boundary value problems
for ordinary differential operators. As $|x| \to \infty$, the coefficients in
equations (3) converge rapidly to those of the following constant
coefficient equation

$$\partial_x(-\partial_x^2 + c)Y = \lambda Y \ . \tag{6}$$

For $\text{Re}\,\lambda > 0$, this equation has solutions $Y(x, \lambda) = e^{\mu_j x}$ for
$j = 1, 2, 3$, where the μ_j, which depend on λ, satisfy

$$\text{Re}\,\mu_1(\lambda) < 0 < \text{Re}\,\mu_j(\lambda) \qquad \text{for } j = 2, 3.$$

Correspondingly, for equation (3) there is a 1-dimensional subspace of
solutions which decay as $x \to \infty$, spanned by a function $Y^+(x, \lambda)$ and
2-dimensional subspace of solutions which decay to zero as $x \to -\infty$,
spanned by functions $Y_1^-(x, \lambda)$ and $Y_2^-(x, \lambda)$. In particular, Y^+ may
be normalized so that

$$Y^+(x, \lambda) \sim e^{\mu_1 x} \qquad \text{as } x \to \infty \ . \tag{7}$$

λ is an eigenvalue when these subspaces meet nontrivially. The angle
between these subspaces may be measured by a Wronskian-like analytic function $D(\lambda)$, named Evans' function by Alexander, Gardner
& Jones [1], after J. W. Evans who pioneered its use in the study
of stability of nerve impulses [8]. In [11, 1], a geometric/topological
approach using Evans' function is developed to study the stability of
traveling waves of singularly perturbed reaction diffusion systems.

We now obtain $D(\lambda)$ in the present application to KdV. Consider,
for $\text{Re}\,\lambda > 0$, the solution $Y^+(x, \lambda)$ as $x \to \infty$. Then $D(\lambda)$ may be
defined by the relation

$$Y^+(x, \lambda) \sim D(\lambda)e^{\mu_1 x} \qquad \text{as } x \to -\infty \ . \tag{8}$$

This interpretation of $D(\lambda)$ as a *transmission* coefficient is exploited
in [23].

In equation (3), for $\text{Re}\,\lambda > 0$, if $D(\lambda)$ vanishes, then λ is an eigenvalue, and conversely. (It also then follows that $-\lambda < 0$ is an eigenvalue, though quite possibly not a zero of $D(\lambda)$.)

What about $\text{Re}\,\lambda = 0$? In this case, it turns out that $\text{Re}\,\mu_2(\lambda) = 0$, and therefore the vanishing of $D(\lambda)$ on the imaginary axis implies the existence of a solution which is exponentially decaying as $x \to +\infty$, but which is merely bounded for $x \to -\infty$ (cf. (8)). In fact one can show, by a perturbation argument about the operator in (6) that the imaginary axis is covered by essential spectrum. Using symmetry properties of gKdV we can in fact show that zeros of $D(\lambda)$ embedded in the essential spectrum are eigenvalues of (3).

Theorem 1

(i) $D(\lambda)$ is defined and is analytic in the half-plane $\Omega = \{\lambda : \text{Re}\,\lambda > -\epsilon\}$, for some $\epsilon > 0$.

(ii) If $\text{Re}\,\lambda \geq 0$, then λ is a zero of $D(\lambda)$ if and only if λ is an eigenvalue for the problem (3). The corresponding eigenfunction decays exponentially as $x \to \pm\infty$.

(iii) $D(\lambda) \to 1$ as $|\lambda| \to \infty$ in Ω.

(iv) $D(\lambda)$ is real for real λ.

It happens naturally that $D(0) = 0$ when linearizing about a traveling wave: for $\lambda = 0$ the function $Y(x) = \partial_x u_c$ satisfies (3). This follows from translation invariance in x.

The crux of our method is that we have new integral formulae for derivatives of $D(\lambda)$. A special case of these formulae which we use is:

Theorem 2 *For all $\lambda \in \Omega$,*

$$D'(\lambda) = \int_{-\infty}^{+\infty} Z^-(x,\lambda)Y^+(x,\lambda) + (\partial\mu_1/\partial\lambda)\, D(\lambda)\, dx \ . \qquad (9)$$

Here, $Y^+(x,\lambda)$ is the solution of (3) satisfying (7), which decays as $x \to +\infty$, and $Z^-(x,\lambda)$ is a solution of the adjoint of (3) which decays as $x \to -\infty$. These solutions are taken with a suitable normalization.

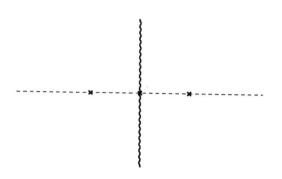

Figure 1: Spectrum of JL when instability holds. Wavy line refers to essential spectrum covering imaginary axis. x's refer to point eigenvalues.

When we apply this result at the eigenvalue $\lambda = 0$, using that $Y^+ = \partial_x u_c/(\mu_1\beta)$ and $Z^- = u_c/(2c\beta)$ for some $\beta > 0$, we find that

$$D(0) = 0, \quad D'(0) = 0 \quad , \quad \mathrm{sgn}\, D''(0) = \mathrm{sgn}\frac{d}{dc}\mathcal{N}[u_c] \ . \tag{10}$$

Thus, if $\frac{d}{dc}\mathcal{N}[u_c] < 0$, it follows that $D(\lambda) < 0$ for small $\lambda > 0$, and since $D(\lambda)$ is continuous with $D(\lambda) \to 1$ as $\lambda \to \infty$, $D(\lambda)$ must vanish for some positive λ. This yields the existence of an unstable eigenvalue for (3).

To summarize, we have

Theorem 3

(i) If $\frac{d}{dc}\mathcal{N}[u_c] > 0$ (p < 4) the spectrum of JL consists of the imaginary axis only.

(ii) If $\frac{d}{dc}\mathcal{N}[u_c] < 0$ (p > 4) the spectrum consists of the imaginary axis, together with two real isolated eigenvalues of opposite sign and equal magnitude. (See Figure 1.)

Always, $\lambda = 0$ is an eigenvalue embedded in the essential spectrum.

4 Transition to Instability; Resonance Poles to Eigenvalues

As noted in section 3, a pair of real eigenvalues appear for $p > 4$, yielding the linearized exponential instability of the solitary wave. What is the origin of these unstable eigenvalues? Is there some trace of them in the stable regime $p < 4$?

For gKdV (and other equations) we study the transition to instability by considering the Taylor expansion of $D(\lambda, p)$ in a neighborhood of the transition point $(\lambda, p) = (0, p_{crit}) = (0, 4)$. (For gKdV with a power nonlinearity the transition point does not depend on the wave speed c, due to a scaling property of the equation.)

We have:

Theorem 4 *The Taylor expansion of $D(\lambda, p)$ at $(0, p_{crit})$ is*

$$D(\lambda, p) = \lambda^2 (\frac{a}{6} \cdot \lambda + \frac{b}{2} \cdot (p - p_{crit}))(1 + O(|\lambda| + |p - p_{crit}|)) ,$$

with $a \equiv \partial_\lambda^3 D(0, p_{crit}) \neq 0$ and $b \equiv \partial_p \partial_\lambda^2 D(0, p_{crit}) \neq 0$.

Therefore, the mechanism for transition from stability to instability may be described as follows: as p varies from below p_{crit} to above p_{crit}, a real root $\lambda_0(p)$ of $D(\lambda, p) = 0$ crosses from the negative real axis $\lambda_0 < 0$ to the positive real axis $\lambda_0 > 0$, with $\lambda(p_{crit}) = 0$. λ_0 is a locally analytic function of p, and $\partial_p \lambda_0(p_{crit}) \neq 0$. Once the root λ_0 is nonnegative it is an eigenvalue of (3). Its existence implies the existence of a symmetrically placed eigenvalue of (3) at $-\lambda_0$.

In finite dimensional Hamiltonian systems the mechanism for the emergence of two real eigenvalues as a parameter varies is quite simple and standard: If r_{crit} denotes the value at which the transition from stability ($r < r_{crit}$) to instability ($r > r_{crit}$), then for $r < r_{crit}$ a pair of pure imaginary eigenvalues exists which coalesce at the origin for $r = r_{crit}$ and branch off symmetrically about the origin on the real axis for $r > r_{crit}$. See figure 2. This scenario is called an "exchange of stability" [7].

By contrast, the transition to instability here does not involve any purely imaginary eigenvalues. An interpretation of $\lambda_0(p)$ when negative is that it corresponds to what is known in quantum scattering

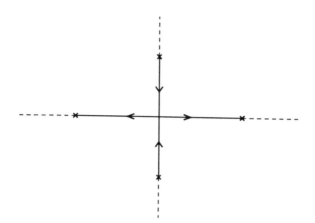

Figure 2: Transition to instability in finite dimensional Hamiltonian systems. Imaginary eigenvalues meet at origin, then branch apart along real axis.

theory as *resonance pole* [18]. The same phenomenon is associated with what is called Landau damping in the Vlasov-Poisson system of plasma physics [5, 6]. To fix ideas, consider the resolvent equation for gKdV in $L^2(R^1)$ written as

$$(JL - \lambda)u = g,$$

where $J = \partial_x$ and $L = -\partial_x^2 + c - u_c^p$. Suppose we are near the transition with $\lambda_0 < 0$, so JL has no eigenvalues off the imaginary axis. We denote the resolvent by $\mathcal{R}_1(\lambda) = (JL - \lambda)^{-1}$. For $\operatorname{Re}\lambda \neq 0$, $\mathcal{R}_1(\lambda)$ is a bounded operator on $L^2(R^1)$. Using the variation of constants formula for ODE's, one can write down an expression for the integral operator defining $\mathcal{R}_1(\lambda)g$ for $\operatorname{Re}\lambda > 0$. As $\operatorname{Re}\lambda \to 0^+$ the operator norm $\| \mathcal{R}_1(\lambda) \|$ becomes singular; the imaginary axis is the essential spectrum. However, for a dense set of $g \in L^2$, namely those which are continuous with compact support, $\mathcal{R}_1(\lambda)g(x)$ (for fixed x) can be analytically continued from the region $\operatorname{Re}\lambda > 0$ across the essential spectrum, to the region $\operatorname{Re}\lambda > -\epsilon$, for some $\epsilon > 0$. This analytic continuation exhibits a pole (called a *resonance pole*) at $\lambda_0(p) < 0$, where $D(\lambda_0) = 0$. During the transition to instability, this resonance pole moves from the negative real axis,

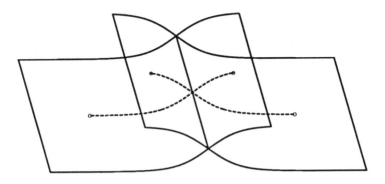

Figure 3: Riemann surface for resolvent. During transition to instability, resonance poles on upper sheet move across the imaginary axis, and become eigenvalues on lower sheet.

across the imaginary axis, to the positive real axis. The emergence of λ_0 as an eigenvalue and its symmetry related eigenvalue $-\lambda_0$, can be understood in terms of the Riemann surface of the resolvent. This is obtained by analytically continuing $\mathcal{R}_1(\lambda)g(x)$ across the imaginary axis to a second sheet over the region $-\epsilon < \operatorname{Re}\lambda < 0$. Using the reflection symmetry of the equation, the second sheet over the region $0 < \operatorname{Re}\lambda < \epsilon$ can similarly be defined [15]. See figure 3.

The arrows in figure 3 indicate the motion of resonance poles (poles of $\mathcal{R}_1(\lambda)g(x)$ on the second sheet) onto the first sheet as p varies from below p_{crit} to above p_{crit}.

5   Evans' Function and Melnikov's Integral

The order of vanishing of $D(\lambda)$ at an eigenvalue can be identified with the eigenvalue's algebraic multiplicity [8, 1, 15]. In the context of the gKdV solitary wave, zero is seen to be an eigenvalue of algebraic multiplicity two (at least). The source of this degeneracy is the translation invariance in space and the existence of solitary waves for a *continuum* of speeds $c > 0$. The integral expression for $D'(\lambda)$

in (9), when evaluated at an eigenvalue λ_0, reduces to

$$D'(\lambda_0) = \int_{-\infty}^{+\infty} Z^-(x, \lambda_0) Y^+(x, \lambda_0) dx \ , \tag{11}$$

and can be viewed as measuring the order of contact or splitting of the *stable* subspace $span\{Y^+\}$ and the *unstable* subspace $span\{Y_1^-, Y_2^-\}$ of the ODE (3) at $\lambda = \lambda_0$. If $\lambda_0 > 0$ is such that $D(\lambda_0) = 0$, and $D'(\lambda_0) \neq 0$, the intersection of these subspaces is transverse, while if $D(\lambda_0) = 0$, and $D'(\lambda_0) = 0$, as is the case for $\lambda_0 = 0$, then we say there is a tangency of the subspaces.

A more general expression for $D'(\lambda_0)$ arises when the ODE (2) is reduced to a first order system $y' = A(x, \lambda)y$ via the standard reduction $y = (Y, Y', Y'')^t$. With $y^+ = (Y^+, Y^{+'}, Y^{+''})^t$ and $z = z^-$ the solution of the adjoint system $z' = -zA(x, \lambda)$ having $z_3^- = Z^-$, we have [15]: $D(\lambda) = z^- \cdot y^+$ for all λ, and

$$D'(\lambda_0) = -\int_{-\infty}^{\infty} z^-(\partial A/\partial \lambda) y^+ \, dx \ . \tag{12}$$

The formula (12) may be regarded as an application of *Melnikov's method*, originally developed to determine the order of contact of the stable and unstable manifolds of a homoclinic point in a periodically perturbed system of autonomous ODEs. To develop the analogy, we describe Melnikov's method following [10]. (For a more general discussion of Melnikov's method and later work, see [13]).

Consider the perturbed Hamiltonian system

$$dx/dt = f(x) + \epsilon g(x, t) \ , \tag{13}$$

where $g(x, t) = g(x, t + T)$, $x \in R^2$, and we assume

$$f(x) = JH'(x) \ , \quad J = \begin{pmatrix} 0 & 1 \\ -1 & 0 \end{pmatrix} \ .$$

For $\epsilon = 0$, we presume that a hyperbolic saddle p_0 exists with a homoclinic orbit $q^0(t)$. When $\epsilon > 0$ is small, for the Poincaré map $P_\epsilon : R^2 \mapsto R^2$ determined by the flow of (13) over one period $[0, T]$, the saddle perturbs to a hyperbolic saddle p_ϵ^0, which determines a

periodic orbit $\gamma_\epsilon^0(t)$ of (13). (For Melnikov's purposes, it was important to consider Poincaré maps for intervals $[t_0, t_0 + T]$, but this is
not important here; we fix $t_0 = 0$.)

The stable and unstable manifolds of p_0, both parameterized by
$q^0(t)$, perturb to solutions $q^s(t, \epsilon)$ and $q^u(t, \epsilon)$, asymptotic to $\gamma_\epsilon^0(t)$ as
$t \to \infty$ and $-\infty$ respectively. As functions of ϵ, $q = q^s$ or q^u has the
form

$$q(t, \epsilon) = q^0(t) + \epsilon q_1(t) + O(\epsilon^2) \,,$$

uniformly on $[0, \infty)$ for q^s, $(-\infty, 0]$ for q^u, where $q_1 = q_1^s$ or q_1^u
satisfies the variational equation

$$dq_1/dt = A(t)q_1 + g(q^0(t), t) \,, \tag{14}$$

where $A(t) = f'(q^0(t))$.

Now, put $z(t) = (f_2(q^0(t)), -f_1(q^0(t))) = (J\dot{q}^0)^t$. Then $z(t)$ is
normal to the homoclinic orbit at $q^0(t)$, and in fact the row vector z
is a solution of the adjoint variational equation $dz/dt = -zA(t)$. A
measure of the separation of the stable and unstable manifolds of p_ϵ^0
is

$$d(\epsilon) = z(0) \cdot (q^u(0, \epsilon) - q^s(0, \epsilon)) = f(q^0(0)) \wedge (q^u(0, \epsilon) - q^s(0, \epsilon)) \,.$$

One has $d(0) = 0$, $d'(0) = z(0) \cdot (q_1^u(0) - q_1^s(0))$. But it follows from
(14) that one can write

$$z \cdot q_1^u(0) - z \cdot q_1^u(t_1) = \int_{t_1}^0 z(t) \cdot g(q^0(t), t) \, dt \,, \tag{15}$$

$$-z \cdot q_1^s(0) + z \cdot q_1^s(t_2) = \int_0^{t_2} z(t) \cdot g(q^0(t), t) \, dt \,. \tag{16}$$

Since $z \cdot q_1^u(t)$ and $z \cdot q_1^s(t)$ approach zero as $t \to -\infty$ and ∞ resp. (as
shown in [10]), it follows that

$$d'(0) = \int_{-\infty}^\infty z(t) \cdot g(q^0(t), t) \, dt = \int_{-\infty}^\infty f(q^0(t)) \wedge g(q^0(t), t) \, dt \,. \tag{17}$$

The analogy with (12) is that z in (17) corresponds to z^- in (12),
while $g(q^0(t), t) = \dot{q}_1 - A(t)q_1$ in (17) corresponds to $(\partial A/\partial \lambda)y^+ =
\dot{y}_\lambda^+ - A(x, \lambda_0)y_\lambda^+$ in (12), where $y_\lambda^+ = \partial y^+/\partial \lambda$. In fact, the proof of
(12) in [15] closely resembles the derivation of (17) sketched above.

Bibliography

[1] Alexander, J., Gardner, R. and Jones, C.K.R.T., *A topological invariant arising in the stability analysis of traveling waves*, J. Reine Angew. Math. **410** (1990) 167–212.

[2] Benjamin, T. B., *The stability of solitary waves*, Proc. Roy. Soc. Lond. **A328** (1972) 153–183.

[3] Bona, J. L., *On the stability of solitary waves*, Proc. R. Soc. Lond. **A344** (1975) 363–374.

[4] Bona, J. L., Souganidis, P. E. and Strauss, W. A., *Stability and instability of solitary waves*, Proc. R. Soc. Lond. **A411** (1987) 395–412.

[5] Crawford, J. D. and Hislop, P. D., *Application of the method of spectral deformation to the Vlasov Poisson system*, Annals of Physics **189** (1989) 265–317.

[6] Crawford, J. D. and Hislop, P. D., *Application of the method of spectral deformation to the Vlasov Poisson system II. Mathematical results*, J. Math. Phys. **30** (1989) 2819–2837.

[7] Drazin, P. G. and Reid, W. H., *Hydrodynamic Stability*, Cambridge University Press, 1989.

[8] Evans, J. W., *Nerve axon equations, IV: The stable and unstable impulse*, Indiana Univ. Math. J. **24** (1975) 1169–1190.

[9] Grillakis, M., Shatah, J. and Strauss, W., *Stability theory of solitary waves in the presence of symmetry, I*, J. Func. Anal **74** (1988) 160–197.

[10] Guckenheimer, J. and Holmes, P., *Nonlinear Oscillations, Dynamical Systems, and Bifurcations of Vector Fields*, Springer-Verlag, New York 1983.

[11] Jones, C.K.R.T., *Stability of the traveling wave solution to the FitzHugh-Hagumo equation*, Transactions Amer. Math. Soc. **286** (1984) 431–469.

[12] Laedke, E. W. and Spatschek, K. H., *Stability theorem for KdV type equations*, J. Plasma Phys. **32** (1984) 263–272.

[13] Palmer, K. J., *Exponential dichotomies and transversal homoclinic points*, J. Diff. Eqns. **55** (1984) 225–256.

[14] Pego, R. L., Smereka, P., and Weinstein, M. I. *Oscillatory instability of traveling waves for a KdV-Burgers equation*, in preparation.

[15] Pego, R. L. and Weinstein, M. I., *Eigenvalues, and instabilities of solitary waves*, Phil. Trans. Roy. Soc. London A340 (1992) 47–94.

[16] Pego, R. L. and Weinstein, M. I., *On asymptotic stability of solitary waves*, Phys. Lett. A **162** (1992) 263-268.

[17] Pego, R. L. and Weinstein, M. I., *Asymptotic stability of solitary waves*, in preparation.

[18] Reed, M. and Simon, B., *Methods of Modern Mathematical Physics IV: Analysis of Operators*, Academic Press, New York-San Francisco-London, 1978.

[19] Souganidis, P. E. and Strauss, W., *Instability of a class of dispersive solitary waves*, Proc. Royal Soc. Edin. **A114** (1990), 643.

[20] Weinstein, M. I., *On the solitary traveling wave of the generalized Korteweg-de Vries equation*, in *Proc. Santa Fe Conf. on Nonlinear PDE, July 1984*, B. Nicolaenko, D. Holm, J. Hyman, eds., Lectures in Appl. Math. **23**, Amer. Math. Soc., 1986.

[21] Weinstein, M. I., *Lyapunov stability of ground states of nonlinear dispersive evolution equations*, Comm. Pure Appl. Math. **39** (1986) 51–68.

[22] Weinstein, M. I., *Existence and dynamic stability of solitary wave solutions of equations arising in long wave propagation*, Comm. P.D.E. **12** (1987) 1133–1173.

[23] Yanagida, E., *Stability of the fast traveling pulse of the FitzHugh-Nagumo equations*, J. Math. Biol. **22** (1985) 81–104.

Ground States of Degenerate Quasilinear Equations

James Serrin and Henghui Zou
Department of Mathematics
University of Minnesota

1   Introduction

In 1989 Chipot and Weissler introduced the interesting quasilinear elliptic equation

$$\Delta u + u^p - |\nabla u|^q = 0, \tag{I}$$

and in particular obtained the existence of ground states when the parameter values p and q satisfy

$$1 < q < \frac{2p}{p+1}, \qquad p > 1.$$

Their study of ground states for (I) was extended to arbitrary exponents $p > 0$ and $q > 0$ in a recent paper of the present authors.

Here we shall show that these considerations can be generalized to the case of the degenerate Laplace operator, that is to the equation

$$\Delta_m u + u^p - |\nabla u|^q = 0, \tag{I$_m$}$$

where $m > 1$ and $\Delta_m u = \operatorname{div}(|\nabla u|^{m-2}\nabla u)$. The interest here lies partly in the fact that the methods for studying ground states for equation (I) do in fact extend to equation (I)$_m$, and also in the somewhat unexpected change in the results for the subcritical parameter

Differential Equations with
Applications to Mathematical
Physics

range, that is, when

$$\frac{(m-1)n}{n-m} < p < \frac{(m-1)n+m}{n-m}, \qquad q > \frac{mp}{p+1}, \qquad (m<n).$$

We recall that a ground state for (I), or for (I)$_m$, is a non-negative non-trivial entire solution. In our previous paper we proved that radial ground states for (I) always exist for the supercritical parameter range $p > (n+2)/(n-2)$ and may or may not exist when p is critical or subcritical, depending on the value of q. Finally we determined a specific bounded range of the parameters p and q, namely

$$\frac{n}{n-2} < p < \frac{n+2}{n-2}, \qquad \frac{2p}{p+1} \le q < \bar{q}(p,n),$$

where existence of ground states could neither be affirmed nor denied using the methods at hand.

Turning to the case of ground states for equation (I)$_m$, we shall show that existence always holds when p is supercritical, that is, when

$$p > l = \frac{(m-1)n+m}{n-m}.$$

For critical p, existence of radial ground states holds if and only if

$$0 < q < q_1 = (m-1)\frac{(m-1)n+m}{(m-1)n-m} \quad \text{if} \quad m > \frac{n}{n-1},$$

and for all $q > 0$ if $m \le n/(n-1)$.

Finally, for subcritical p, existence holds when

$$q < \frac{mp}{p+1},$$

while we prove non-existence of radial ground states when

$$q > \frac{mp}{p+1}, \qquad p \le l_1 = \frac{(m-1)n}{n-m}.$$

For the remaining parameter range, namely

$$q > \frac{mp}{p+1}, \qquad l_1 < p < l, \tag{1.1}$$

the results are somewhat complicated, and at the same time not complete. When $n \geq 2$ and $m < n$, we have the following non-existence ranges:

(1) $q \geq \bar{q}$, $\quad l_1 < p < \bar{p}$, $\quad m < n/(n-1)$;

(2) $q \geq \bar{q}$, $\quad l_1 < p < l$, $\quad m = n/(n-1)$;

(3) $q \geq \bar{q}$, $\quad l_1 < p \leq l$, $\quad m > n/(n-1)$;

here $\bar{q}$ ($> mp/(p+1)$) satisfies the quadratic equation

$$D(t) = [2(m-1)t + (m-2)pt - m(m-1)p]^2$$
$$- \frac{4(m-1)(n-m)}{m(n-1)}(l-p)[(p+1)t^2 - mpt] = 0$$

while $\bar{p}$ is the unique root of the equation

$$P(s) = [2(m-1) - (2-m)s]^2 - \frac{4(m-1)(n-m)}{m(n-1)}(l-s)(s+1) = 0$$

in the interval $l_1 < s < l$. When $m < n < 2$, we show that radial ground states cannot exist for the parameter range

(4) $q \geq \bar{q}$, $\quad l_1 < p \leq \bar{p}$, $\quad 4(n-m) - m(m-1)(n-2)^2 > 0$,

see Theorem C'. Only case (3) corresponds directly with the results in our earlier paper.

Regions of existence and non-existence for radial ground states in the various cases (1)-(3) are shown in Figure 1 on the following page, for particular values of m and n.

For values of (p, q) satisfying (1.1) but not covered by the cases (1)-(4) above, we have not been able to determine the existence or the non-existence of ground states. This problem certainly deserves further study.

2 Preliminary Results

In this section we consider some preliminary results for positive radial ground states $u(r)$ of $(I)_m$, where $r = |x|$ is the radius. Obviously we

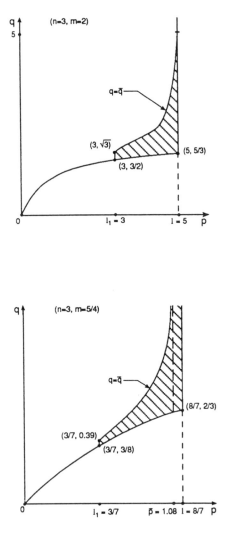

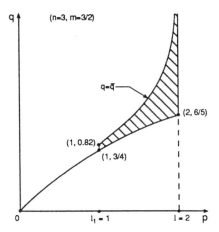

Figure 1. Existence and non-existence domains for radial non-negative ground states of equation $(I)_m$. Existence holds for (p, q) in the domain below the curve $q = mp/(p+1)$ and to the right of the critical line $p = l = [(m-1)n + m]/(n-m)$. Non-existence holds elsewhere, except in the shaded region (see discussion in the text).

Case (a) $n = 3$, $m = 2$; Case (b) $n = 3$, $m = 3/2$; Case (c) $n = 3$, $m = 5/4$.

can consider $u(r)$ as a solution, for some $\xi > 0$, of the initial value problem

$$(|u'|^{m-2}u'(r))' + \frac{n-1}{r}|u'|^{m-2}u' + u^p - |u'|^q = 0,$$
$$u(0) = \xi > 0, \qquad u'(0) = 0 \tag{IVP}_m$$

with

$$u(r) > 0 \qquad \text{for all } r > 0. \tag{I$'_m$}$$

It is also important to consider solutions of $(\text{IVP})_m$ which do not satisfy $(\text{I})'_m$. These solutions, which we continue to denote by u, thus satisfy

$$u(r) > 0 \quad \text{for } 0 \le r < R, \quad u(R) = 0. \tag{II$'_m$}$$

The local existence of C^1-solutions of $(\text{IVP})_m$ is assured by standard theory. Moreover, regularity theory shows that u is twice continuously differentiable wherever $u' \ne 0$, and can be continued so that it satisfies either $(\text{I})'_m$ or $(\text{II})'_m$. The following results are similar to those for the Laplace operator, see reference [2] for details.

Lemma 1 *Suppose $u(r)$ is a solution of $(IVP)_m$ satisfying either $(I)'_m$ or $(II)'_m$. Then $u'(r) < 0$ when $u > 0$. Moreover*

$$|u'(r)| < \left(\frac{m}{(m-1)(p+1)}\right)^{1/m} \xi^{(p+1)/m} \tag{2.1}$$

for $0 < r < R$. Here we define $R = \infty$ in case $(I)'_m$ holds.

Proof. From $(\text{IVP})_m$, the quantity $|u'|^{m-2}u'$ is decreasing whenever $u' = 0$ and $u > 0$. At $r = 0$, in particular, $(|u'|^{m-2}u')' = -\xi^p/n$. Hence $u' < 0$ for $r > 0$ near zero, and even more u' cannot return to zero as long as $u > 0$.

To obtain (2.1), we define

$$H_m(r) = \frac{m-1}{m}|u'|^m(r) + \frac{u^{p+1}}{p+1}, \tag{2.2}$$

and observe by direct computation that

$$H'_m(r) = -\frac{n-1}{r}|u'|^m(r) - |u'|^{q+1} < 0 \tag{2.3}$$

whenever $u' < 0$. Hence for $0 < r < R$ we have

$$H_m(r) < \frac{\xi^{p+1}}{p+1},$$

which obviously yields (2.1).

Lemma 2 *Let $u(r)$ be a solution of $(IVP)_m$ satisfying $(I)'_m$. Then necessarily*

$$u(r) \to 0, \quad u'(r) \to 0 \tag{2.4}$$

as $r \to \infty$.

For real α, we define the modified "energy" function (see [2])

$$G_m(r) = H_m(r) + \frac{\alpha u u' |u'|^{m-2}}{r}, \tag{2.5}$$

for which the following identity holds.

Lemma 3 *Let $u(r)$ be a solution of $(IVP)_m$ and let k be a real number. Then*

$$
(r^k G_m(r))' = r^{k-1} \left\{ \left(\frac{m-1}{m} k + \alpha - n + 1 \right) |u'|^m - r |u'|^{q+1} \right.
$$
$$
\left. + \alpha(n-k) r^{-1} u |u'|^{m-1} \left(\frac{k}{p+1} - \alpha \right) u^{p+1} + \alpha u |u'|^q \right\} \tag{2.6}
$$

for $0 < r \leq R$.

When p is subcritical, i.e., when

$$0 < p < l = \frac{(m-1)n + 2}{n - m}, \quad m < n, \tag{2.7}$$

we have the following existence result, proved exactly as in [2].

Theorem 1 *Suppose that (2.7) holds. Then equation $(I)_m$ admits infinitely many positive ground states provided that*

$$q \leq p, \quad p < m - 1 \quad or \quad q < \frac{mp}{p+1}, \quad m - 1 \leq p < l.$$

3 Existence of Ground States

Here we obtain the existence of ground states for equation $(I)_m$ in the supercritical case. Since $(I)_m$ has a radially symmetric structure, we seek in particular the existence of radial ground states. As is customary, we consider the initial value problem $(IVP)_m$ and use an ODE shooting argument to find the desired solutions.

Theorem A *(i) If*

$$p \geq l \quad and \quad 0 < q \leq \frac{m(n-1)p}{mn + (m-1)(p+1)},$$

then for each $\xi > 0$ equation $(I)_m$ has a unique positive radial ground state $u(r)$ with central value $u(0) = \xi$ (that is, $|u|_{L^\infty} = \xi$).

(ii) If

$$p > l \quad and \quad q > \frac{m(n-1)p}{mn + (m-1)(p+1)},$$

then for all sufficiently small $\xi > 0$, equation $(I)_m$ has a unique positive radial ground state $u(r)$ with central value $u(0) = \xi$.

The proof of Theorem A is almost the same as in [2], the following PPS (Pohozaev-Pucci-Serrin) type identity for solutions of $(IVP)_m$ being the main tool for the argument.

Proposition 1 *Let u be a solution of $(IVP)_m$ satisfying $(II)'_m$. Then for all real numbers α and for $k > 0$ we have*

$$
\frac{m-1}{m}|u'|^m R^k = \left(\frac{k}{p+1} - \alpha\right) \int_0^R |u|^{p+1} r^{k-1} + \alpha \int_0^R u|u'|^q r^{k-1}
$$
$$
-\alpha(k-n) \int_0^R u|u'|^{m-1} r^{k-2} - \int_0^R |u'|^{q+1} r^k
$$
$$
+ \left(\frac{m-1}{m}k + \alpha - n + 1\right) \int_0^R |u'|^m r^{k-1} \quad (3.1)
$$

In order to apply (3.1) we shall also need the following generalization of the well-known Poincaré inequality (see [2], Proposition 2).

Proposition 2 *Let $\Omega \subset \mathbf{R}^n$ be bounded with smooth boundary. Suppose that $u \in C^1(\Omega) \cap C^0(\bar{\Omega})$ with $u = 0$ on $\partial\Omega$. Then for $0 < q < k$,*

$$\int_\Omega |u||Du|^q r^{k-n} \leq \frac{q+1}{k-q} \int_\Omega |Du|^{q+1} r^{k-n+1}. \tag{3.2}$$

Finally, see [2], Lemma 4, we have

Lemma 4 *Suppose that u is a solution of $(IVP)_m$ satisfying $(II)'_m$, with $q > mp/(p+1)$. Then, provided that $\xi \leq 1$, we have*

$$\int_0^R u|u'|^q r^{n-1} \leq m\xi^s \int_0^R |u'|^m r^{n-1}, \tag{3.3}$$

where $s = [q(p+1) - mp]/2 > 0$.

Sketch of proof of Theorem A. Consider solutions of $(IVP)_m$. Our goal is to show that for appropriate initial values $\xi > 0$ a solution $u(r)$ can never reach zero at a finite value of r, i.e., $u(r)$ exists and stays positive for all $r > 0$. If this is done then the theorem will be proved, for by Lemma 2 the function $u(r)$ must tend to zero as r goes to infinity and consequently u must be a positive ground state with central value ξ.

The proof is by contradiction. If $u(r)$ reaches zero at a finite point R, then choosing

$$k = \frac{m(n-1)(p+1)}{m+(m-1)(p+1)}, \qquad \alpha = \frac{k}{p+1},$$

we obtain a direct contradiction with (3.1) and (3.2) in case (i), while in case (ii) we choose $k = n$ and $\alpha = k/(p+1)$ and use (3.1) and (3.3). For details, the reader is referred to [2].

4   A Fundamental Identity

In this section we prove an identity for solutions of $(IVP)_m$ which will be crucial for the remaining results in the paper. Let $u(r)$ be a solution of $(IVP)_m$ and $R = R(\xi)$ the first positive zero of u (if $u > 0$ for all r, then $R = \infty$). For $k, \alpha, \beta \in \mathbf{R}$, consider the function

$$Z_m(r) = r^k G_m(r) - \beta r^k u |u'|^q, \qquad 0 < r < R, \qquad (4.1)$$

where $G_m(r)$ is given by (2.5).

Lemma 5 *Suppose that $u(r)$ is a solution of $(IVP)_m$ and let*

$$k = \frac{m(n-1)(p+1)}{m + (m-1)(p+1)}, \quad \alpha = \frac{k}{p+1}, \quad \beta = \frac{m}{m + q(p+1)}. \quad (4.2)$$

Then $Z_m(r)$ satisfies the initial value problem

$$Z_m'(r) + \theta |u'|^{q-1} Z_m(r) = K(r), \qquad r \geq 0,$$
$$Z_m(0) = 0, \qquad\qquad\qquad\qquad\qquad (4.3)$$

where $\theta = \beta q (p+1)/(m-1)$ and

$$K(r) = C r^{k-2} u |u'|^{m-1} (a + bX + cX^2) \qquad (4.4)$$

with

$$
\begin{aligned}
a &= (m-1)[mn - (n-m)(p+1)], \\
b &= [2(m-1) + (m-2)p]q - m(m-1)p, \\
c &= [(p+1)q^2 - mpq]/m(n-1),
\end{aligned}
$$

and

$$
\begin{aligned}
C &= \frac{m}{m-1} \frac{n-1}{[m + (m-1)(p+1)]^2}, \\
X &= \frac{m + (m-1)(p+1)}{m + q(p+1)} r |u'|^{q+1-m}.
\end{aligned}
$$

In particular

$$Z_m(r) = e^{-\theta \int_0^r |u'|^{q+1-m} ds} \int_0^r K(s) e^{\theta \int_0^s |u'|^{q+1-m} dt} ds. \qquad (4.5)$$

Proof. Equation (4.3), with (4.4) and (4.5), was proved in [2] for $m = 2$. The proof for arbitrary $m > 1$ is based on the following identity for any $k, \alpha, \beta, \theta \in \mathbf{R}$

$$Z'_m(r) + \theta|u'|^{q+1-m} Z_m(r) =$$

$$\left(\tfrac{m-1}{m}\theta + \beta - 1\right) r^k |u'|^{q+1}$$

$$+\beta \left(\tfrac{q}{m-1} - \theta\right) r^k u|u'|^{2q+1-m}$$

$$+ \left(\tfrac{m-1}{m}k + \alpha - n + 1\right) r^{k-1}|u'|^m$$

$$+ \left(\tfrac{\theta}{p+1} - \tfrac{q\beta}{m-1}\right) r^k u^{p+1}|u'|^{q+1-m}$$

$$+ \left(\alpha(1-\theta) + \tfrac{(n-1)}{m-1}q\beta - \beta k\right) r^{k-1} u|u'|^q$$

$$+\alpha(n-k)r^{k-2}u|u'|^{m-1} + \left(\tfrac{k}{p+1} - \alpha\right) r^{k-1} u^{p+1},$$

this being a direct extension of (2.6).

5 Asymptotic Estimates

The goal of this section is to determine the asymptotic behavior at infinity of solutions of (IVP)$_m$ satisfying (I)$'_m$, as well as uniform estimates for solutions satisfying (II)$'_m$. This will be important for the results of Sections 6 and 7. Throughout the section we assume that

$$q > p, \quad 0 < p < m - 1 \quad \text{or} \quad q \geq \frac{mp}{p+1}, \quad p \geq m - 1. \quad (5.1)$$

First, we establish the uniform estimates, depending on solution only through its initial value, and valid for all $r \in (0, R)$, but at the same time depending on p, q, n and m.

Theorem 2 *Let $u(r)$ be a solution of (IVP)$_m$ satisfying (I)$'_m$ or (II)$'_m$. Suppose that (5.1) holds with $p > n(m-1)/(n-m)$. Then*

$$u \leq C_1 r^{-m/(p+1-m)}, \qquad |u'|^{q+1-m} \leq C_2 r^{-1-\epsilon} \quad (5.2)$$

for $0 < r < R$, where C_1 and C_2 are positive constants depending only on p, q, n, m and ξ, and

$$\epsilon = \frac{(p+1)q - mp}{p+1-m} \geq 0.$$

Moreover, C_1 and C_2 remain bounded as $\xi \to 0$.

The estimate $(5.2)_1$ follows directly from a series of lemmas corresponding to those for the case $m = 2$ in [2], Section 5.

To prove $(5.2)_2$, we choose $k > m(p + 1)/(p + 1 - m) > m - 1$, multiply $(\text{IVP})_m$ by r^k and integrate from 0 to r to obtain

$$
\begin{aligned}
|u'|^{m-1} r^k &= \int_0^r u^p s^k - \int_0^r |u'|^q s^k + (k - n + 1) \int_0^r |u'|^{m-1} s^{k-1} \\
&\leq \int_0^r u^p s^k + (k - n + 1) \int_0^r |u'|^{m-1} s^{k-1}.
\end{aligned}
\tag{5.3}
$$

It is clear that

$$
\int_0^r u^p s^k = O(r^{k-(m-1)p/(p+1-m)})
\tag{5.4}
$$

since $k > m(p+1)/(p+1-m)$ and $u = O(r^{-m/(p+1-m)})$. To estimate the second integral in (5.3), observe that

$$
\begin{aligned}
\int_0^r |u'|^m s^k &= -\int_0^r u'|u'|^{m-1} s^k \\
&= k \int_0^r u|u'|^{m-1} s^{k-1} + \int_0^r u(|u'|^{m-1})' s^k - u|u'|^{m-1} r^k \\
&\leq \frac{m-1}{m} \int_0^r |u'|^m s^k + \int_0^r u^{p+1} s^k + \frac{k^m}{m} \int_0^r u^m s^{k-m}
\end{aligned}
$$

by Young's inequality together with the differential equation itself. It follows that

$$
\begin{aligned}
\int_0^r |u'|^m s^k &\leq k^m \int_0^r u^m s^{k-m} + m \int_0^r u^{p+1} s^k \\
&= O(r^{k+1-m(p+1)/(p+1-m)}).
\end{aligned}
$$

Therefore

$$
\begin{aligned}
\int_0^r |u'|^{m-1} s^{k-1} &\leq \left(\int_0^r |u'|^m s^k \right)^{(m-1)/m} \left(\int_0^r s^{k-m} \right)^{1/m} \\
&= O(r^{k-(m-1)(p+1)/(p+1-m)}).
\end{aligned}
\tag{5.5}
$$

Combining (5.3), (5.4) and (5.5) immediately yields

$$
|u'|^{m-1} r^k = O(r^{k-(m-1)(p+1)/(p+1-m)}),
$$

which gives $(5.2)_2$.

Theorem 3 *Suppose that $u(r)$ is a solution of (IVP) satisfying $(I)'_m$, and that (5.1) holds. Then*

(a) If $p < m - 1$, there is no solution;

(b) If $p = m - 1$, then $u = O(e^{-r})$, $u' = O(e^{-r})$ as $r \to \infty$;

(c) If $p > m - 1$, then

$$u = O(r^{-m/(p+1-m)}), \qquad u' = O(r^{-(p+1)/(p+1-m)})$$

as $r \to \infty$.

The proof is essentially the same as for Theorem 2, cf. [2], Section 6. As a consequence of Theorem 3, we also have the following lower asymptotic estimate for ground states.

Corollary 1 *Suppose $p \geq m - 1$ and $q > mp/(p + 1)$. Then*

$$\int_0^\infty |u'|^{q+1-m} < \infty. \tag{5.6}$$

Moreover, there exists a constant $\rho > 0$ such that

$$u \geq cr^{(m-n)/(m-1)}, \qquad r \geq \rho, \tag{5.7}$$

where c is a positive constant.

Proof. (5.6) is a direct consequence of Theorem 3. To prove (5.7), we notice that from (IVP)

$$\left(|u'|^{m-2}u'(r)r^{n-1}e^{\int_0^r |u'|^{q+1-m}}\right)' = -u^p r^{n-1}e^{\int_0^r |u'|^{q+1-m}} < 0.$$

It follows that the function $|u'|^{m-2}u'(r)r^{n-1}e^{\int_0^r |u'|^{q+1-m}}$ is decreasing. In turn $|u'|^{m-2}u'(r)r^{n-1}$ tends to a negative (possibly infinite) limit by (5.6), since $u' < 0$. We then infer (5.7) by integration.

6 Existence of Ground States: The Critical Case

Here we extend the existence theory of Section 3 to the range

$$q > \frac{(m-1)n + m}{n}, \quad p = l = \frac{(m-1)n + m}{n - m}, \quad n > m. \quad (6.1)$$

The situation for the degenerate operator is somewhat different than for the Laplace operator. To be precise, it will be shown that there exist ground states for the *entire range* (6.1) when $n \leq m/(m-1)$, while for $n > m/(m-1)$, existence holds only if

$$q < q_1 = (m-1)\frac{(m-1)n + m}{(m-1)n - m}.$$

On the other hand, Theorem A (i) already gives the existence of ground states for the values

$$p = l = \frac{(m-1)n + m}{n - m}, \quad q \leq \frac{mp}{p+1} = \frac{(m-1)n + m}{n}.$$

It follows that existence holds on the critical line exactly for the range

$$0 < q < \begin{cases} q_1, & \text{if } n > m/(m-1) \\ \infty, & \text{if } n \leq m/(m-1). \end{cases} \quad (6.2)$$

Note that the first case can happen only if $n > 2$.

Theorem B *Suppose that (6.2) holds and $p = l$. Then for all sufficiently small values $\xi > 0$, depending only on q, m and n, equation $(I)_m$ has a unique positive radial ground state $u(r)$ with central value ξ.*

Theorem B is proved by combining the uniform estimate (5.2) and the identity (4.3) for $Z_m(r)$, see [2], Section 5. Indeed, setting $p = l$ in (4.4) gives

$$K(r) = \hat{C} r^{n-1} u|u'|^q (\hat{a} r |u'|^{q+1-m} - \hat{b})$$

with

$$\hat{a} = (nq - [(m-1)n + m])q,$$

$$\hat{b} = \frac{n-m}{m}([(m-1)n - m]q - (m-1)[(m-1)n + m]),$$

$$\hat{C} = \frac{1}{(m-1)(nq + n - m)}.$$

By (5.2) and Lemma 1 we have for suitably small ξ (depending on q, n and m)

$$\hat{a}r|u'|^{q+1-m} < \hat{b}, \qquad 0 < r < R,$$

which implies $K(r) < 0$ for $0 < r < R$. If $R < \infty$, it follows that $Z_m(R) < 0$. But also $u(R) = 0$, so that $Z_m(R) > 0$, a contradiction. Hence $R = \infty$ and the theorem is proved.

7 Non-existence of Ground States

If p is strictly greater than the critical exponent, there always exists at least one ground state for $(I)_m$. However, if p is subcritical we know so far only that $(I)_m$ admits ground states if $q < mp/(p+1)$.

In this section we shall use the main lemma established in Section 4 to prove non-existence for suitable pairs (p,q) in the remaining part of the subcritical region.

For convenience, we denote by l_1 and l the two critical values

$$l_1 = \frac{(m-1)n}{n-m}, \qquad l = \frac{(m-1)n + m}{n-m}, \qquad m < n.$$

As in the case $m = 2$ (see [2], Section 7), it is important to study the roots of the quadratic equation

$$\begin{aligned} D(t) &= [2(m-1)t + (m-2)pt - m(m-1)p]^2 \\ &\quad - \frac{4(m-1)(n-m)}{m(n-1)}(l-p)[(p+1)t^2 - mpt] = 0 \quad (7.1) \end{aligned}$$

for fixed values of p in $(l_1, l]$.

When $n > m/(m-1)$, so that of course also $n > 2$, the situation is simple: for $p \in (l_1, l]$ equation (7.1) has a unique root $\bar{q}$ in the interval

$$\left(\frac{mp}{p+1}, \frac{m(m-1)p}{2(m-1)+(m-2)p} \right). \tag{7.2}$$

The case $2 \le n < m/(m-1)$ is more delicate. In this case $m < 2$, and moreover there is a unique root $\bar{p}$ of the quadratic equation

$$
\begin{aligned}
P(s) \;=\;& [2(m-1)-(2-m)s]^2 \\
& -\frac{4(m-1)(n-m)}{m(n-1)}(l-s)(s+1) = 0 \tag{7.3}
\end{aligned}
$$

in the interval $l_1 < s < l$; to prove the existence of $\bar{p}$, note that

$$P(l_1) = \frac{m(m-1)}{(n-m)^2}[m(m-1)(n-2)^2 - 4(n-m)] < 0,$$

$$P(l) = [2(m-1)-(2-m)l]^2 > 0$$

and that $P(s)$ is quadratic in s.

Now consider (7.1) when $l_1 < p < \bar{p}$. Observe that $P(l_1) < 0$ and $P(l) > 0$ while $P(\bar{p}) = 0$, so $P(p) < 0$ for $l_1 < p < \bar{p}$. Since $P(p)$ is the coefficient of t^2 in $D(t)$ we now have

$$D(mp/(p+1)) > 0, \qquad D(t) \to -\infty \quad \text{as } t \to \infty.$$

Hence (7.1) has exactly one root $\bar{q}$ in the interval $(mp/(p+1), \infty)$ since $D(t)$ is quadratic in t. Clearly $\bar{p} \to l$ as $n \to m/(m-1)$ and $\bar{q} \to \infty$ as $p \to \bar{p}$.

When $n = m/(m-1)$ then $m < 2$ and $n > 2$, and

$$\frac{2(m-1)}{2-m} = l, \qquad P(l) = 0.$$

In turn, for $l_1 < p < l$ there is again exactly one root $\bar{q}$ of (7.1) in the interval (7.2).

Theorem C *Suppose $m < n$ and $n \ge 2$. Then equation $(I)_m$ admits no positive radial ground states if any of the following conditions holds:*

(i) $q > p$, $0 < p < m - 1$;

(ii) $q > mp/(p + 1)$, $m - 1 \leq p \leq l_1$;

(iii) $q \geq \bar{q}$, $l_1 < p < \bar{p}$, $m < n/(n - 1)$;

(iv) $q \geq \bar{q}$, $l_1 < p < l$, $m = n/(n - 1)$;

(v) $q \geq \bar{q}$, $l_1 < p \leq l$, $m > n/(n - 1)$.

Here $\bar{q}$ ($> mp/(p + 1)$) satisfies (7.1), while $\bar{p}$ is the unique root of (7.3) in the interval $l_1 < s < l$.

If $m < n < 2$ then $P(l) > 0$. Clearly cases (iv) and (v) cannot occur, but otherwise the results continue to apply when $P(l_1) < 0$. On the other hand, if $P(l_1) \geq 0$, that is if

$$m(m - 1)(n - 2)^2 \geq 4(n - m),$$

then equation (7.3) has no root in $(l_1, l]$. In this case part (iii) of the above theorem also does not apply, and the whole strip

$$q > \frac{mp}{p + 1}, \quad l_1 < p < l$$

is left undetermined. We state this as Theorem C'.

Theorem C' *Suppose $m < n < 2$, and let*

$$\mu = 4(n - m) - m(m - 1)(n - 2)^2.$$

If $\mu > 0$ then the first three parts of Theorem C continue to apply, while if $\mu \leq 0$ just the first two parts of Theorem C are valid.

The proof of the first two parts of Theorem C only involves the asymptotic estimates given in Section 5. To prove the remaining parts we use the identity in Section 4 and the following lower estimate for ground states.

Lemma 6 *Suppose that $u(r)$ is a solution of $(IVP)_m$ satisfying $(I)'_m$. Let*

$$k = \frac{m(n - 1)(p + 1)}{m + (m - 1)(p + 1)}, \qquad \alpha = \frac{k}{p + 1}, \qquad (7.5)$$

and assume that any of the conditions (iii), (iv) or (v) holds. Then

$$\liminf_{r \to \infty} r^k G(r) > 0. \tag{7.6}$$

The proof of this lemma and of parts (iii)-(v) of the theorem is exactly parallel to that for the case $m = 2$ in Section 7 of [2].

Remark. When $m - 1 < p \le l_1$ and $q > mp/(p + 1)$, equation $(I)_m$ does not even admit singular radial ground states, that is, nonnegative solutions of $(I)_m$ on $\mathbf{R}^n \setminus 0$ which tend to infinity at the origin. Indeed, for this range the argument only depends on the asymptotic behavior of solutions at infinity, having nothing to do with their behavior at the origin. However, when

$$q = \frac{n - m}{n - 1} \, p, \qquad 0 < p < \infty,$$

$(I)_m$ does admit singular solutions of the form $cr^{-\alpha}$.

The nonexistence results above do not cover the supercritical range $mp/(p + 1) < q < \bar{q}$ if $l_1 < p < \bar{p}$ and $mp/(p + 1) < q$ if $\bar{p} \le p < l$, since $D(t, p) > 0$ in this region and the proof does not apply. Indeed we do not know whether or not existence holds for these parameter values. However, we can show that any radial ground state with (p, q) in this range must have a suitably large central value (depending only on p, q, n and m).

Theorem 4 *Suppose that either*

$$\frac{mp}{p + 1} < q < \bar{q}, \quad l_1 < p < \bar{p} \quad or \quad \frac{mp}{p + 1} < q, \quad \bar{p} \le p < l. \tag{7.7}$$

Then there exists a constant $\xi_0 = \xi_0(p, q, n, m)$ such that if $u = u(r)$ is a radial ground state of $(I)_m$, then necessarily

$$u(0) > \xi_0(p, q, n).$$

When $n \le m$, both the critical exponents l_1 and l are infinity so that every p is subcritical. In this case the following non-existence theorem holds, a simple extension of Theorem C.

Theorem D *Suppose $n \leq m$ and*

$$q > \min \{p, \frac{mp}{p+1}\}, \quad 0 < p < \infty.$$

Then equation $(I)_m$ admits no positive radial ground states.

Theorem D involves only the first two cases of Theorem C, the proof of which only used the asymptotic behavior of solutions. Since these estimates apply for all $m > 1$ and $n > 1$, Theorem D follows directly from Theorem 3 and the corollary in Section 5.

8   Compact Support Ground States

We have concentrated on *positive* radial ground states in the earlier results of the paper. However, equation $(I)_m$ can admit *compact support* ground states, these being positive for $r < a$ $(0 < a < \infty)$ and identically zero for $r \geq a$. This section contains two results about radial compact support ground states, the first concerning existence, the second non-existence.

We first note, as shown in the following lemma, that radial compact support ground states can only exist when $q < m - 1$ and $p < l$.

Lemma 7 *Suppose that $q \geq m - 1$ or $p \geq l$. Then equation $(I)_m$ cannot admit any radial compact support ground state.*

Proof. The case $p \geq l$ follows from the proof in Section 3. In the case $q \geq m - 1$, suppose for contradiction that $(I)_m$ admits radial compact support ground states and let u be such a solution. Then there exists a finite number $R > 0$ such that

$$u(r) > 0, \quad r \in [0, R); \qquad u(R) = u'(R) = 0. \tag{8.1}$$

Obviously

$$\int_0^R |u'|^{q+1-m} < \infty, \tag{8.2}$$

since $q \geq m - 1$. Using $(IVP)_m$, (8.1) and (8.2), we have

$$|u'|^{m-2} u' r^{n-1} e^{\int_0^r |u'|^{q+1-m}} = \int_r^R u^p s^{n-1} e^{\int_0^s |u'|^{q+1-m}} \tag{8.3}$$

for $r \in (0, R)$, since $u' < 0$ by Lemma 1. Clearly this is impossible since the left side is negative, while the right positive. This completes the proof.

Theorem 5 *Suppose $p < q < mp/(p + 1)$, $0 < p < m - 1$. Then equation $(I)_m$ admits infinitely many compact support ground states.*

The proof is a combination of those for Theorem 1 and Theorem C. Indeed by Theorem C we have $R < \infty$. Then $u'(R) = 0$ if the initial value ξ is small enough (see Theorem 5 of [2]). Extending u by defining $u \equiv 0$ for $r \geq R$ gives the desired compact support ground state.

Theorem 6 *Suppose $q \geq mp/(p + 1)$ and $p > 0$. Then equation $(I)_m$ admits no radial compact support ground states.*

The proof is exactly parallel to that for Theorem 6 in [2].

Bibliography

[1] M. Chipot and F. Weissler, *Some blow up results for nonlinear evolution equations with a gradient term.* SIAM J. Math. Anal. 20 (1989), p. 886-907.

[2] J. Serrin and H. Zou, *Existence and non-existence for ground states of quasilinear elliptic equations.* Arch. Rational Mech. Anal., to appear.

Gradient Estimates, Rearrangements and Symmetries

Giorgio Talenti
Department of Mathematics
University of Florence
viale Morgagni 67A, 50134
Florence, Italy

Consider a Dirichlet boundary value problem for a second–order partial differential equation of elliptic type and suppose that estimates of the gradient of relevant solutions are in demand. By way of example, consider the following archetype

$$-\sum_{i,j=1}^{n} \frac{\partial}{\partial x_i}\left\{ a_{ij}(x)\frac{\partial u}{\partial x_j}\right\} = f(x) \quad \text{in } G, \tag{1a}$$

$$u = 0 \qquad \text{on the boundary of } G \tag{1b}$$

Here G is an open subset of euclidean n–dimensional space $\mathbb{R}^n$; coefficients a_{ij} are real–valued, measurable and bounded; ellipticity reads

$$\sum_{i,j=1}^{n} a_{ij}(x)\xi_i\xi_j \geq \xi_1^2 + \cdots + \xi_n^2 \tag{2}$$

for every x in G and every ξ from $\mathbb{R}^n$; f is real–valued and belongs to an appropriate Lebesgue space. If u is a *weak* or *variational* solution to problem (1), then the following inequality

$$\left(\int_G |\text{grad } u|^p dx \right)^{1/p} \leq C \left(\int_G |f|^q dx \right)^{1/q} \tag{3a}$$

Differential Equations with
Applications to Mathematical
Physics

holds, where

$$C = \frac{p^{-1/p}}{\sqrt{\pi}n} \left(\frac{q}{q-1}\right)^{1/q} \left\{\frac{\Gamma(n)\Gamma(n/2)}{2\Gamma(n/q)\Gamma(n-n/q)}\right\}^{1/n}, \qquad (3b)$$

$q = np/(n+p)$ and $n/(n-1) \leq p \leq 2$. The inequality is *sharp*: some coefficients a_{ij}, some domain G, and special f and u exist which render (3a) an equality. Inequality (3) is an easy corollary of the following theorem.

Theorem 1 *Consider the variational problem:*

$$\int_G |\text{grad } u|^p dx = \text{maximum},$$

where u is a weak solution to problem (1), and domain G, coefficients a_{ij}, and right-hand side f are the competing variables. Assume the measure of G is given; coefficients a_{ij} satisfy ellipticity condition (2); f belongs to $L^{2n/(n+2)}(G)$ and is equidistributed with a given function. In other words, the collection of those points x from G such that $|f(x)|$ exceeds t — a level set of f — has a prescribed measure for every nonnegative t. (There are some minor alterations if the dimension, n, is 2.) Assume $0 < p \leq 2$. Then the maximum in question is acheived when G is a ball — centered at the origin, say; $a_{ij} = \delta_{ij}$, the coefficients of Laplace operator; f is nonnegative, spherically symmetric — i.e., invariant under rotations about the origin — and radially decreasing.

Inequality (3) and Theorem 1 appeared in [6]. Related results and a bibliography are presented in [7]. Further advances are in [1].

Now let a nonnegative nonincreasing function μ be given, assume f obeys

$$\text{measure of } \{x \in \mathbb{R}^n : |f(x)| > t\} = \mu(t) \qquad (4)$$

for every nonnegative t and belongs to appropriate Lebesgue spaces, and consider the following problem

$$-\Delta u = f(x) \qquad \text{in } \mathbb{R}^n, \qquad (5a)$$

$$u(x) = o(|x|) \qquad \text{as } |x| \to \infty. \tag{5b}$$

Question: If u is a weak solution to problem (5), which f renders

$$\left(\int_{\mathbb{R}^n} |\text{grad } u|^p dx \right)^{1/p}$$

a maximum?

If $0 < p \leq 2$, Theorem 1 settles the question: a typical maximizer is the function f defined by $f \geq 0$ and

$$\{x \in \mathbb{R}^n : f(x) \geq t\} = \left\{ x \in \mathbb{R}^n : \pi^{n/2} |x|^n \leq \Gamma\left(\frac{n}{2} + 1\right) \mu(t) \right\} \tag{6}$$

for every nonnegative t. Thus, the *symmetry about a point* governs the affairs.

Studies in progress, which cannot be detailed here, indicate that the symmetry about a point definitely *breaks down if p is large*. If $p = \infty$, the question in hand is settled by the following theorem, showing that the *symmetry about a line* prevails.

Theorem 2 *If f obeys condition (4) and u is a weak solution to problem (5), then the following inequality*

$$\sup\{|\text{grad } u(x)| : x \in \mathbb{R}^n\} \leq C \int_0^\infty \mu(t)^{1/n} dt \tag{7a}$$

holds, where

$$C = \frac{1}{2} \pi^{1/(2n)-1} n^{1/n} (n-1)^{1/n-1} \Gamma\left(\frac{n}{2}\right) \left\{ \frac{\Gamma\left(\frac{1}{2n-2}\right)}{\Gamma\left(\frac{n^2}{2n-2}\right)} \right\}^{1-1/n}. \tag{7b}$$

Inequality (7) is sharp. Indeed, define f in the following way. Firstly, let

$$A = \frac{\pi^{(n-1)/2} \Gamma\left(\frac{1}{2n-2}\right)}{n(n-1)\Gamma\left(\frac{n^2}{2n-2}\right)}; \tag{8a}$$

secondly, let the absolute value of f be specified by

$$\{x \in \mathbb{R}^n : |f(x)| \geq t\} =$$
$$\left\{ x \in \mathbb{R}^n : |x_1| \leq (\mu(t)/A)^{1/n}, \right.$$
$$\left. x_2^2 + \cdots + x_n^2 \leq (\mu(t)A)^{2(1-1/n)/n} x_1^{2/n} - x_1^2 \right\} \tag{8b}$$

for every nonnegative t; thirdly, let the sign of f be specified by

$f(x)$ *is positive if* x_1 *is positive, negative otherwise.*   (8c)

Then equation (4) holds and any weak solution u to problem (5) satisfies

$$-\frac{\partial u}{\partial x_1}(0) = \text{the right-hand side of (7a).} \qquad (9)$$

Theorem 2 appears in [2], together with variants and refinements. Notice the following corollary of Theorems 1 and 2. Let E be a 3–dimensional concentration of electric charges; suppose E has a given volume V and the density of charge takes the values $+1$ and -1 only. Assertions: (i) the *total energy* of the electric field generated by E is a maximum if E is a ball and the charges have all the same sign; (ii) the *value at a given point* — the origin, say — of the electric field in question turns out to point towards a given direction — the direction $(-1, 0, \ldots, 0)$, say — and simultaneously take its largest absolute value if E is the set — symmetric about a line — defined by

$$-\left(\frac{15}{8\pi}V\right)^{1/3} \leq x_1 \leq \left(\frac{15}{8\pi}V\right)^{1/3} \text{ and } x_2^2+x_3^2 \leq \left(\frac{15}{8\pi}V\right)^{4/9} x_1^{2/3}-x_1^2,$$

moreover the positive charges are concentrated in the subset of E where $x_1 \geq 0$ and the negative charges lie in the remaining part .of E.

Proof of Theorem 2, outlined. Standard properties of harmonic functions and Poisson equation, equation (5a) and condition (5b) imply

$$\text{grad } u(x) = \frac{\Gamma\left(\frac{n}{2}\right)}{2\pi^{n/2}} \int_{\mathbb{R}^n} f(y)\frac{x-y}{|x-y|^n}\, dy, \qquad (10a)$$

in particular

$$-\frac{\partial u}{\partial x_1}(0) = \frac{\Gamma\left(\frac{n}{2}\right)}{2\pi^{n/2}} \int_{\mathbb{R}^n} f(x)\frac{x_1}{|x|^n}\, dx. \qquad (10b)$$

Let rearrangements à la Hardy and Littlewood come into play. Recall that the *decreasing rearrangement*, f^*, of f is defined by

$$f^*(s) = \sup\{t \geq 0 : \mu(t) > s\} \qquad (11)$$

if $0 \leq s < \mu(0+)$, and $f^*(s) = 0$ if $s \geq \mu(0+)$ — here μ stands for the *distribution function* of f, i.e., use is made of equation (4). (We refer to [3], [4], [5] and [7] for more information.)

Define h by

$$h(x) = \frac{1}{2}\,\pi^{-n/2}\Gamma\left(\frac{n}{2}\right)x_1|x|^{-n} \tag{12a}$$

and compute the decreasing rearrangement of h. As is easy to check,

$$h^*(s) = \frac{C}{n}s^{-1+1/n} \tag{12b}$$

for every nonnegative s — constant C is given by (7b). (Observe incidentally that h is *hamonic*.)

A theorem by Hardy and Littlewood says

$$\left|\int_{\mathbb{R}^n} f(x)h(x)dx\right| \leq \int_0^{+\infty} f^*(s)h^*(s)ds. \tag{13}$$

Formulas (10b), (12a), (12b) and (13) give

$$\left|\frac{\partial u}{\partial x_1}(0)\right| \leq \frac{C}{n}\int_0^{+\infty} f^*(s)s^{-1+1/n}ds. \tag{14a}$$

As equation (5a) and condition (5b) are invariant under translations and rotations, we conclude that

$$|\text{grad } u(x)| \leq \frac{C}{n}\int_0^{+\infty} f^*(s)s^{-1+1/n}ds \tag{14b}$$

for every x in $\mathbb{R}^n$. Inequality (7) follows, since

$$\int_0^{+\infty} f^*(s)s^{-1+1/n}ds = n\int_0^{+\infty} [\mu(t)]^{1/n}dt \tag{15}$$

thanks to (11).

Suppose f is defined by (8). An inspection shows that such a f satisfies (4). Importantly, f has the same sign as h and any level set of f is a level set of h. It can be shown that equality holds in Hardy and Littlewood inequality (13) if f and h have the same sign and any level set of the former is a level set of the latter. The last part of Theorem 2 follows. □

Bibliography

[1] A. Alvino, P. L. Lions, and G. Trombetti, *Comparison results for elliptic and parabolic equations via Schwarz symmetrization,* Ann. Inst. Henri Poincaré **7** (1990) 37–65.

[2] A. Cianchi, *Maximizing the L^∞ norm of the gradient of solutions to Poisson equation,* Journal of Geometric Anal., to appear.

[3] G. H. Hardy, J. E. Littlewood, and G. Pólya, *Inequalities,* Cambridge University Press, 1965.

[4] K. Kawohl, *Rearrangements and Convexity of Level Sets in PDE,* Lecture Notes in Math., no. 1150, Springer–Verlag, 1985.

[5] G. Pólya and G. Szegö, *Isoperimetric Inequalities in Mathematical Physics,* Princeton University Press, 1951.

[6] G. Talenti, *Elliptic equations and rearrangements,* Ann. Scuola Norm. Sup. Pisa (4) **3** (1976) 697–718.

[7] G. Talenti, *Rearrangements and partial differential equations,* Inequalities (edited by W. N. Everitt), Lecture Notes in Pure and Appl. Math., no. 129, pp. 211–230, Dekker, 1991.

Purely Nonlinear Norm Spectra and Multidimensional Solitary Waves

Henry A. Warchall
Department of Mathematics
University of North Texas
Denton, TX 76203-5116

Abstract

Semilinear elliptic equations on R^N with appropriate nonlinearities have countable sets of localized classical solutions, both spherical
and nonspherical. In explicit examples, norms of these solutions exhibit interesting and unexplained patterns. These norm spectra are
purely the result of nonlinearity, having no analog in corresponding
linearized problems. We present the examples and discuss their relation to the theory of bifurcation of solutions to semilinear elliptic
equations on bounded domains.

Interest in localized solutions to semilinear elliptic equations on
R^N stems in part from the role of such solutions as the spatial profiles
of standing solitary wave solutions to nonlinear wave equations in
$(N + 1)$ spacetime dimensions. A nonspherical solitary wave carries
nonzero (classical) angular momentum in its center-of-momentum
frame, and thus represents a "spinning" excitation. The spectra of
masses and spins of such solitary waves are determined by the spectra
of norms of solutions to the associated elliptic equations.

Differential Equations with
Applications to Mathematical
Physics

1 Introduction

The purpose of this note is to raise a question, motivated by study of model equations in physics, that could lead to new results in nonlinear spectral theory. We first state the mathematical question, then discuss its relation to bifurcation diagrams for solutions of semilinear equations on bounded domains. We conclude by presenting explicit examples of purely nonlinear norm spectra, and indicating their relationship to some quantities with physical interpretations.

2 The Question

Consider the semilinear elliptic equation

$$-\Delta v = f(|v|)\frac{v}{|v|} \qquad \text{(NLE)}$$

where $v : R^N \to C$, and $f : R \to R$ is a continuous odd function. This special kind of nonlinearity arises in some physical models and is particularly amenable to study. Here we are interested in localized classical solutions, for which $v \in C^2$ with $|v(x)| \to 0$ as $|x| \to \infty$. Conditions on f that guarantee the existence of such solutions are spelled out in [1]-[5]. Roughly, it is required that $f'(0) < 0$ and $F(s) \equiv \int_0^s f(t)\, dt > 0$ for some $s > 0$. It is known that for such f, (NLE) has infinite families of localized classical solutions, of which there are at least two types:

A. Spherically symmetric real-valued solutions $v(x) = w(|x|)$, where the function $w : [0, \infty) \to R$ satisfies the radial ordinary differential equation $w'' + \frac{N-1}{r}w' + f(w) = 0$ with $r \equiv |x|$. Generically, there is such a radial solution with each prescribed number of nodes. (If r is interpreted as time, these solutions may be visualized as describing one-dimensional motion in a potential well F with time-dependent damping.)

B. Nonspherical complex-valued solutions, constructed as follows. If N is even $(N = 2n)$, group the coordinates of $x \in R^N$ into n pairs: $(x_1, x_2), (x_3, x_4), \ldots, (x_{2n-1}, x_{2n})$. If N is odd $(N = 2n + 1)$, group the first $N - 1$ coordinates into n pairs $(x_1, x_2), (x_3, x_4), \ldots,$

(x_{2n-1}, x_{2n}), and let $z \equiv x_{2n+1}$. Let r_j and θ_j be polar coordinates for the pair (x_{2j-1}, x_{2j}). Then we can find solutions of the form $v(x) = e^{i(m_1\theta_1 + \cdots + m_n\theta_n)} w(r_1, \ldots, r_n [, z])$, where $m_1, \ldots, m_n$ are integers; the function w satisfies the reduced equation

$$-\Delta w + \left(\frac{m_1^2}{r_1^2} + \cdots + \frac{m_n^2}{r_n^2} \right) w = f(|w|) \frac{w}{|w|}.$$

For example, in $N = 2$ dimensions, $v(r, \theta) = e^{im\theta} w(r)$, where $w : [0, \infty) \to R$ satisfies $w'' + \frac{1}{r}w' - \frac{m^2}{r^2}w + f(w) = 0$. Again, generically there is a solution for each number of nodes.

Each of these countable infinity of distinct localized solutions v has $\|v\|_2$ and $\|\nabla v\|_2$ finite. The set of these solution norms, which might be regarded as a "spectrum," is a signature of the nonlinearity. Note that these numbers arise naturally; no eigenvalue parameter is inserted by hand. Furthermore, there is no linear analog: the corresponding linearized autonomous equation on R^N has no nontrivial solutions with finite L^2 norm; it would require non-autonomous terms or boundary conditions on a bounded domain (both of which serve to set the scale by hand) to generate a discrete structure to the solution set.

We will call the set of solution norms the "purely nonlinear spectrum" of the nonlinearity. This set appears to be a natural mathematical object that reflects the interplay between the Laplacian and the nonlinearity f. Natural questions about the nonlinear norm spectrum include: Given an appropriate semilinearity f, what are the properties of the norm spectrum? For example, does knowledge of the general properties of the nonlinearity determine the asymptotics of the norm spectrum? Conversely, given the norm spectrum, what about f is determined?

3 Relation to Bifurcation Diagrams for Problems on Bounded Domains

Since relatively more is known about semilinear elliptic equations on bounded domains than is known about such equations on R^N, we reformulate the purely nonlinear norm spectrum in terms of quantities associated with problems on bounded domains. Let B be a ball of radius one in R^N. We define $g : C \to C$ by $g(v) \equiv f(|v|)\frac{v}{|v|}$, and consider the "standard problem"

$$\begin{cases} -\Delta u_\lambda = \lambda\, g(u_\lambda) & \text{in } B \\ u_\lambda = 0 & \text{on } \partial B \end{cases}$$

parametrized by real parameter λ.

We will compare solutions to this standard problem with those to the "unbounded problem"

$$\begin{cases} -\Delta v = f(v) & \text{on } R^N \\ v \to 0 & \text{as } |x| \to \infty \end{cases}$$

Given a nontrivial branch of solutions $\{u_\lambda \mid \lambda \in (0, \infty)\}$ to the standard problem, set $w_\rho(x) \equiv u_\rho\left(\frac{x}{\sqrt{\rho}}\right)$. Then w_ρ satisfies $-\Delta w_\rho = f(w_\rho)$, and $w_\rho(x) = 0$ for $|x| = \sqrt{\rho}$. It follows that, as $\rho \to \infty$, w_ρ converges to a nontrivial solution v of the unbounded problem.

Since $\|w_\rho\|_2 = \rho^{N/4}\|u_\rho\|_2$, we have $\lim_{\rho \to \infty} \rho^{N/4}\|u_\rho\|_2 = \|v\|_2$, with a similar result for the norm of the gradient. Thus the norms of solutions to the unbounded problem determine the asymptotic behavior of the bifurcation diagram for the standard problem, as indicated in Figure 1.

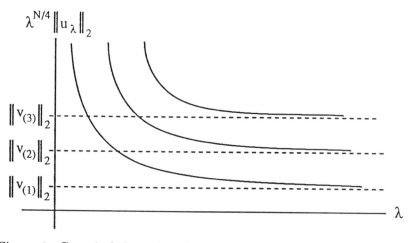

Figure 1. Generic bifurcation diagram for the standard problem.

That is, the purely nonlinear norm spectrum is determined by the set of asymptotic values of $\lambda^{N/4} \|u_\lambda\|$ for the standard problem (on a fixed bounded domain) in the limit $\lambda \to \infty$.

4 Example

To allow explicit computation of solutions, we may take f to be a piecewise linear function. For specificity we take $f = f_\sigma$ with f_σ as shown in Figure 2.

We may compute explicitly solutions of both types mentioned above.

A. Real-valued spherically symmetric solutions. Reference [6] gives an explicit construction of all spherically symmetric solutions in $N = 3$ spatial dimensions. (Similar results hold for any $N > 1$.) Substitution of the radial ansatz $v(x) = w(|x|)$ into (NLE) gives the ordinary differential equation $-w'' - \frac{2}{r}w' = f_\sigma(w)$ for w. Since f_σ is piecewise linear, we have explicit solutions in each w-amplitude region:

$$-1 \le w \le 1: \quad w'' + \frac{2}{r}w' = \sigma^2 w \quad \Rightarrow \quad w(r) = \frac{a}{r}e^{\sigma r} + \frac{b}{r}e^{-\sigma r}$$

$$\left.\begin{array}{c} w \ge 1 \\ w \le -1 \end{array}\right\} : \quad w'' + \frac{2}{r}w' = -w \pm \left(1 + \sigma^2\right)$$

$$\Rightarrow \quad w(r) = \frac{c}{r}\sin r + \frac{d}{r}\cos r \pm \left(1 + \sigma^2\right)$$

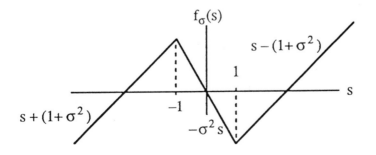

Figure 2. Piecewise linear nonlinearity

The unknown parameters that characterize a solution are the values of coefficients a, b, c, d in various amplitude regions, and the values R of radii at which the solution crosses amplitude threshold ± 1. We can determine these unknowns from the transcendental equations that insist w is continuously differentiable [then w is also C^2 by virtue of the ordinary differential equation] and w is localized. In particular, we match the values of w and of w' across amplitude threshold crossings, and impose appropriate limits at $r = 0$ and at $r = \infty$. A typical solution is shown schematically in Figure 3.

We find, in explicit form, all spherical solutions with these piecewise linear nonlinearities; they are indexed by number of nodes in the radial profile.

B. Complex-valued nonspherical solutions. Reference [7] gives an explicit construction of nonspherical solutions in dimension $N = 2$. Substitution of the ansatz $v(r, \theta) \equiv e^{i m \theta} w(r)$ into (NLE) gives the ordinary differential equation $-w'' - \frac{1}{r} w' + \frac{m^2}{r^2} w = f_\sigma(w)$ for w. Again we have explicit solutions in each w-amplitude region:

$$-1 \le w \le 1 : w'' + \frac{1}{r} w' - \frac{m^2}{r^2} w = \sigma^2 w \Rightarrow w(r) = a\, I_m(\sigma\, r) + b\, K_m(\sigma\, r)$$

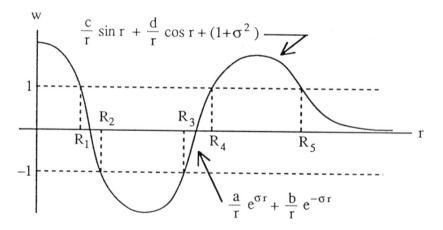

Figure 3. Patching together local solutions to form solution w.

(Here I_m and K_m are modified Bessel functions.)

$$\left.\begin{array}{c} w \geq 1 \\ w \leq -1 \end{array}\right\} : \quad w'' + \frac{1}{r}w' - \frac{m^2}{r^2}w = -w \pm \left(1 + \sigma^2\right) \Rightarrow$$

$$w(r) = c\,J_m(r) + d\,Y_m(r) \pm F_m(r)$$

Here J_m and Y_m are Bessel functions, and F_m is a particular solution. Specifically, $F_0(r) = 1 + \sigma^2$, and $F_1(r) = \frac{\pi}{2}\left(1 + \sigma^2\right) H_1(r)$, where H_1 is a Struve function; for general m, F_m is built from Bessel and Struve functions.

Unknowns are again the values of coefficients a, b, c, d in various amplitude regions, and the values R of radii at which the solution crosses amplitude threshold ± 1. These are determined by the transcendental equations that insist w is continuously differentiable and localized. Solutions are parametrized by σ, m, and "excitation number" n.

The norms $\|v\|_2$ and $\|\nabla v\|_2$ for forty-two solutions in dimension $N = 2$ for a fixed value of σ are shown in Figures 4 through 7. We observe that, for fixed excitation number n, the quantities $\|v\|_2^2$ and $\|\nabla v\|_2^2$ are each approximately linear in spin m. Careful analysis shows, however, that the relationship is not exactly linear. Similarly, for fixed spin m, the quantities $\|v\|_2^2$ and $\|\nabla v\|_2^2$ are each approximately quadratic in excitation number n, but not exactly quadratic. The precise behavior of these norm spectra and their relationship to f_σ is not currently known.

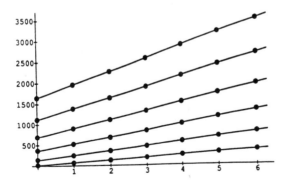

Figure 4. $\|v\|_2^2$ versus m for $1 \leq n \leq 6$. Lines are least-squares linear fits through families with fixed n.

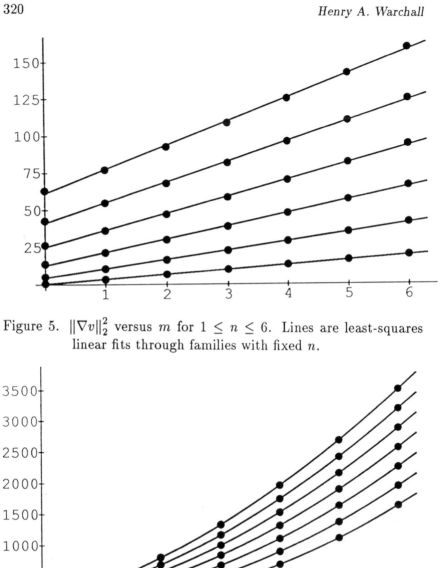

Figure 5. $\|\nabla v\|_2^2$ versus m for $1 \leq n \leq 6$. Lines are least-squares linear fits through families with fixed n.

Figure 6. $\|v\|_2^2$ versus n for $0 \leq m \leq 6$. Curves are least-squares quadratic fits through families with fixed m.

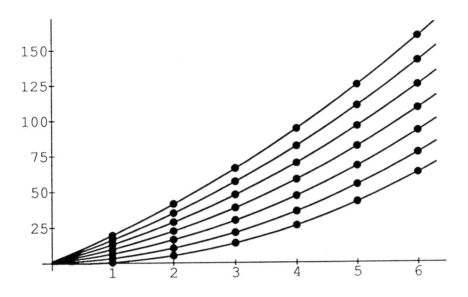

Figure 7. $\|\nabla v\|_2^2$ versus n for $0 \le m \le 6$. Curves are least-squares quadratic fits through families with fixed m.

5 Quantities of Physical Interest

Elliptic equations of the type considered here govern the spatial profiles of stationary multidimensional solitary wave solutions to nonlinear wave equations of the form $u_{tt} - \Delta u = g(u)$ where $u : R^{N+1} \to C$ and $g : C \to C$ with $g(r\,e^{i\theta}) = g(r)e^{i\theta}$. Special solutions of the form $u(x,t) = e^{i\omega t} v(x)$ with ω real and $v : R^N \to C$ are called standing waves. Substitution of this standing wave form for u into the nonlinear wave equation gives the semilinear elliptic equation $-\Delta v = f_\omega(|v|) \frac{v}{|v|}$ for v, where $f_\omega(s) \equiv g(s) + \omega^2 s$.

An interesting class of solitary wave solutions to the nonlinear wave equation is obtained by applying Lorentz boosts to standing waves that are localized (for which $|v(x)| \to 0$ as $|x| \to \infty$). Thus the search for solitary wave solutions of nonlinear wave equations motivates the study of localized solutions to nonlinear elliptic equations on R^N.

Among the conserved quantities for solutions of the nonlinear wave equation are energy

$$E[u] \equiv \int_{R^N} \left\{ \tfrac{1}{2} |u_t|^2 + \tfrac{1}{2} |\nabla u|^2 - H(|u|) \right\} dx,$$

charge $Q[u] \equiv \int_{R^N} \mathrm{Im}\,[-\bar{u}_t u]\, dx$, and angular momentum $L[u] \equiv \int_{R^N} \left\{ x \times \mathrm{Re}\,[-\bar{u}_t \nabla u] \right\} dx$. The values of these functionals evaluated on standing wave solutions are interpreted as the (rest-frame) mass, charge, and spin, respectively, of the associated solitary wave. Explicitly, these physically meaningful quantities are related to the purely nonlinear norm spectrum of the elliptic equation as follows. The solitary wave mass is $E = \omega^2 \|v\|_2^2 + \tfrac{1}{2} \|\nabla v\|_2^2$. The solitary wave charge is $Q = \omega \|v\|_2^2$. The solitary wave spin is $|L| = |m\omega| \|v\|_2^2$. (We note that nonspherical solitary waves necessarily carry nonzero angular momentum in the rest frame.)

The purely nonlinear norm spectrum of the elliptic equation is thus reflected in the discrete set of allowed values of solitary-wave masses, charges, and spins. Thorough understanding of the relationship between the nonlinearity f and its norm spectrum will shed light on the relationship between the dynamics of solitary waves and the structure of the solitary wave family in nonlinear wave equations.

Bibliography

[1] H. Berestycki & P.-L. Lions, *Nonlinear scalar field equations, I and II.* Arch. Rat. Mech. Anal. 82 (1983) 313-375.

[2] M. Berger, *Nonlinearity and Functional Analysis.* Academic Press, New York (1977).

[3] C. Jones & T. Küpper, *On the infinitely many solutions of a semilinear elliptic equation.* SIAM J. Math. Anal. 17 (1986) 803-835.

[4] P.-L. Lions, *Solutions complexes d'equations elliptiques semilineaires dans R^N.* C.R. Acad. Sc. Paris 302 (#19) (1986) 673-676.

[5] W. Strauss, *Existence of solitary waves in higher dimensions.* Comm. Math. Phys. 55 (1977) 149-162.

[6] E. Deumens & H. Warchall, *Explicit construction of all spherically symmetric solitary waves for a nonlinear wave equation in multiple dimensions.* Nonlin. Anal. T. M. A. 12 (1988) 419-447.

[7] G. King, *Explicit multidimensional solitary waves.* University of North Texas Master's Thesis, August 1990.

On Gelfand-Dickey Systems and Inelastic Solitons

Rudi Weikard
Department of Mathematics
University of Alabama at Birmingham
Birmingham, AL 35294, USA

1 Introduction

The Boussinesq equation

$$u_{tt} = (u^2)_{xx} + u_{xxxx}$$

is known to have so called N-soliton solutions, i.e., solutions that exhibit asymptotically (as $t \to \pm\infty$) N solitary waves of the typical sech^2-form (see Hirota [5]). Here I am mainly interested in (a scaled version of) the Boussinesq equation in imaginary time, specifically,

$$u_{tt} = -\frac{2}{3}(u^2)_{xx} - \frac{1}{3}u_{xxxx}. \tag{1}$$

This equation renders "inelastic solitons", i.e., solitary waves of the sech^2-form which may stick together after interaction thus forming a new sech^2-wave (see Figure 1).

These inelastic solitons can be obtained via an auto-Bäcklund transformation for the Gelfand-Dickey system associated with the

Differential Equations with
Applications to Mathematical
Physics

Boussinesq-type equation.* In the following I will define what Gelfand-Dickey systems and their "modified" counterparts, the Drinfeld-Sokolov systems, are. Section 2 then reviews the above mentioned auto-Bäcklund transformation (see [3] and [4]). Section 3 describes briefly (details will appear elsewhere) how the inelastic solitons are constructed.

Gelfand-Dickey systems are most easily defined in terms of Lax pairs. By a Lax pair is meant a pair of two ordinary differential expressions

$$
\begin{aligned}
L &= \partial_x^n + q_{n-2}\partial_x^{n-2} + \dots + q_0, \\
P &= \partial_x^r + p_{r-2}\partial_x^{r-2} + \dots + p_0,
\end{aligned}
$$

which are almost commuting, i.e., their commutator $[P, L]$ is a differential expression of order $n - 2$ only. Under an additional homogeneity condition it is always possible to find uniquely coefficients p_j, $j = 0, ..., r - 2$ such that this holds (Wilson [7]). This distinguishes between n and r and causes the two operators to play very different roles. The Lax equation

$$
\frac{dL}{dt} = [P, L]
$$

is then equivalent to a system of nonlinear evolution equations which is called a Gelfand-Dickey system. In particular the well-known KdV equation is recovered in the case $n = 2$ and $r = 3$, while the case $n = 3$ and $r = 2$ yields the Boussinesq-type equation (1).

An important ingredient in the construction of the auto-Bäcklund transformation is another system of evolution equations, the Drinfeld-Sokolov system which is defined as follows: Given functions $\phi_i(x, t)$, $i = 1, ..., n$ such that their sum is identically equal to zero,

---

*As I realized only after finishing this work these solutions were obtained earlier by Tajiri and Nishitani (*J. Phys. Soc. J.*, 51:3720–3723, 1982) and by Lambert, Musette and Kesteloot (*Inv. Prob.*, 3:275–288, 1987) using different methods. However, the construction of these solutions in the present context should be viewed as an illustration of how the auto-Bäcklund transformation of Section 2 works.

construct the matrix

$$
M = \begin{pmatrix}
0 & \cdots & 0 & \partial_x + \phi_n \\
\partial_x + \phi_1 & \cdots & 0 & 0 \\
0 & \cdots & 0 & 0 \\
\vdots & \ddots & \vdots & \vdots \\
0 & \cdots & \partial_x + \phi_{n-1} & 0
\end{pmatrix}.
$$

Then $M^n = \operatorname{diag}(L_1, ..., L_n)$ where each L_j has the form of the above L:

$$
\begin{aligned}
L_j &= (\partial_x + \phi_{n+j-1})...(\partial_x + \phi_j) \\
&= \partial_x^n + q_{j,n-2}\partial_x^{n-2} + ... + q_{j,1}\partial_x + q_{j,0}.
\end{aligned}
\tag{2}
$$

Note that L_j is obtained from L_{j-1} by commuting the first $n-1$ factors with the last one. This basic idea of commutation goes back to Darboux and was used by Deift [1] to construct the N-soliton solution of the KdV equation.

Now let $Q = \operatorname{diag}(P_1, P_2, ..., P_n)$ where P_j, $j = 1, ..., n$ is the uniquely defined differential expression of order r that almost commutes with L_j. Then

$$
\frac{dM}{dt} = [Q, M]
$$

is equivalent to a system of $n-1$ nonlinear evolution equations, called the Drinfeld-Sokolov system or modified Gelfand-Dickey system.

2   An Auto-Bäcklund Transformation

Given a solution of the Drinfeld-Sokolov system, i.e., a set of ϕ_j, $j = 1, ..., n$ such that $dM/dt = [Q, M]$ then it is easy to see that this implies $d(M^n)/dt = [Q, M^n]$, which is equivalent to

$$
\frac{dL_j}{dt} = [P_j, L_j], \quad j = 1, ..., n.
$$

This means one has found n solutions of the associated Gelfand-Dickey system. This observation is due to Sokolov and Shabat [6]. Now the following question arises: Is it possible to reverse this process and to construct a solution ϕ_j, $j = 1, ..., n$ of the Drinfeld-Sokolov system given a solution of the Gelfand-Dickey system? If so then one has immediately $n - 1$ new solutions of the Gelfand-Dickey system. It is precisely this question which was answered affirmatively by Gesztesy and Simon in [2] in the case of the KdV equation and by Gesztesy, Race and myself in [3] in the case of a Boussinesq-type equation.

The answer in the general case was given in [4]. The method there allows the coefficients of L to be matrices with entries in some commutative algebra with two independent derivations. For simplicity, however, I give in the following the scalar version using just functions of x and t as coefficients of L.

Theorem 1 (Gesztesy, Race, Unterkofler, W.) *Suppose that* $(q_{n-2}, ..., q_0)$ *is a real-valued solution of the Gelfand-Dickey system. Also assume that the* q_i *and their* x*-derivatives up to order* $r + i$ *are continuous functions in* $\mathbb{R}^2$. *Let* $\psi_1, ..., \psi_n$ *be a fundamental system of solutions of* $L\psi = 0$ *and* $\psi_t = P\psi$ *and define* $\phi_1, ..., \phi_n$ *according to*

$$\phi_k = -\frac{\partial}{\partial x} \log \left| \frac{W_k}{W_{k-1}} \right|$$

where $W_0 = 1$ *and* $W_k = W(\psi_1, ..., \psi_k)$, *the Wronskian of* $\psi_1, ..., \psi_k$ *for* $k = 1, ..., n$. *(Note that this implies that* $\sum_{k=1}^{n} \phi_k = 0$.) *Then* $(\phi_1, ..., \phi_n)$ *satisfies the Drinfeld-Sokolov system. Furthermore define* $q_{k,i}$, $k = 1, ..., n$, $i = 0, ..., n - 2$ *through (2). Then each tuple* $(q_{k,n-2}, ..., q_{k,0})$ *satisfies the Gelfand-Dickey system. In particular* $(q_{1,n-2}, ..., q_{1,0}) = (q_{n-2}, ..., q_0)$.

One can allow for an "energy" parameter λ and consider $L\psi = \lambda\psi$ instead of $L\psi = 0$. The method can now be applied repeatedly to construct new solutions in each step, i.e., new operators $L_{j,1} = L_{j-1,2}$ starting from a given $L_{0,1}$. This way one may derive the following formula

$$q_{j,1,n-2} = q_{0,1,n-2} + n(\log W(\psi_{1,1}, ..., \psi_{j,1}))_{xx}, \qquad (3)$$

where $q_{i,1,n-2}$ is the leading non-trivial coefficient in $L_{i,1}$ and $\psi_{i,1}$ is a solution of $L_{0,1}\psi = \lambda_i\psi$ and $P_{0,1}\psi = \psi_t$.

In general the solutions constructed by the method described above may have singularities since the Wronskians used may have zeros. In the KdV case as well as in the Boussinesq-type case it is possible to show that under certain conditions the new solutions inherit some properties from the original solution.

Theorem 2 (Gesztesy, Race, W.) *Let* (q_1, q_0) *be such that the Gelfand-Dickey system for* $n = 3$, $r = 2$ *is satisfied. Furthermore assume that* $q_i, ..., q_i^{(3+i)}$ *are in* $C^0(\mathbb{R}^2) \cap L^\infty(\mathbb{R}^2)$ *and that* $L\psi = 0$ *is disconjugate at time* t_0.

Then $L\psi = 0$ *is disconjugate at all times. Moreover, for a suitable choice of a solution system* $(\psi_1, ..., \psi_n)$, *the solutions constructed in Theorem 1 satisfy the same smoothness and boundedness conditions as the original one, in particular there are no local singularities.*

A similar result was proven by Gesztesy and Simon [2] for the KdV case.

3   Inelastic Solitons

A solution of the Gelfand-Dickey system for $n = 3$ and $r = 2$

$$q_{1t} = 2q_{0x} - q_{1xx}, \quad q_{0t} = q_{0xx} - \frac{2}{3}(q_{1xxx} + q_1 q_{1x})$$

yields at once a solution of the Boussinesq-type equation

$$u_{tt} = -\frac{1}{6}a(u^2)_{xx} + bu_{xx} - \frac{1}{3}u_{xxxx}$$

upon letting $u = (4q_1 + 3b)/a$ ($a \neq 0$). The Lax pair associated to this Gelfand-Dickey system is

$$L = \partial_x^3 + q_1 \partial_x + q_0, \quad P = \partial_x^2 + \frac{2}{3}q_1.$$

Starting now from the trivial solution where both coefficients $q_1 = q_{1,1}$ and $q_0 = q_{1,0}$ of $L = L_1$ are constant, new nontrivial solutions of the Boussinesq-type equation are constructed. The coefficient $q_{2,1}$ of L_2 is given in terms of one solution ψ_1 of $L\psi = 0$ and $P\psi = \psi_t$ as

$$q_{2,1} = q_{1,1} + 3(\log \psi_1)_{xx}.$$

A fundamental system of solutions of $L\psi = 0$ and $P\psi = \psi_t$ is of course given by a set of exponential functions. If ψ_1 is now chosen to be one of these exponential functions then $q_{2,1} = q_{1,1}$, i.e., no new solution is constructed. If ψ_1 is chosen to be a linear combination of two of these exponentials then one obtains a one-soliton solution, i.e., a sech^2-wave. This solution, however, involves two parameters instead of one in the Boussinesq case.

However if one linearly combines all three of the exponentials then something unexpected happens: initially there are two solitons well separated moving with constant velocity towards each other. When they eventually get into the same region they collide inelastically, i.e., one soliton only emerges after the interaction. This situation is shown in Figure 1, where $q_{2,1} - q_{1,1}$ is plotted as a function of x for five different t. Defining the mass of a soliton to be the product of height and width then mass as well as momentum are conserved during this collision but (kinetic) energy gets destroyed.

Considering $q_{3,1}$ instead of $q_{2,1}$ or performing the transformation $t \to -t$ shows that one can also have the reverse situation, namely a single soliton moving along that all of a sudden decays into two different solitons under conservation of mass and momentum but producing kinetic energy while it decays.

Finally using the method of repeated commutation, i.e., formula (3) one can construct other interesting solutions. In the case $j = 2$ one gets according to the different possibilities of linearly combining $\psi_{1,1}$ and $\psi_{2,1}$ out of appropriate exponential functions besides the already known two further phenomena:

- Two elastically interacting solitons moving towards each other or following each other. In contrast to the Boussinesq case the smaller one is here the faster one. This situation is shown in Figure 2.

- Three solitons two of which collide inelastically forming one soliton after the collision while the third interacts elastically with both of the other two. This situation is shown in Figure 3.

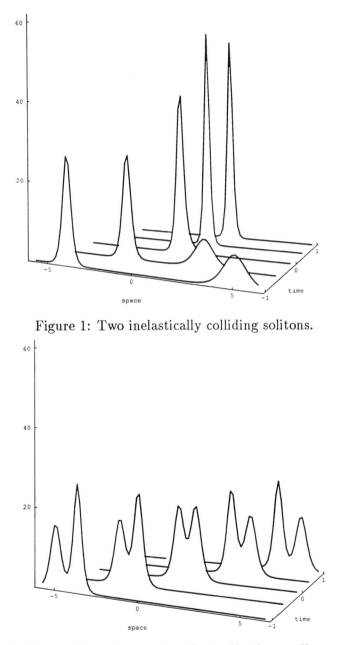

Figure 1: Two inelastically colliding solitons.

Figure 2: Two solitons interacting elastically the smaller one being faster than the bigger one.

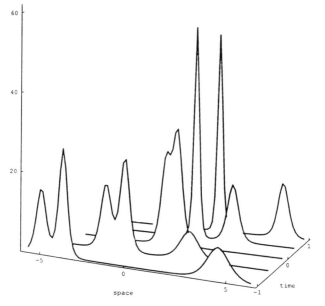

Figure 3: Three solitons, two of which collide inelastically while the third one is interacting elastically with both of the others.

Acknowledgement

It is a pleasure to thank F. Gesztesy for introducing me into this exciting field as well as for many helpful discussions.

Bibliography

[1] P. A. Deift. *Application of a commutation formula.* Duke Math. J., 45:267–310, 1978.

[2] F. Gesztesy and B. Simon. *Constructing solutions of the mKdV-equation.* J. Funct. Anal., 89:53–60, 1990.

[3] F. Gesztesy, D. Race, and R. Weikard. *On (modified) Boussinesq-type systems and factorizations of associated linear differential expressions.* J. London Math. Soc., to appear.

[4] F. Gesztesy, D. Race, K. Unterkofler, and R. Weikard. *On Gelfand-Dikii and Drinfeld-Sokolov systems.* In preparation.

[5] R. Hirota. *Exact N-soliton solutions of the wave equation of long waves in shallow-water and in nonlinear lattices.* J. Math. Phys., 14:810–814, 1972.

[6] V. V. Sokolov and A. B. Shabat. *(L, A)-pairs and a Ricatti type substitution.* Funct. Anal. Appl., 14:148–150, 1980.

[7] G. Wilson. *Algebraic curves and soliton equations.* In E. Arbarello, C. Procesi, and E. Strickland, editors, *Geometry Today*, International Conference, Rome, June, 1984, Birkhäuser, Boston etc., 1985.

Inertial Manifolds and Stabilization in Nonlinear Elastic Systems with Structural Damping

Yuncheng You
Department of Mathematics
University of South Florida
Tampa, FL 33620

Abstract

Consider a nonlinear extensible flexural beam equation with Balakrishnan-Taylor damping and a structural damping, which serves as a model of deflection and control of nonlinear aerospace structures. To solve the "spillover" problem in stabilization of the relevant vibration, which is significant in achieving system stability and performance with controllers involving only finitely many modes, this work provides a new approach by proving the existence of inertial manifolds for the uncontrolled nonlinear equation. The results show that, based on the information of inertial forms, the system is exponentially stabilizable by a linear finite-dimensional feedback control which is robust with respect to the uncertainty of parameters.

1 Introduction

The objective of this paper is to study the following initial-boundary value problem of a nonlinear beam equation, cf. Bass and Zes (1991),

Differential Equations with
Applications to Mathematical
Physics

$$u_{tt} + \alpha u_{xxxx} - \delta u_{xxt}$$

$$- \left[a + b \int_0^1 |u_x(t,\xi)|^2 \, d\xi + q \left(\int_0^1 (u_x u_{xt})(t,\xi) \, d\xi \right) \right] u_{xx} = f$$

$$\text{for } (t,x) \in R^+ \times (0,1),$$

$$u(t, 0) = u_{xx}(t, 0) = u(t, 1) = u_{xx}(t, 1) = 0, \text{ for } t \geq 0, \qquad (1)$$

$$u(0, x) = u_0(x), u_t(0, x) = u_1(x), \quad \text{for } x \in [0, 1].$$

Here $u(t, x)$ is the dynamical transverse deflection of the beam. All the parameters α, δ, b, and q are assumed to be positive constants but $a \in R$. The term $-\delta u_{xxt}$ represents the *structural damping*, $\left[a + b \|u_x\|^2 \right] u_{xx}$ is the tension from the extensibility, and the last term at the left-hand side stands for Balakrishnan-Taylor damping. $f = f(t, x)$ is an external input which in this paper is a control function. We consider the hinged boundary condition. For the cantilevel boundary condition, the notion of "comparable" fractional power operators will be involved, and we shall deal with it separately. Consider the *spillover* problem which concerns whether it is possible and how to design a control involving only finitely many modes and achieving a high performance, for instance, in terms of robust stabilizing the system of some parameter uncertainty at a uniform rate.

2   Formulation as Abstract Evolution Equation

First formulate the initial-boundary value problem of the uncontrolled equation

$$u_{tt} + \alpha u_{xxxx} - \delta u_{xxt} \qquad (2)$$

$$- \left[a + b \int_0^1 |u_x(t,\xi)|^2 \, d\xi + q \left(\int_0^1 (u_x u_{xt})(t,\xi) \, d\xi \right) \right] u_{xx} = 0,$$

as an abstract semilinear evolution equation and study the existence and properties of local solutions in this section. Denote by $H = L^2(0, 1)$ with its norm $|.|$ and inner-product $\langle,\rangle$. Define a linear operator $A : D(A) \longrightarrow H$ by

$$A\varphi = \frac{d^4\varphi}{dx^4} \text{ (in the distribution sense)}, \forall \varphi \in D(A), \qquad (3)$$

$$D(A) = \left\{ \varphi \in H^4(0, 1) : \varphi(0) = \varphi''(0) = \varphi(1) = \varphi''(1) = 0 \right\}.$$

The operator A is densely defined, self-adjoint, and coercively positive, with compact resolvent A^{-1}. The spectrum $\sigma(A)$ consists of only eigenvalues $\{\lambda_k = k^4\pi^4 : k = 1, 2, \ldots\}$ of multiplicity one, with the eigenvectors $\left\{ e_k = \sqrt{2}\sin(k\pi x) : k = 1, 2, \ldots \right\}$. Note that $A^{1/2}\varphi = -d^2\varphi/dx^2$ and $|d\varphi/dx|^2 = \left|A^{1/4}\varphi\right|^2$. Thus the original equation (2) can be formulated as

$$\frac{d^2u}{dt^2} + \alpha Au + \delta A^{1/2}\frac{du}{dt}$$
$$+ \left[a + b\left|A^{1/4}u\right|^2 + q\left\langle A^{1/2}u, u_t \right\rangle \right] A^{1/2}u = 0, \qquad t > 0,$$
$$u(0) = u_0, \quad \frac{du}{dt}(0) = u_1, \qquad (4)$$

Denote by $V = D\left(A^{1/2}\right)$ with the norm $\|v\| = \left|A^{1/2}v\right|$. Define a product real Hilbert space $E = V \times H$. Define a linear operator

$$G = \begin{pmatrix} 0 & I \\ -\alpha A & -\delta A^{1/2} \end{pmatrix} : D(G)\left(= D(A) \times D\left(A^{1/2}\right)\right) \longrightarrow E, \qquad (5)$$

in which I is the identity on V. Also define a nonlinear mapping g by

$$g\begin{pmatrix} \varphi \\ \psi \end{pmatrix} = \begin{pmatrix} 0 \\ -\left[a + b\left|A^{1/4}\varphi\right|^2 + q\left\langle A^{1/4}\varphi, A^{1/4}\psi \right\rangle \right] A^{1/2}\varphi \end{pmatrix}. \qquad (6)$$

Then the equation (4) can be forumlated as a first-order evolution equation:

$$\frac{d}{dt}\begin{pmatrix} u \\ v \end{pmatrix} = G\begin{pmatrix} u(t) \\ v(t) \end{pmatrix} + g\begin{pmatrix} u(t) \\ v(t) \end{pmatrix}, \quad t \geq 0,$$

$$\begin{pmatrix} u(0) \\ v(0) \end{pmatrix} = \begin{pmatrix} u_0 \\ v_0 \end{pmatrix} \in E,$$

or, let $w(t) = \begin{pmatrix} u(t) \\ v(t) \end{pmatrix}$ and $w_0 = \begin{pmatrix} u_0 \\ v_0 \end{pmatrix}$,

$$\frac{d}{dt}w = Gw + g(w), \quad t \geq 0, \quad w_0 \in E. \tag{7}$$

It can be shown that the operator $-G$ is sectorial, G generates an analytic semigroup of contraction, denoted by $\{T(t), t \geq 0\}$, and G has compact resolvent. Denote by $E^1 = D\left((-G)^{1/2}\right)$ and $E^2 = D(A) \times D\left(A^{1/2}\right)$ with the graph norms. The nonlinear mapping $g : E \longrightarrow E$ (resp. $g : E^1 \longrightarrow E^1$) is locally Lipschitz continuous and maps any bounded set of E to a bounded set of E (resp. for E^1). The proof of the following existence and regularity of local solutions is omitted.

Lemma 1 *For any $w_0 \in E$, there is a $\tau = \tau(w_0) > 0$ such that the mild solution of the equation (7) with the initial condition $w(0) = w_0$ exists uniquely for $t \in [0,\tau]$, and $w \in C([0,\tau]; E) \cap C^1((0,\tau); E) \cap C((0,\tau); E^2)$. If $w_0 \in E^2$, then this mild solution is a classical solution of (7) for $t \in [0,\tau]$.*

3   Dissipation of the Semiflow

In this section we will prove simultaneously the global existence of mild solutions of the equation (7) and the dissipation property of the generated semiflow in terms of the existence of absorbing sets in E and in E^1. As a result, there exists a global attractor in E and in E^1 respectively.

Lemma 2 *For any $w_0 \in E$, there exists a unique global mild solution $w(t)$, $t \in [0,\infty)$, of the equation (7), which has the regularity as described in Lemma 1. The generated semiflow ϑ is dissipative, i.e. absorbing sets exist in E and in E^1 respectively.*

Proof. Take the inner-product in H of the equation (2) with $2u_t$ and with ϵu respectively and then add up, by choosing the undetermined constant $\epsilon > 0$ sufficiently small, we obtain

$$\frac{d}{dt} L(t) + \frac{\epsilon}{2} L(t) \le \frac{\epsilon a^2}{2b}, \tag{8}$$

for $t \in I_{\max}$, where

$$L(t) = |u_t|^2 + \alpha |u_{xx}|^2 + \frac{1}{2b} \left(a + b |u_x|^2 \right)^2$$
$$+ \epsilon \langle u_t, u \rangle + \frac{\epsilon \delta}{2} |u_x|^2 + \frac{\epsilon q}{4} |u_x|^4.$$

By the integration of (8) and the usual denseness-approximation argument, it follows that

$$\frac{1}{2} \min \{1, \alpha\} \|(u(t), u_t(t))\|_E^2 \le (1 - \epsilon) |u_t|^2 + (\alpha - \epsilon) |u_{xx}|^2 \le L(t)$$

$$\le (0) \exp \left(-\frac{\epsilon}{2} t \right) + \frac{a^2}{b}, \tag{9}$$

for $t \in I_{\max}$. Thus the mild solution will exist globally over $[0, \infty)$, and the closed ball $B_R = \{y \in E : \|y\|_E \le R\}$, with $R = \left[2a^2 b^{-1} \min \{1, \alpha\}^{-1} + 1 \right]^{1/2}$, is an absorbing set for the semiflow ϑ. The existence of absorbing set in E^1 can be shown by more *a priori* estimates but the same approach. ∎

By the basic theorem on the existence of global attractors, we can prove:

Lemma 3 *There exists a global attractor in E and in E^1 resp. for the semiflow ϑ.*

4   The Existence of Inertial Manifolds

We refer the definition of an inertial manifold to Foias-Sell-Temam (1988). Let $H_m = Span\{e_1, \ldots, e_m\}$. Denote by $P_m : H \longrightarrow H_m$ the orthogonal projection and $Q_m = I_H - P_m$. Denote by $\Pi_m = \begin{pmatrix} P_m & 0 \\ 0 & P_m \end{pmatrix} : E \longrightarrow H_m \times H_m$ and $\Theta_m = I_E - \Pi_m$. We have decompositions $H = P_m H \oplus Q_m H$ and $E = \Pi_m E \oplus (\Theta_m E)$. The H-valued function $u(t)$ has a corresponding decomposition $u(t) = p(t) \oplus h(t)$. The second-order evolution equation (4) is decomposed as follows:

$$\frac{d^2 p}{dt^2} + \alpha A p + \delta A^{1/2} \frac{dp}{dt} + J_u(t) A^{1/2} p = 0, \tag{10}$$

$$\frac{d^2 h}{dt^2} + \alpha A h + \delta A^{1/2} \frac{dh}{dt} + J_u(t) A^{1/2} h = 0. \tag{11}$$

where $J_u(t) = a + b \left| A^{1/4} u \right|^2 + q \left\langle A^{1/2} u, u_t \right\rangle$.

Theorem 1 *There exists a flat inertial manifold $M = H_m \times H_m$ for the semiflow ϑ generated by the equation (7), where $m > 0$ is suitably large.*

Proof. Since that M is a positively invariant can be easily shown due to the commutivity between $A^{1/2}$ and P_m, it remains to prove that M has the exponential attracting property. Take the inner-product in H of the equation (11) with $2h_t + \xi h$ to get

$$\frac{d}{dt}\left\{ |h_t|^2 + \alpha|h_{xx}|^2 + \xi \langle h_t, h \rangle + (\xi\delta/2)|h_x|^2 \right\}$$
$$+ \left\{ 2\delta |h_{xt}|^2 - \xi|h_t|^2 + \xi\alpha|h_{xx}|^2 \right\}$$
$$+ \left\{ 2J_u(t)\langle h_x, h_{xt} \rangle + \xi J_u(t)|h_x|^2 \right\} = 0. \tag{12}$$

By the absorbing property, for every given bounded set Z in E and for any initial point $w_0 \in Z$, the solution trajectory $w(t; w_0)$ will enter a fixed absorbing ball B_R at a universal exponential decay rate

$\epsilon/2$ and after a transient period $[0, t_0]$ with $t_0 = t_0(Z)$. First we consider the trajectories already in the absorbing ball B_R. Hence,

$$\begin{aligned}
|J_u(t)| &\leq \left| a + b|u_x|^2 - q\langle u_{xx},\, u_t \rangle \right| \\
&\leq |a| + (b+q)R^2, \qquad \text{for } t \geq t_0, \tag{13}
\end{aligned}$$

and

$$\begin{aligned}
&\left| 2J_u(t)\langle h_x,\, h_{xt} \rangle + \xi J_u(t)|h_x|^2 \right| \\
&\leq \left(|a| + (b+q)R^2 \right)^2 \delta^{-1}|h_x|^2 + \delta|h_{xt}|^2 \\
&\quad + \xi\left(|a| + (b+q)R^2 \right)|h_x|^2 \\
&\leq \frac{\left(|a| + (b+q)R^2 \right)^2 \delta^{-1} + \xi\left(|a| + (b+q)R^2 \right)}{\sqrt{\lambda_{m+1}}}|h_{xx}|^2 + \delta|h_{xt}|^2 \\
&\leq K(R,\xi)(m+1)^{-2}\pi^{-2}|h_{xx}|^2 + \delta|h_{xt}|^2 \tag{14}
\end{aligned}$$

where $K(R,\xi) = \left(|a| + (b+q)R^2 \right)^2 \delta^{-1} + \xi\left(|a| + (b+q)R^2 \right)$. Substitute (14) into (12) to get

$$\frac{d}{dt}Y(t) + \frac{\xi}{2}Y(t) \leq 0, \quad \text{for } t \geq t_0, \tag{15}$$

where $Y(t) = |h_t|^2 + \alpha|h_{xx}|^2 + \xi\langle h_t,\, h \rangle + (\xi\delta/2)|h_x|^2$, ξ is a constant satisfying $0 \leq \xi \leq \min\left\{ 1,\, \alpha(1+\delta)^{-1},\, \delta/2 \right\}$, and $(m+1)^2 \geq (\xi\alpha\pi^2)^{-1}(\xi + 2K(R,\xi))$. It follows that

$$\begin{aligned}
\frac{1}{2}\min\{1,\alpha\}\left\| \begin{pmatrix} h(t) \\ h_t(t) \end{pmatrix} \right\|_E^2 &\leq Y(t) \leq Y(t_0)\exp\left(-\frac{\xi}{2}(t - t_0) \right) \\
&\leq [1 + \alpha + \xi + (\xi\delta/2)]\left\| \begin{pmatrix} u(t_0) \\ u_t(t_0) \end{pmatrix} \right\|_E^2 \exp\left(-\frac{\xi}{2}(t - t_0) \right) \\
&\leq (2 + \alpha + \delta)R^2 \exp\left(-\frac{\xi}{2}(t - t_0) \right) \tag{16}
\end{aligned}$$

so that

$$\|\theta(t)\|_E^2 \leq 2\min\{1,\alpha\}^{-1}(2 + \alpha + \delta)R^2 \exp\left(-\frac{\xi}{2}(t - t_0) \right), \tag{17}$$

$t \geq t_0$, for any solutions with initial data in E. Note that ξ only depends on the system parameters α and δ. Finally, (17) and (9) imply that

$$\text{dist}_E\left(S\left(t\right)w_0,\, M\right)$$

$$\leq 2\min\left\{1,\,\alpha\right\}^{-1}\left\{\left(2+\alpha+\delta\right)\exp\left(-\frac{\xi}{2}\left(t-t_0\right)\right)\right.$$

$$\left.\cdot\left[L\left(0\right)\exp\left(-\frac{\epsilon}{2}t_0\right)+\frac{a^2}{b}\right]\right\},$$

where $L\left(0\right)$ is a functional of w_0. For any given bounded set Z in E, let $K_1\left(Z\right)=\sup\left\{L\left(0\right):w_0\in Z\right\}$. Then it follows that

$$\text{dist}_E\left(S\left(t\right)w_0,\, M\right)\leq 2\min\left\{1,\,\alpha\right\}^{-1}\left(2+\alpha+\delta\right)\cdot$$

$$\cdot\left\{K_1\left(Z\right)\exp\left(-\frac{\epsilon}{2}t_0\right)+\frac{a^2}{b}\right\}\exp\left(-\frac{\xi}{2}\left(t-t_0\right)\right),$$

for $t \geq t_0$. To include the behavior in the transient period, denote by $\nu = \frac{1}{2}\min\left\{\epsilon,\,\xi\right\}$, and $K_2\left(Z,\,t_0\left(Z\right)\right)=2\min\left\{1,\,\alpha\right\}^{-1}\left(2+\alpha+\delta\right)$ $\left\{K_1\left(Z\right)+\frac{a^2}{b}\exp\left(\frac{\epsilon}{2}t_0\right)\right\}$ then we have the following exponential attraction expression,

$$\text{dist}_E\left(S\left(t\right)w_0,\, M\right)\leq K_2\left(Z,\,t_0\left(Z\right)\right)\exp\left(-\nu t\right),\quad\text{for }t\geq 0.$$

Thus $M = H_m \times H_m$ is an inertial manifold for the semiflow ϑ. ∎

As a consequence implied by the intermediate steps of the above proof, we have a lower bound of the dimension of the inertial manifold M.

Corollary 1 *Let m be the smallest positive integer which satisfies*

$$m > -1 + \frac{1}{\pi\alpha^{1/2}}\times$$

$$\sqrt{1+2\rho\left(\alpha,a,b,q\right)\left[1+\delta^{-1}\rho\left(\alpha,a,b,q\right)\max\left\{1,\alpha^{-1}\left(1+\delta\right),2\delta^{-1}\right\}\right]},$$

where $\rho\left(\alpha,\,a,\,b,\,q\right)=\left|a\right|+2a^2\left(1+q/b\right)\max\left\{1,\,a^{-1}\right\}$, then there exists an inertial manifold $M = H_m \times H_m$ with $\dim M = 2m$.

The governing equation of the subflow on an inertial manifold is called the *inertial form*, which is a system of ordinary differential equations.

Corollary 2 *For the inertial manifold M, the inertial form is the following equation in the subspace H_m,*

$$\frac{d^2 p}{dt^2} + \alpha A p(t) + \delta A^{1/2} \frac{dp}{dt}$$
$$+ \left[a + b \left| A^{1/4} p(t) \right|^2 + q \left\langle A^{1/2} p(t), \frac{dp}{dt} \right\rangle \right] A^{1/2} p(t) = 0,$$
$$p(0) = p_0 \in H_m, p_t(0) = p_1 \in H_m. \tag{18}$$

5 Robust Stabilization by Finite-Mode Feedback Control

In this section, the *spillover problem* is solved based on the existence of inertial manifolds. Now consider the full equation (1) with control function $f(t, x)$ on the right-hand side.

Theorem 2 *The control system* (1) *is exponentially stabilizable by a finite-dimensional linear feedback control*

$$f(t) = a A^{1/2} P_m u(t), t \geq 0, \tag{19}$$

where $P_m : H \longrightarrow H_m$ is the orthogonal projection, and H_m is the factor subspace associated with the inertial manifold $M = H_m \times H_m$ for the uncontrolled equation (2) *or* (7).

Proof. Apply this feedback control in the equation (1) and decompose it into two component equations in accordance with the decomposition of $H = P_m H \oplus Q_m H$, we get

$$p_{tt} + \alpha p_{xxxx} - \delta p_{xxt} - (J_u(t) - a) p_{xx} = 0, \tag{20}$$

$$h_{tt} + \alpha h_{xxxx} - \delta h_{xxt} - J_u(t) h_{xx} = 0, \tag{21}$$

and $u(t) = p(t) + h(t)$ is a solution of the closed-loop equation. An easy adaption ensures that Lemma 2 remains valid and the afore-mentioned ball B_R is still an absorbing set for the new closed-loop equation (20)–(21). Since this linear feedback (19) does not change the h-component equation at all, the argument in the proof of Theorem 1 in showing the exponential attraction (within the absorbing ball B_R) of the manifold M remains true. Hence,

$$\left\| \begin{pmatrix} h(t) \\ h_t(t) \end{pmatrix} \right\|_E^2 \leq K_2(Z, t_0(Z)) \exp(-\nu t), t \geq 0, \qquad (22)$$

where $h(t) = Q_m u(t)$, and the constants K_2 and ν are the same as above. Now we need only to handle the p-component equation (20). We want to prove that the component $p(t) = P_m u(t)$ of the closed-loop solution $u(t)$ also converges to zero at a uniform exponential decay rate. Taking the inner-product of the equation (20) in H with $2p_t + \kappa p$, we have

$$\frac{d}{dt}\Big\{ |p_t|^2 + \alpha |p_{xx}|^2 + \kappa \langle p_t, p \rangle + (\kappa\delta/2)|p_x|^2$$
$$+ (b/2)|p_x|^4 + (\kappa q/4)|p_x|^4 \Big\}$$
$$+ \Big\{ 2\delta |p_{xt}|^2 - \kappa |p_t|^2 + 2q|\langle p_x, p_{xt}\rangle|^2 + \kappa\alpha |p_{xx}|^2 + \kappa b |p_x|^4 \Big\}$$
$$+ \Big\{ 2b\langle p_{xx}, p_t\rangle |h_x|^2 + 2q\langle p_{xx}, p_t\rangle \langle h_{xx}, h_t\rangle$$
$$+ \kappa b |p_x|^2 |h_x|^2 - \kappa q |p_x|^2 \langle h_{xx}, h_t\rangle \Big\} = 0. \qquad (23)$$

Denote by

$$\Gamma(t) = |p_t|^2 + \alpha |p_{xx}|^2 + \kappa \langle p_t, p\rangle + (\kappa\delta/2)|p_x|^2 + [(b/2)+(\kappa q/4)]|p_x|^4,$$

$$\Delta(t) = 2\delta |p_{xt}|^2 - \kappa |p_t|^2 + 2q|\langle p_x, p_{xt}\rangle|^2 + \kappa\alpha |p_{xx}|^2 + \kappa b |p_x|^4.$$

Then we have

$$\Delta(t) - \frac{\kappa}{2}\Gamma(t) \geq \left(2\delta - \frac{3\kappa}{2} - \frac{\kappa^2}{2}\right)|p_t|^2$$
$$+ \kappa\left(\frac{\alpha}{2} - \frac{\kappa}{2} - \frac{\kappa\delta}{4}\right)|p_{xx}|^2 + \frac{\kappa}{4}\left(3b - \frac{\kappa q}{2}\right)|p_x|^4 \geq 0,$$

$t \geq 0$, if we choose $\kappa > 0$ sufficiently small. For the mixed terms in (23), we have the following estimate valid within the absorbing ball B_R.

$$\left| 2b \langle p_{xx}, p_t \rangle |h_x|^2 + 2q \langle p_{xx}, p_t \rangle \langle h_{xx}, h_t \rangle + \kappa b |p_x|^2 |h_x|^2 \right.$$
$$- \kappa q |p_x|^2 \langle h_{xx}, h_t \rangle \bigg|$$
$$\leq (2 + \kappa)(b + q) R^2 K_2 (Z, t_0(Z)) \exp(-\nu t),$$

for $t \geq t_0$. Now substitute these into (23), we get

$$\frac{d}{dt} \Gamma(t) + \frac{\kappa}{2} \Gamma(t) \leq K_3 (Z, t_0(Z)) \exp(-\nu t), t \geq t_0,$$

where $K_3 (Z, t_0(Z)) = (2 + \kappa)(b + q) R^2 K_2 (Z, t_0(Z))$. Therefore,

$$\frac{1}{2} \min\{1, \alpha\} \left\| \begin{pmatrix} p(t) \\ p_t(t) \end{pmatrix} \right\|_E^2 \leq \Gamma(t) \leq \Gamma(t_0) \exp\left(-\frac{\kappa}{2}(t - t_0) \right)$$
$$+ \frac{K_3 (Z, t_0(Z))}{\left| \frac{\kappa}{2} - v \right|} \exp\left(-\min\left\{ \frac{\kappa}{2}, v \right\} (t - t_0) \right), \quad t \geq t_0. \quad (24)$$

Note that $\Gamma(t_0) \leq \left(3 + \frac{\delta}{2} \right) R^2 + \left(\frac{b}{2} + \frac{q}{4} \right) R^4$, where $R^2 = 2a^2 b^{-1} \min\{1, \alpha\}^{-1} + 1$ as shown before. Denote by $\mu = (1/2) \min\{\kappa, \epsilon, \xi\}$ and

$$K_4 (Z, t_0(Z)) = 2 \max\{1, \alpha^{-1}\} \left[\left(3 + \frac{\delta}{2} \right) R^2 + \left(\frac{b}{2} + \frac{q}{4} \right) R^4 \right.$$
$$+ \frac{K_3 (Z, t_0(z))}{\left| \frac{\kappa}{2} - v \right|} \right] \exp(-\mu t_0).$$

Then from (24) and the exponential decay during the transient period $[0, t_0]$ we get

$$\left\| \begin{pmatrix} p(t) \\ p_t(t) \end{pmatrix} \right\|_E^2 \leq K_4 (Z, t_0(Z)) \exp(-\mu t), \quad t \geq 0, \quad (25)$$

Finally combine (25) with the result for h–component, we obtain

$$\left\| \begin{pmatrix} u(t) \\ u_t(t) \end{pmatrix} \right\|_E^2 \leq K_5 (Z, t_0(Z)) \exp(-\mu t), \quad t \geq 0,$$

for some constant $K_5(Z, t_0(Z))$. ∎

Remark. If we replace the parameters appearing in the dimension bound formula by their conservative bounds of uncertainty, then Theorem 1 and Theorem 2 become the robust existence of inertial manifolds and the robust stabilization, respectively.

Bibliography

[1] R. W. Bass and D. Zes (1991): *Spillover, nonlinearity, and flexible structures,* The 4th NASA Workshop on Computational Control of Flexible Aerospace Systems, NASA Conference Publication 10065, compiled by L. W. Taylor, Jr., Hampton, VA, pp. 1-14.

[2] C. Foias, G. R. Sell and R. Temam (1988): *Inertial manifolds for evolutionary equations,* J. Diff. Eqns., Vol. 73, pp. 309-353.

Index

Mathematics in Science and Engineering

Edited by Willliam F. Ames, *Georgia Institute of Technology*

ISBN 0-12-056740-7

90018